THE MAKING OF MEMORY

Steven Rose is Professor of Biology and Director of the Brain and Behaviour Research Group at the Open University and Visiting Professor in the Department of Anatomy and Developmental Biology at University College London. His books include *The Chemistry of Life*, *The Conscious Brain*, *Molecules and Minds*, *Lifelines* and, most recently, *Alas, Poor Darwin: Arguments Against Evolutionary Psychology* (co-editor, with Hilary Rose).

Steven Rose

THE MAKING OF MEMORY

From Molecules to Mind

VINTAGE

Published by Vintage 2003

2 4 6 8 10 9 7 5 3 1

Copyright © Steven Rose 1992, 2003

Steven Rose has asserted his right under the Copyright,
Designs and Patents Act 1988 to be identified as the
author of this work

First published in Great Britain in 1992 by Bantam

Revised edition first published in Great Britain in 2003
by Vintage

Random House, 20 Vauxhall Bridge Road,
London SW1V 2SA

Random House Australia (Pty) Limited
20 Alfred Street, Milsons Point, Sydney
New South Wales 2061, Australia

Random House New Zealand Limited
18 Poland Road, Glenfield,
Auckland 10, New Zealand

Random House (Pty) Limited
Endulini, 5A Jubilee Road, Parktown 2193,
South Africa

The Random House Group Limited Reg. No. 954009
www.randomhouse.co.uk

A CIP catalogue record for this book
is available from the British Library

ISBN 0 099 44998 6

Papers used by Random House are natural, recyclable
products made from wood grown in sustainable forests.
The manufacturing processes conform to the environmental regulations of the country of origin

Printed and bound in Great Britain by
Cox & Wyman Limited, Reading, Berkshire

Contents

Preface to the Vintage Edition

As I wrote in the preface to the first edition of *The Making of Memory*, a decade ago now, I've wanted to write this book – or at least a book like this – for many years. The ways in which my many non-scientist friends and colleagues often regard me as a laboratory scientist – with incomprehension and awe, tinged, I sometimes feel, with faint patronage – engendered in me the idea of a sort of apologia for laboratory life. Could I explain what I did day by day in the laboratory in a way which could give a sense of this arcane activity? Furthermore, could I make it clear why I believe these minuscule observations help cast light on what for most of my working life I have seen as one of the most challenging of all biological and human phenomena, that of memory? And could I in doing so avoid the naïve positivism, which is how most of us as working scientists go about our day-to-day labours, but instead set this account of my own laboratory practice into the richer and more complex context which the present-day philosophy, politics and sociology of science have revealed as framing scientific theory and experiment?

Well, I tried, and the book seemed to work – somewhat to my surprise and pleasure, it won the annual Science Book Prize in 1993. But especially in the context of fast moving science, books these days tend to have a rather short shelf life – sometimes I wonder if they ought not to be stamped 'best before . . .' like supermarket food. In any event, when my current publisher, Will Sulkin at Random House suggested a new edition I was delighted. In many respects the first edition of *The Making of Memory* was a sort of detective story, as I traced the history of my own then twenty years of work exploring the molecular mechanisms of memory in the young chick. Ten years on, some of the paradoxes and problems still unsolved a decade ago had been resolved, only for new ones to emerge. But perhaps most importantly, what had seemed then to me as a quest for basic scientific understanding has begun to yield clues as to how to treat that most desperate and intractable of diseases – Alzheimer's Disease. So this new, and fully revised and rewritten edition now ends with a hope for therapy that I couldn't have guessed at a decade back.

But to return to the text. Memory is a rich area of interpretation and research, and the very term means many things to many people. So, apologies, from the start, to those readers who have picked up this book in the expectation of an engagement with literary criticism, psychoanalysis, neuropsychology or cognitive neuroscience, or, for that matter, a textbook approach to my theme. What I am trying to do draws on all of these disciplines, and several more besides, but it is at once more and less ambitious. I will attempt no further summary or apologia here; the chapters that follow must speak for themselves.

There is no doubt as to who the main influence on me has been as I have tried to develop the synthesis in the following pages: forty years of living and working – sometimes writing – with the feminist sociologist of science Hilary Rose have been a continuous dialectic (a word she herself abjures) whose traces are apparent throughout this book. No book whose central characters are young chicks could be complete without a tribute to Pat Bateson, who was responsible, back in the 1960s, for my first blind date with what became a continuing obsession, and who since then has continued to sharpen the experimental wits of one whose first training was, after all, in the most reductionist of the neurosciences. The experimental programme that forms the core of the book has involved the collaboration of many scores of colleagues, visitors from five continents, students and technicians within the Brain and Behaviour Research Group in the years since we established it in 1969. Some of them are mentioned in the text, others in the references to particular pieces of research.

Four not specifically mentioned there I would like to pay special tribute to: Arun Sinha, who joined me first as technician, later as student and then postdoctoral colleague, and was with me through all the early years at Imperial College and the Open University; and John Hambley, an early graduate student at the OU, working on imprinting, a great lovable shambling bear of a man, and a loyal comrade, who died tragically young in Sydney in 1990. For more than twenty years my laboratory life has been stimulated by what were first occasional visits and finally permanent collaboration with Radmila (Buca) Mileusnic, once a Belgrade Yugoslav, now a UK citizen. Without her, and our colleague Chris Lancashire, the work on cell adhesion molecules and the approach to restoring memory destroyed in Alzheimer's Disease would never have occurred. Others who have worked in the group but are not mentioned by name here may feel that I have missed their own special contribution to the research I describe; my apologies for any omissions: I have valued working with you all over the years. Thanks especially, though, to four people who are too often invisible supports but without whose constant and committed backup it would be impossible for us to maintain a laboratory, rear animals, run experiments or manage people, grants, budgets and the multifarious activities of a university department. For the chicks, the subjects/objects of my research (and who deserve their own dedication), I thank Dawn Sadler and Steve Walters. For much else besides, Heather Holden and Les Pearce.

Wanting to write this book and finding the space and energy to do so are two separate things. What made the first edition possible was a precious year away from the responsibilities of running a busy university department. What made the new edition even thinkable was retirement from that burden so as to be able to focus uninterruptedly on research.

Because this book is in no way intended to provide a comprehensive introduction to memory, but rather to chronicle an adventure in research, illustrate the nature of doing science, and reflect on a theory of mind, I have been a little concerned as to how to deal with the need to explain certain vital aspects of brain structure and biochemistry, necessary to follow the arguments, without formally 'teaching' it. I've opted to introduce key ideas about the brain, neurons, proteins and so on where they are needed for the story line, rather than systematically at the beginning. I hope this will enable a reader, however unfamiliar with the brain, to follow the argument without having to learn too many new bits of biologese. The illustrations were especially prepared for the first edition of the book by Debra Woodward and I hope will help illuminate difficult concepts as well as being pleasing in their own right. I've adopted a similar approach to the conventional academic apparatus of references and credits for research, where I've made a sort of compromise. Some reflections or deepenings of issues raised in the main body of the text will be found in footnotes. Key references are indicated by number and collected at the back of the book. Some chapters are only lightly referenced, and colleagues in the memory mafia whose work is not referred to directly (and I know that the first thing many of us do when we see a new book is to check the references in case we have been cited!) will I hope forgive me. The key 'experimental' chapters, 8–12 are, however, more conventionally densely referenced.

In addition to Debra Woodward's drawings I would like to thank Mike Stewart for permission to use the light and electron micrographs of Figures 3.2, 3.3, 10.2, 10.6, and Larry Squire and the *Journal of Neuroscience* for the NMR photograph of Figure 5.4.

Steven Rose, London, April 2003

CHAPTER 1

The search for the
Rosetta stone

MEMORIES ARE OUR MOST ENDURING CHARACTERISTIC. IN OLD AGE WE can remember our childhood eighty or more years ago; a chance remark can conjure up a face, a name, a vision of sea or mountains once seen and apparently long forgotten. Memory defines who we are and shapes the way we act more closely than any other single aspect of our personhood. All of life is a trajectory from experienced past to unknown future, illuminated only during the always receding instant that we call the present, the moment of our actual conscious experience. Yet our present appears continuous with our past, grows out of it, is shaped by it, because of our capacity for memory. It is this which prevents the past from being lost, as unknowable as the future. It is memory which thus provides time with its arrow.

For each of us, our memories are unique. You can lose a limb, have plastic surgery, a kidney transplant or a sex-change operation, yet you are still in an important sense recognisably yourself so long as your memories persist. We know who we are, and who other people are, in terms of memory. Lose your memory and you, as you, cease to exist, which is why clinical cases of amnesia are so endlessly fascinating and frightening. Advocates of cryonics, that Californian fantasy of quick-freezing the dead until future advances in medical technology can bring them back to life, recognise this; they propose a computer backup store for the frozen corpse's memories which may somehow be read into the revivified body at a future time. But our own human memories are not embedded in a computer, they are encoded in the brain, in the hundred thousand million nerve cells that comprise the human cerebrum – and the hundred million million connections and pathways between those cells. Memories are living processes, which become transformed, imbued with new meanings, each time we recall them.

Most of us worry that we have a poor memory, that we forget names, faces, vital appointments. Yet the scale and extent of what any one of us can remember is prodigious. Imagine sitting down and looking at a photograph for a few seconds. Then another, then another, then another . . . Suppose that a week later I show you the photographs again, each accompanied by a new, different one, and ask you to say which you had seen before. How many photographs do you think you could identify correctly before your memory runs out or you become confused? When I asked my colleagues in the lab, their guesses ranged from twenty to fifty. Yet when the experiment is done in reality most people can identify accurately at least *ten thousand* different photographs without showing any signs of 'running out' of memory capacity.

Do we then really forget at all? Are all our past experiences, as some schools of psychoanalysis maintain, encoded in some way within our brains so that, if only we could find the key to accessing them, every detail of our past would become as transparent to us as is the present moment of our consciousness? Or is forgetting functional, so that we record and remember only those things which we have reason to believe are important for our future survival? If that were so then to have a perfect memory would not be a help but a hindrance in our day-to-day existence, and the long search for techniques or drugs to improve our memory – a search which goes back far into antiquity – would be at best a chimera.

Above all, how do we remember at all? How can the subtleties of our day-to-day experiences, the joys and humiliations of childhood, the trivia of last night's supper or the random digits of a passing car's numberplate become represented within the mix of molecules, of ions, proteins and lipids that make up the hundred billion nerve cells of our brains? If it is hard to envisage such a great number of cells, it is enough to note that each human brain contains getting on for twenty times as many nerve cells as there are people alive on the earth today, and that if you were to begin counting the connections between them at the rate of one every second, it would take you anything from three to thirty million years to complete your tally. Enough here perhaps to store the memories of a lifetime . . .

And yet there is a problem. During a human lifetime every molecule of our body is replaced many times over, cells die and are replaced, the connections between them are made and broken thousands, perhaps millions of times. Yet despite this great flux which constitutes our biological existence, memories remain. No memory within a computer could survive such a complete turnover of all the machine's constituent parts. Somehow, just as the shapes of our bodies persist despite the ceaseless ebb and flow of their molecular components, so do our memories, embedded in the structure and processes of the brain.

It is this central paradox which dominates the dramatic progress that neuroscience – brain research – has made over the past decades. In the United States the 1990s were called the 'decade of the brain'; and many have argued that the first decade of this new century became 'the decade of the mind'. This paradox is one that has dominated my own thoughts and laboratory work ever since, some forty years ago, I first felt able to call myself a neuroscientist. When we talk about memory in our day-to-day lives, we refer to it as a feature of our minds, our sensations, thoughts and emotions. But in this book I will mainly be talking not about the mind but about the brain. Neuroscientists are committed to the view that not only is it possible to explore the workings of the mind in all its many dimensions by the methods of science, but also that those workings can be described in terms of the properties, structures and processes of the brain.

There are some who would view such a statement as either sacrilegious or absurd. The methods of science, or at least of biological science, they would maintain, cannot provide understanding of the mind either because the mind is fundamentally inaccessible to materialist investigation, or because our techniques, while maybe applicable to understanding animal brains and behaviour, fail when confronted with the complexities of human thought, speech and social existence. Or maybe we are simply asking the wrong questions; to try to understand the mind and its memories by understanding the brain is like trying to understand how a computer and its programs work by analysing the chemical constituents of the machine and its disks. But when I talk about 'the methods of science' in this somewhat formal way I certainly don't mean 'the methods of nineteenth-century physics' as if there were only one science – as if a slightly old-fashioned view of physics, actively propagated by traditional philosophers of science and virtually all school teaching, was what every different science – from chemistry to psychology and economics – aimed to become.

What I mean by science and its methods is something a good deal broader and less restrictive; a commitment to a unitary, materialist view of the world, a world capable of exploration by methods of rational enquiry and experiment. If I can achieve what I am setting out to do in this book, then what that description of science means in practice, and why I believe it can be applied to the study of memory, will become clear.

My task as a neuroscientist, I believe, has been to try to put flesh on the bones of this statement of faith. The workings of the mind, I repeat, are to be described in terms of the properties, structures and processes of the brain and body, and describing them in this way will help us understand some of the fundamental questions that all of us as humans ask about our own existence, of who we are and why we are as we are. Please note I do not write that the workings of the mind 'are to be explained'. To use the term explain might be interpreted as implying that if I could describe precisely the molecular and cellular components of the brain, the complex organising relationships between them and their evolutionary and developmental history, then I would have said all there was to be said of the mind; I would have emptied the word of any meaning at all, reduced it to nothing but an assemblage of brain processes. I do not mean this; describing mind in terms of brain and body is not the same as explaining mind away – which is why I referred above to brain and body being embedded in culture. I am not planning, as did some of the schools of psychology which flourished in the twentieth century, and some sociobiologists today, to try to abolish mind terms from any account of how and why we are what we are; why we do what we do; why I write and you read these sentences. Let me try an analogy.

Pass through the massive neoclassical entrance to the British Museum in London, turn left through the shop and pick your way through the throngs of tourists tramping the Egyptian and Assyrian galleries. A knot of Japanese lean over a slab of black stone mounted at a slight angle to the floor. If you can interpose your body between the tourists and their miniature camcorders, you will see that the flat surface of the stone is divided into three sections, each covered with white marks. The marks in the top third are ancient Egyptian hieroglyphs; those in the middle are in a cursive script, demotic Egyptian; and if you had what used to be regarded as a 'sound classical education' or have been to Greece on holiday, you will recognise the writing in the lower third as Greek. You are looking at the Rosetta stone, the text of a decree passed by a general council of Egyptian priests who assembled at Memphis on the Nile on the first anniversary of the coronation of King Ptolemy in 196 BCE. 'Discovered' (in the sense that Europeans talk of artefacts of which they were previously unaware, irrespective of what the local population might have known of them) by a Lieutenant of Engineers in Napoleon's Egyptian expeditionary force in 1799, the stone became British booty of war with the French defeat, and was brought back to London and placed ritually amongst the great heap of spoils of ancient empires with which the British aggrandised themselves during their own century of imperial domination.

But the importance of the Rosetta stone lies not in its symbolism of the rise and fall of empires (even the Greek portion of its three scripts indicated that at the time it was carved Egyptian power was in slow decline, and the rise of European pre-eminence was beginning). The fact that its three scripts each carry the same message, and that nineteenth-century scholars could read the Greek, meant that they could begin the task of deciphering the hitherto incomprehensible hieroglyphs that formed ancient Egypt's written language. The simultaneous translation offered by the Rosetta stone became a code-breaking device, and for me it is a metaphor for the task of translation that we face in understanding the relationships between mind and brain.

As humans trying to understand and act upon the world we inhabit, we work with several languages. Speaking of our own experience we talk personally, subjectively. The classical goal of science has been to eliminate this personal subjective quality of language and replace it with a public voice of claimed objectivity. Yet this is easier to do if we are dealing with physics or chemistry than if we are concerned with the biological and psychological worlds. Dealing with psychological, mental experience, describing what we and others do, why we do it and how we feel about it, we have available at least two alternative languages, each of which makes claims to objectivity – that of mind and that of brain. Brain language has many dialects, spoken by many different sorts of biologists – physiologists, biochemists, anatomists – and handles its claims to objectivity with confidence. Mind language can be – generally is – subjective, the language of everyday life, or of the poet or novelist. But in the hands of psychology, it too aspires towards objectivity. One of the tasks of the new breed of neuroscientist that was born in the 1960s, grew through the 1980s and has matured into the new and scientifically glorious (if socially and politically somewhat inglorious!) decades of the 1990s and 2000s, is to learn how to translate between the two languages of mind and brain. To help that translation we need a Rosetta stone, some inscription in which the two languages, the Greek of mind and the hieroglyphs of brain, can be read in parallel and the interpretation rules deciphered. Deciphering translation rules is not the same as reducing one language to the other. The Greek is never replaced by the Egyptian; the mind is never replaced by the brain. Instead, we have at least two distinct and legitimate languages, each describing the same unitary phenomena of the material world.

The separate histories of these languages as they have developed over the past century have hitherto made them sometimes rivals, sometimes allies. But the prospect of unity, of healing old divisions and of learning the translation rules, has never seemed brighter. I believe, for reasons that this book will I hope make clear, that the study of memory gives us the best chance of learning these rules, that memory will turn out to be the brain's Rosetta stone.

Is this claim arrogant? When I began work on the first edition of this book, I felt reasonably sure that it was not. Today, I am less satisfied with my own metaphor. At least one of the reasons derives from the findings of our own and other labs, some of which I will come back to in chapter 12. For it turns out, as should anyway have been obvious, that mind and memory are not solely 'in the brain' but are profoundly dependent on body states too – circulating hormones, physiological processes, the immune system, all interact reciprocally with the brain in ways that mean that brain and body – and hence mind and body – cannot be separated. But there are more profound reasons too.

Memory is not only – or even, some would say, primarily – the terrain of the sciences of brain and behaviour. To achieve my translation, I have abolished the subjective at the stroke of a computer key, and offered instead merely the objective language of psychology. Yet for each of us our personal memories are profoundly subjective. Day-by-day in the lab I explore the biology of memory in experiments with young chicks and in the evening I go home to a world richly inhabited by my own personal memories. How do these two halves of my life relate? To heal the split in our own lives between subjective and objective surely requires more than just translation between two equally objective languages, but the bridging of a much more profound chasm that has developed within the fragmented culture of Western industrialised society, a chasm that the very power and professed objectivity of science is seen by some as deepening. Can we be at peace with ourselves if we recognise that our deepest, most sacred feelings, of love for others and awe at the universe in which we find ourselves, are at the same time in some way represented inside our own bodies and brains by patterns of connections between nerve cells and the electrical flux between them, the circulation of hormones, the synthesis of particular proteins and the breakdown of others? I believe that we have to learn how to integrate these separate knowledges and feelings if we are to achieve the potential that our very humanity, our own evolved brains and societies, offer us. However, although this has been my life's credo, I am less certain than ever at the end of my research career how it might be achieved. The best perhaps, is to agree with the philosopher Mary Midgley that although we live in one world, it is a very big one. (I've tried to write about this elsewhere, especially in my book *Lifelines*, which followed the first edition of this book on memory, and will return to it again in the book I'm working on now, on the future of the brain.)

But to return to memory. Every culture has offered its own analogy for human memory. For the Greeks, it was inscriptions on wax tablets, for the medievals a complex system of hydraulics, of pipes and valves. At the birth of modern Western science in the seventeenth century, clockwork mechanisms seemed the appropriate metaphor; clockwork was replaced in the nineteenth century by electricity, and in the late twentieth, by computers. I will argue that none of these analogies captures the richness of real human memory, whose understanding lies in the biology of the brain itself, the dynamics of the structural, chemical and electrical interactions between its molecules and cells, and yet which cannot be reduced to 'merely' these. Yet I have to accept the limits of neuroscience, to concede that it has so far been left to the other half of our fragmented culture, the terrain traditionally inhabited by poets and novelists, to try to explore the subjective meanings of memory.

Memory has fascinated philosophers since ancient times; Aristotle devoted an entire book to the topic. Like most such writing it focused on the question, not of how we learn, but of how we remember. Memory, for Aristotle, was about the presence in our minds of mental images of past events. Memory also serves to link the present to the past, and to enable a sense of elapsed time.[1] Describing memory in terms of images is, as will become apparent in later chapters, a recurrent theme in early writings. No one ever put it better than St Augustine, in his *Confessions*, written some 1600 years ago.[2]

Memory, for Augustine, is like a 'spacious palace, a storehouse for countless images'. But memory is capricious. Some things come 'spilling from the memory' unwanted, whilst others are forthcoming only after a delay (p. 214). Memory enables one to envisage colours even in the dark, to taste in the absence of food, to hear in the absence of sound. 'All this goes on inside me in the vast cloisters of my memory' (p. 215). But memory also contains 'all that I have ever learnt of the liberal sciences, except what I have forgotten . . . innumerable principles and laws of numbers and dimensions . . . my feelings, not in the same way as they are present to the mind when it experiences them, but in a quite different way . . .' (pp. 216-20), and things too, such as false arguments, which are known not to be true. Further, when one remembers something, one can later remember that one has remembered it. St Augustine's perceptions and puzzles are those of this book too, and indeed of all memory researchers, even today, albeit we fail to write about them so clearly and elegantly as he did all those years ago.

Memory pervades ancient ballads and modern novels alike. Especially in the twentieth century, from James Joyce and Marcel Proust to the writing of Margaret Atwood, of J. G. Ballard, Toni Morrison, Salman Rushdie and Alice Walker, the theme of personal memory, of the constant examination, interpretation and reinterpretation of lived experience has been central. Can I integrate the minuscule observations of the behaviour of the chicks I work with and the chemistry of their brains with such richness of evocation? Or are we doomed to live always in the divided worlds of subjectivity and objectivity, with no translation possible between these languages?

By offering merely to translate between the languages of brain and mind, biology and psychology, I have ignored agency and intentionality, the fact that humans are not isolated monads, existing trapped inside their own heads, but are profoundly social beings, continually interacting with the outside worlds of things and people. Humans, their minds and brains, are not closed but open systems. The very experiments I conduct could not even have been conceived of without my own sense of agency and intentionality. To understand ourselves demands a recognition of this openness, and that the sciences which can account for its consequences are no longer those of individual psychology or neuroscience, but of the collective of individuals who comprise human society. Ecology and ethology, sociology and economics are such sciences – languages – of the collective, and they cannot (despite the imperialising claims of some sociobiologists and neuroscientists) simply be collapsed into sciences of the individual.

In these sciences of the collective individual memories too become subordinate, to evolution and to history. Although this book is primarily about individual memories and their biology, in writing it I have found myself continually confronted with the phenomena of such collective memory, and you will find them forming a subtext through many chapters. Especially in a society enmeshed in its own cultural artefacts, with a history which transcends any individual experience and memory, but is recorded in texts and images, memory has burst the confines of the individual, the personal, and has become collective. Where once memories were bounded by an animal's – or a person's – own life history and begun afresh with the conception and development of each new life, technology now means that we share as a society memories none of us has ever had personally. Who of my generation can fail to have scarred on their own memories the images of the bulldozed piles of bodies recovered from the Nazi death camps in 1945, or the screaming, napalm-flamed girl-child running down the road from her village in Vietnam, the massacres in Bosnia, the human debris left by the suicide bombers and collective reprisals in Israel and Palestine? Ancient tribal memories which stretch back to generations long gone have engulfed whole regions of eastern and southern Europe in bitter and deadly conflict. No book on memory can avoid crossing the boundary between the individual and the collective, any more than it can avoid facing the divide between objective and subjective.

This book has a further purpose, integral to its every chapter. Each experiment we conduct within our laboratories is dependent for its meaning on the cultural and ideological assumptions of the world which surrounds the lab, just as the lab could not exist without the technological underpinnings of machinery, chemicals, power and money which are omitted from the conventional accounts of science. And no act we make within the lab is a mere passive contemplation of nature; the products of our work themselves generate new technologies just as certainly as they generate new understanding. Laboratories have become the ideological and technological power-houses of modern society. Yet within and because of the fragmented culture of modern society, what goes on inside the laboratory seems arcane, mysterious. 'Science' is represented in the media – especially in television and film – as two-faced, Janus-like. On the one hand there is technological gee-whizzery, in which incomprehensible 'breakthroughs' promise more and more technological and medical marvels, from computers the size of matchboxes to smart drugs which will restore the memories of old age. On the other, there are the mad scientists, the modern Frankensteins, threatening to destroy the world by inadvertence, technological overconfidence, or an insane lust for power. On the whole, scientists don't help demystify this process. We rejoice in our white-coated expertocratic objectivity, called upon to offer certainty in an uncertain world. We tell our own history in triumphalist terms, as a steady pushing back of the frontiers of ignorance and darkness, as an account of inevitable progress, the sort that modern historians dismiss as Whiggish. By and large the media believe us, accepting our press releases and sometimes self-serving triumphalisms; scientific journalists hold up a handsome reflecting mirror to science, disfigured only by episodes of inves-tigative journalism cataloguing seamy stories of cheating or priority disputes. Faced with such charges the 'scientific community' tends to close ranks; by searching out and excluding the occasional rotten apple, the rest of the barrel hopes to retain its virtue.

The result is that outside the charmed circle that constitutes 'the scientific community' the real life of the lab remains unknown territory. As lay people we may have a shrewd idea of how many of our fellows pass their working day; in factories or offices, as house workers, health workers or teachers, even as artists, writers or politicians. But *scientists*? What does it mean to design or perform an experiment, to make an inference, to raise the money to make such experiments possible, or to write the papers to persuade our scientific peers that what we say about the world, what we maintain we have discovered, is true – or at least provisionally true, for this is all we are allowed these days to claim?

Over the past decades, philosophers, sociologists, even anthropologists, have tried to strip away some of the mystery of the natural sciences, to cut our claims to objectivity down to size, to subject our methods of arguing, our assertions that we can tell the truth about the natural world, to searching criticism. Philosophers have questioned our claims to be able to achieve certain knowledge about the material world; realism has been in retreat. Sociologists have laid bare the culture- and ideology-bound nature of the preconceptions with which we attempt to interpret the pointer-readings of the instruments with which we observe the world, the social relations of a science which is not above society but has grown up historically as men's work within the framework of Western, white capitalism. Anthropologists have abandoned their traditional enthusiasm for studying remote tribespeople to sit instead in our laboratories, recording our chatter as we battle with recalcitrant instruments and even more stubborn data.[3]

This sustained theoretical criticism from outside the laboratory has largely passed working scientists by. What C. P. Snow described in the 1950s as the two cultures, the science/arts divide, is still alive and well amongst us, and scientists remain confident that they are what Snow once called them – 'men [sic, and in the main actually so!] with the future in their bones'. Because of this most scientists find it hard to comprehend the popular reaction against science, a hostility which sees us as more likely to be bringing radioactivity than the future to the genetically engineered bones, and not just of men but women and children too. We can feel, even if we do not understand, the popular anger which brings animal rights activists to the doors of our laboratories and environmentalists to reject our technological optimisms, our claims to be able to control and manipulate physical and biological nature. We may feel superior to those who prefer astrology and tarot cards to astronomy and statistics, but it is a superiority tinged with anxiety.

The simplest way to write this book would have been to ignore the philosophical and social issues and to tell a straight story. Look, I would say, the start of this new millennium is a marvellous time to be a neuro-scientist; we have the tools in our hands to understand brain processes, at many levels. It will mean discarding many shibboleths, the naïve molecular reductionism of the biochemists, the arid behaviourism of the psychologists, but we can see the goal clearly. Memory may be our Rosetta stone in obtaining such an understanding. And with the new knowledge may come new techniques that can potentially transform the quality of our lives – at least those of us living in relative abundance in technologically advanced societies – from cradle to grave.

But I want to do more than that: I want to describe what it feels like to be a neuroscientist, to design experiments, to train animals, study their biology, build – and reject – theories, to demystify the workings of my sort of science. I am not writing this book as an observer from outside, nor yet as a textbook or state-of-the art review. I want to share with you, as the reader, something of the excitement and frustrations of how I have spent the last forty years of my life in the lab. And in doing so I want also to go some way towards bridging the gap in my own life, between would-be objective observer in the laboratory, and subjective human outside. Bridging this gap, I maintain, is essential if we are to move beyond our fragmented culture towards a new synthesis which transcends both the ruthless reduc-tionism of a science indifferent to human values and a subjectivism for which truth is but one story amongst many of equal worth. Such a synthesis will represent the real rationality towards which we must strive, both inside and outside the laboratory. Of course, I cannot pretend to be blind to the social and philosophical environment of my work and to the debates which have been raging around the labo-ratory and its claims to truth. As one of the radical critics of reduc-tionist science in the last decades, I have taken my own part in these debates, and I have lived the best part of my life with a feminist sociologist of science whose searching exposure of the nature of a masculinist and largely white science as it is practised in Western capitalist societies has revealed the weaknesses of any uncritical defence of a science that refuses to recognise its limitations.[4]

Yet without knowledge of the life of the laboratory, those outside can have few ways of judging its claims or controlling its products. Knowledge is power, wrote Francis Bacon, that prophet, philosopher and politician *extraordinaire*, at the birth of modern science. Democracy is about the control of power. I am sufficiently a political product of the 1960s to continue to believe that if knowledge is not democratised, power can never be – and human survival itself is placed in jeopardy. Such knowledge can never be democratised while culture is fragmented, while the laboratory is as much a closed door as its findings are a closed book. Can this present text help open that door?

These then are the goals which inform the pages that follow. But I reach my target by a circuitous route. In the next chapter I ask you simply to come with me through a day in the life of the lab, as I go through the routine tasks of experimentation, training chicks, dissecting their brains, measuring their biochemical constituents in quantities of thousandths of a milligram, and trying to extract meaning from the tables of figures that these measurements produce. The experiment I describe is the one I was starting as I began writing the first edition of this book early in the 1990s; where it fits into the general scheme of my work and theories will not become apparent until many chapters later, because to make sense of the experiment you need to understand something both about me as a memory researcher, and about the current state of memory research. Chapter 3 therefore describes my own, subjective memories, and my formation as a neuroscientist. Following this, it is time to consider theories of memory, beginning with the obsessions with memory metaphors, from the Greek and Roman theatres to today's computers. Just why brains are not computers, and why computer memory is a poor substitute for human brain memory, will then become apparent. So what can one learn of human memory by studying humans – especially the disorders of memory resulting from disease or accident – what neuropsychologists describe as inferring function from dysfunction? Some of these findings, together with the fascinating strengths and limitations of human memory, form the subject of chapter 5. This tells us what it is that needs to be explained if we are to develop an adequate biological theory of memory.

Chapters 6 and 7 begin the task of constructing that theory. The starting point here is no longer humans, but animals. Does it even make sense to speak of such quintessentially human activity of learning and remembering in animals? When do memory-like properties appear in the evolution of nervous systems and brains? Studying animal brains and memory, I will show, provides us with a theory, based on the properties and actions of individual nerve cells and their modification by experience.

Chapter 8 enters the era of modern memory research. It does so, however, by way of a curious aberration which overtook memory researchers in the 1960s, when bizarre experiments, and even more bizarre theories, led for a while to the belief that there were specific 'memory molecules' which could be transferred, along with the memories they carried, from animal to animal and even across species. The reasons why this error took hold, and the fallacious experiments on which it was based, make an instructive – and sobering – episode, not much approved of in the traditional Whig accounts of scientific progress, but one with important lessons still. Chapter 9 brings us forward into the new millennium with its discussion of modern memory research, the search for 'God's organism', the ideal species in which to study memory – which in some United States researchers' hands is a sea slug, and in many Europeans' view a sea-horse – or rather, a part of the mammalian brain which is named for its resemblance to that pretty animal. At last, in chapters 10, 11 and 12 we return to my own version of 'God's organism', the chick, and how, in the last decades, my own research has revealed something about the molecular, electrical and morphological pattern of events occurring during learning and memory. Here, we see where the experiment of chapter 2 fits into what, by the end of chapter 11, begins to look like a theory of learning. Chapters 10 and 11 were the core of the first edition and, rather than do more than minor updating, I have left them largely intact. But over the intervening decade, the findings from this research have taken new and wholly unexpected turns, leading me towards – and perhaps some would say not before time – a potential practical outcome from my decades of 'pure scientific' enquiry. For it turns out that my chicks might provide a clue as to a potential treatment for that most devastating of diseases of ageing, the loss of memory that goes with Alzheimer's disease. I must be cautious; these experiments are still under way as I write, in 2002, and this may be fool's gold, but on this I combine a necessary pessimism of the intellect with an optimism of the will. By the time this new edition appears, the answers may be clearer.

But in these last chapters I seem to have abandoned the problematic social and political context of science, even my own subjective world outside the lab, for almost straight scientific storytelling. Chapter 13 is the time to reproblematise that account, to look at it with the eyes of the sociologists and the science policy-makers who know that laboratories are not enough. Can I meet my final goal, that of synthesising the diverse elements in my story? Chapter 14 is, at the least, my best try at doing so. Now, read on . . .

Chapter 2

Reading the record

MOST DAYS, DEPENDING ON THE TRAINS, I GET IN ABOUT NINE O'CLOCK. I dump my coat and bag in the office, pull on a white coat and head for the laboratory. The white coat is important; it has a strong symbolic value. First, the fact that I am wearing it at all means that today is a lab day: I'm running an experiment, not to be disrupted with meetings and administrative queries. Second, the white coat singles me out, separates me from the rest of the world of non-researchers, who wear no coats. It begins to envelop me in the aura of being 'a scientist', ready to perform mysterious – even almost priestly – labours. But not all lab coats are identical; there is a subtle hierarchy amongst them in which their primary function – to keep chemical and biological goo from polluting one's day-clothes – has become overladen with symbolism. You don't have to be long in the lab to notice it. First, protective coats come in several colours; if you are a workshop technician, the coat is likely to be blue, if a porter, brown. There's no real reason why a researcher shouldn't wear a blue or brown coat, but the colour difference, like an army uniform, is an unspoken indication of rank. If you are a lab technician, you'll wear a white coat too, but with a certain difference in the cut – it will button differently at the neck. What's more, you'll be likely to keep it properly fastened at the front. We researchers, on the other hand, leave ours more casually open, swinging round us as we sprint down the corridor from office to lab, though the tradition is fading a bit now as biologists spend less and less time amongst chemical reagents and living things and more in the computer room watching complex multicoloured displays on the screens of the image analysis gear.

But the hierarchy of the coat doesn't simply end there. Even if they are given them free, graduate students won't wear lab coats unless they are actually doing something rather hazardous – handling radioactivity say, when the regulations get quite severe – or a bit bloody, like a minor piece of dissection. The jeans, open collars or tee-shirts that the students wear may also be saying: my clothes are too poor to be troubled by pollution and anyhow I don't want to be fenced in by regulations. At the other end of the hierarchy the rules shift again; the more senior the technician, the less likely to wear a lab coat; here the message is that I have graduated from lab to office, away from manual and towards mental labour.

Not true for the senior researchers; many who may never have been near the lab in years nonetheless put on a spotless white coat each morning as they arrive, sit behind their office desk in it, take it off only to go to the refectory for lunch and when they leave for home at 5.30. They'll even go to committee meetings in their lab coats. Again the message is pretty clear; I would like to be down in the lab actually doing experiments with all you guys, I'm just too busy at the moment with a paper, a grant, a committee, but I haven't lost touch with what really matters, and I can still make a scientific contribution, not just an administrative one.

Today I'm going first to the animal house, where to control for pollution I have to change into another, far from spotless coat – the chicks I was working with yesterday have already seen to that. Today's will add to the mess, till I look as if I'm the starting point for a Persil ad. These days entry into the animal house is getting more and more like getting into a United States airbase. Since the animal rights activists started blowing up buildings and 'liberating' lab animals a few years ago most laboratories have stepped up security – both for the animals and for their carers. A television monitor broods over the heavy door, and I have to feed a plastic card into the lock and punch in a personal number before the door unbolts itself, letting me into an airlock and a second door before I get into the animal house proper.

As I pass through the doors I find myself singing. I don't sing often, for I am, I must make clear, practically tone-deaf; listening to my singing is, I am told, an excruciating experience. But leaving the office for the lab always cheers me; first formless hums and finally snatches of mixed song, odd verses which clutter the recesses of my memory like undisposed of garbage, break out. Catching myself singing, I am amused. I know what it signifies; it is a day for real experiments, not just paper shuffling.

A central fluorescent-lit corridor, doors off into rooms for rabbits, rats and mice, and then my own set, the chick rooms. In one, large incubators, like massive domestic ovens, with racks of eggs, held for eighteen days in carefully controlled temperature and humidity, rocking gently. For the last couple of days before the chicks are due to hatch, batches of fifty or so eggs are moved by the animal technicians, Steve or Dawn, into the brooders in the next room, whose lights are timed to go on and off to simulate night and day outside the brooders – shallow temperature-controlled boxes each covered with a clear plastic dome. There the chicks chip away at their shells and pull themselves, sticky, wet and tired, into their first contact with the external world. By the time I collect them from the brooders the next morning they are dry and twittering, the yellow 50 g bundles of fluff of Easter cards and children's toys – impossible not to find them charming. Leave them a few days though and the yellow fluff will go and white feathers begin to appear; within three weeks they will be strutting adolescents and by twelve they will have put on enough weight, if they were reared in farm or battery, to appear plucked and oven-ready on the super-market shelves. For these are chickens bred as a broiler strain, the eggs bought in by us in tiny quantities from the hundreds of thousands of the weekly production of the poultry breeders. Mass-produced they may be, but these chicks are still biological creatures, not little machines, and it is their biology which makes them important to me.

My experimental design today needs sixteen birds – but I'll take a few more to be on the safe side. I pick up a plastic box, sprinkle some sawdust on its floor to stop the chicks scrabbling, open the brooder dome, count out twenty-four and take them into the next room. This is where my experiment will start. A small, warm room, about three metres square; a bench and sink at one end; shelves down each side and on each shelf a set of small aluminium pens, each about 20 by 25 cm in area and 20 cm high. Above each pen, a 25-watt red light. With the room lights off and the red pen lights on, the room looks oddly comforting, friendly. I put two birds into each pen, marking the back of one of each pair with a felt pen so that I can tell it from its companion, and scatter some crumbs of food onto the floor of the pen. Day-old chicks don't actually need to eat, because they still have plenty of reserve food in a yolk sac, which they won't use up for the next day or so, but they do like to practise pecking and play with the crumbs.

I keep the birds in pairs in the pens because chicks are rather companionable – if they are by themselves they get a bit lonely and shout. 'Peep' is the technical term for a chick's distress call – one of the few occasions when we haven't managed to make our language so pompous that only the initiates can understand it. Chicks will peep if they are alone, if they get cold, if the lights go out suddenly; sometimes a roomful of the birds will start peeping loudly, each setting the other off, until the noise is deafening, like an infant-school classroom when the teacher is out. If I shush loudly they will calm and quieten down – for a few minutes at least. I'm told you can do it by playing a tape of a clucking hen. By contrast, if the chicks are content the only sounds they make are low chirrups ('twitters'). In some behavioural experiments you can actually count the peeps and twitters to check on the birds' wellbeing. If I needed to, I would check the sex of the birds – a procedure that we all used to think very difficult without special training until someone told us that at a day old you could tell males from females by looking at their wing feathers; females are slightly more developed than males and have a double row of feathers to the males' single – it isn't foolproof, but a rough and ready guide.

Now the birds are set up for the experiment I can relax. I'll leave them there for an hour or so to get used to their new environment ('equilibrate') – time to take the lab coat off briefly, pick up a coffee, open the post, talk with Heather in the office, and check the details of the biochemical bit of today's experiment with Reza. It will probably be about 10.15 before I go back into the chick room to start the training.

I've made a chart on a fresh page of my lab book, a stiff-covered A4 notebook; down the side I've written the number of each of the twelve pens I am using, with a symbol for the marked and for the unmarked bird in each. Next to each pen I've listed a series of times, to remind me of what I have to do and when I have to do it. At the top of the page, the date and a title for the experiment: 'To check double wave of fucose incorporation; repeat of experiment of page 34.'

These books are amongst the few items that probably still look much the same in a modern lab as they did a century ago. Go to the Science Museum and you will find lots of such old notebooks preserved; page after page of hieroglyphic records, sketches, quick calculations, pasted-in data from someone else in the lab . . . the raw material of observation from which we construct order out of the chaos of the natural world we study – as revealing in many ways as painters' sketchbooks or writers' manuscripts. The only difference between modern lab notebooks and those you'll find in the museums is that the old foolscap and quarto have been Euro-standardised to A4, and our notes are more likely to be scribbled with a Pentel than a Parker.* But there is a comfortable continuity – and even a continuity with school science, when we were instructed to write at the top of the page 'Experiment 37; to show that water is composed of one part of oxygen and two parts of hydrogen'.

Not experiments at all, of course, these school exercises, but demonstrations of known facts. If we don't show that water is made of oxygen and hydrogen in the proportions 1:2, no one will believe that we have found out something new about the world, only that we did something wrong, miscalculated. Our teachers will then instruct us as to what should have happened, and we will write it down in place of what we actually saw – a triumph of authority over

---

*In a bizarre case in the United States a few years ago, the FBI seized and analysed reams of such lab notes to detect allegedly faked data.[1]

observation just as much as in medieval times. (I know school science is supposed to have grown out of doing such pseudo-'experiments' but there's precious little evidence that it has actually done so.)

Today though, my experiment is a real one. I don't know how it will turn out. I have a theory – if you can dignify a half-baked hunch that I have brooded on evenings and weekends over the past weeks with such a grandiose title. There have been hints in the literature that I might possibly find the phenomenon I am hunting for, and a few weeks ago I ran a different sort of experiment, quick, relatively cheap and using as few animals as I could get away with, that made me feel it was worth trying this one, which will be much more expensive and time-consuming. I am going to have to compare a particular set of biochemical processes in the brains of chicks that have been treated in several different ways. But I know that any differences I find will be small and only detectable by statistical analysis and that I will end up needing at least sixteen birds in each of the four conditions I intend to study.

Why so many? The problem is straightforward. On average, the difference I can expect to find between birds in the measure I am making for each of the conditions I am studying is likely to be around 20 per cent. But if I take a group of birds at random from the hatch, without doing anything different to them at all, and simply carry out the same biochemical procedure on each, the measurements I make will not be quite identical. This is because each bird is unique; they all differ ever so slightly genetically, and in how their egg was treated during incubation, and in just when they struggled out of their egg and into the external world of the brooder – were they the first of the hatch, or were there many other chicks already surrounding them; did they crack their egg easily or only with difficulty? These tiny differences in life history are part of the many sources of variation that make each living organism individual and not in any simply predictable way merely the product of its genes and environment; they are one of the many reasons why biology is inherently a more complex science than physics or chemistry.

More to the present point, though, these small differences in history may all affect the biochemical measures I am going to make. So too will any small differences in the way I treat each chick, and in processing the samples for the biochemical analysis I am going to make of its brain. So at best, if I treat all the birds as similarly as possible, I must expect that the measurements I make will differ from one another by as much as 10 per cent simply by chance variation – by which I mean variation whose causes do not interest me in the context of my present experiment and whose consequences I need to eliminate from any interpretation of the results I get. To be reasonably sure – say 95 per cent sure, the standard safety measure that has become the scientific convention – that, if I find a difference of 20 per cent between groups, it is not simply due to these chance variations, I will need at least the sixteen birds I plan to use in each condition, and I will need to do some statistical analysis of the results. Of course, I might be wrong in my hunch; the effects might be much larger than I suspect, in which case I will need fewer birds. If the effects are much smaller though, the whole experiment may turn out inconclusive; I might never be able to be sure that the effect I suspect is there.

I'll end up cross with myself – a waste of time, a waste of money, a waste of birds – and all because I can't design a decent experiment, or my hunch is not even half-baked, just mistaken. And unlike the school exercise, I won't have an instructor to tell me to write down what should have happened; the stubbornness of facts that simply won't turn out as predicted will have triumphed over my pretty theory. Except that it is so pretty that I will go on turning it over in my mind just to check – could anything have gone wrong, could there be another way of testing it? It isn't that hard to think of experiments to do – I invent half a dozen new ones every evening as I digest the day's results. It is much harder to decide which experiments *not* to do because they won't give conclusive results, or will provide merely trivial knowledge, or are a distraction from the main line of the work. What distinguishes an inspired from a dull researcher is having a feel for the right experiment.

Theories die hard, and I suspect I will struggle to save the one I am testing now if I have to. But all that is way down the line. I won't know the answer for weeks yet. Because each time I run the experiment (replicate it, in lab jargon) I can only conveniently handle sixteen birds, I will need to run it four times to get the numbers I need, and there's no point in even trying to calculate the results until I have the whole set to study. Today's is the second time I've run the experiment, it will take a week to process the data, and it will be a month, if nothing screws up, before I can sit down with all the figures. Only then will I know whether my hunch looks even remotely sensible.

Meantime, I must not look beyond today's programme. Amongst the instruments and debris on the bench with the sink I pick out a piece of stiff wire, about 20 cm long. Stuck on the end of the wire is a small white bead. I offer the bead to each chick in each pen in turn. Mostly the chicks eye the bead cautiously for a few seconds, then peck at it briefly; sometimes they peck at it repetitively for several seconds, sometimes just once before going about more interesting business. I note the behaviour of each chick in the book with a simple code – Peck; Ignore; or Actively avoid the bead. I do the same thing for each bird through all twelve pens – and then repeat the same exercise twice more. Each chick has had three goes at pecking the little bead, and most will have done so on at least two occasions. Anyone who doesn't gets excluded from the rest of the experiment. I call this process 'pretraining'.

Now comes the time for the training proper. I pick up two more beads with wire handles – this time they are larger, about 4 mm in diameter, and made of bright chrome; they are identical except that one has a coloured tape round its handle so I can tell it apart. I fill two small glass beakers, one with tapwater, the other from a little brown glass bottle on the bench. The liquid from this bottle is treacly yellow with a rather pungent smell. The bottle is labelled 'Methylanthranilate'.

I dip one bead in the tapwater and offer it to the two chicks in the first pen – they peck at it excitedly, and come back to peck again. I note the pecks in the lab book, wait six minutes till I am at the time I've already marked down for Pen 2 in the book, then dip the second bead in the yellow liquid and offer it to the two chicks in the second pen. Each pecks enthusiastically, then stops short. They shake their heads fiercely, stoop down and wipe their bills on the floor of the pen. If they were human, I'd say they had just tasted something bitter, had a bad taste in their mouths and were pretty disgusted with the whole affair. Why shouldn't I anthropomorphise? It's considered bad form, especially in psychology in the Anglo-American tradition, to attribute such human-like sensations to non-humans; I am supposed to record 'objectively' what happens. I write down in the notebook by each chick the code letters P–S for peck-and-shake. But I don't have any real doubts about how they experience the taste – I have licked the methylanthranilate myself on the end of a bead, and I know it is bitter and slightly burny, like vinegar and chilli. But the sensation doesn't last long; within a few seconds I couldn't taste it any more, and as for the chicks, they are soon wandering unconcernedly around their pen again. I proceed down the line of pens, training all the birds on either the water or the methylanthranilate-coated bead at six-minute intervals.

Meanwhile, Reza has joined me; he is carrying a polystyrene box filled with ice. Bedded in the ice is a metal rack and in the rack a minuscule plastic test-tube. He is wearing thin rubber gloves as well as his lab coat, for the liquid in the tube is ever so slightly radioactive. From the bench in front of us he picks up a tiny syringe fitted with a long, extremely thin needle. He draws fluid up from the tube into the syringe; the glass body of the syringe is graduated with black lines; each division corresponds to a tiny volume – two microlitres – that is, two millionths of a litre. It used to be impossible even to think seriously about handling such minuscule quantities; now I take them for granted and use them routinely in the lab. The syringe holds in all 10 microlitres; I am going to inject 5 microlitres into each side of each chick's brain.

Now I put on gloves too; there's almost no radioactivity there really, but the new health and safety regulations are very strict, and it's on my mind that if I am sloppy about this myself how can I expect the graduate students to behave themselves properly?

Five minutes after I have trained the chicks in the first pen, I pick each out in turn. Through the yellow down on its head I can make out the shape of the skull, the midline dividing the two hemispheres of the brain, and other guidemarks I have grown as familiar with as a walker reading a map of a familiar route. Reza hands me the syringe, and I rapidly inject the solution first into the left, then the right side of the brain, through the skull. The sharp syringe needle has no difficulty entering, and we've fitted it with a thin plastic sleeve to 4 mm from the end, so it cannot go too deep. There is no blood, the chick seems wholly unconcerned by the procedure, and within twenty seconds both injections have been made and it is back in the pen.

As there is a six-minute gap between training times, up to twelve pens to be trained if any problem arises in the birds' responses, and five minutes between training and injection, I am soon busy doing both almost simultaneously, timing them against my notebook and my Casio set to its stopwatch mode. Once we have the sixteen birds, eight trained on the bitter yellow liquid and eight on the water, I can relax again. I am going to inject four birds in each group now, just after training, and another four in each group five hours later.

The solution I have injected contains a type of sugar, fucose, which the brain uses to synthesise some of the vital molecular components of its nerve cell membranes. The sugar will enter the nerve cells and, over the next couple of hours, begin to be built into the cell membranes. The more membrane that is built, the more sugar will be added into it. So if one group of chicks is making more nerve cell membrane, on average, than the chicks of another group, there will be more of the injected radioactive sugar in its membranes at the end of two hours. All I need is to have a way of separating the various brain regions, purifying the membranes and measuring how much sugar has gone into them. That is why the sugar I have injected is mildly radioactive – some of its normal carbon atoms have been replaced by a radioactive form of carbon. The radioactivity is acting like a coloured dye, a tracer (the same principle, but using very short-lived radioactive substances, is used in the brain and body scanners which are increasingly part of the routine equipment of big hospitals). If I then measure how much radioactivity is in the membranes – a relatively straightforward task with modern equipment – I will be able to calculate how much sugar has been built into them. By making the membranes from different brain regions, and by choosing two times to make this measurement, one beginning just after training and one five hours later, I will be able to see in which parts of the brain the changes occur – if they do – and how long any differences between the groups last. That at any rate is the principle of what I am trying to do in this experiment.

Another gap, now I have completed the injections; time to talk to the graduate students, check how some of last week's experiments are running, make a quick phone call. An hour after the chicks have been trained, it is time to test them. I return to the pens, pick up the bright chrome bead again, but this time dry, without water or methylanthranilate on it, and offer it to each chick in turn. Those that pecked it before when it was covered in water approach it enthusiastically and peck again. Most of those which pecked it when it tasted bitter an hour back now eye the bead distinctly cautiously; some back away from it, others merely turn their head aloofly aside. They have learned, and now remember, that the bead has a bad taste, and their behaviour has changed accordingly. I record the differences in behaviour in the lab book.

This little difference, to peck or not to peck at a bead depending on your past experience, is the starting point for my experimental programme; it has dominated my working life for the past twenty years – and it is what much of this book is about. What goes on inside the chick's head so that its behaviour changes so decisively? I call it learning, memory, remembering.

I can hear the sceptical voices almost as I write these words: 'what a crazy way to spend a life.' (When are you going to stop studying and get a real job, my aunts and uncles asked me in my twenties. They shut up when I became a professor – that was serious stuff, even if they remained uncertain about what it implied. 'My son the doctor' everyone understands; but 'my son the professor'?). You get chicks to peck a bead and call that memory? Memory is our lives, recalling our childhood, recognising a friend's voice on the phone, picking up an old photo and reliving a holiday from twenty years ago, reciting our times table or the team that won the Cup in 1985.What have chicks pecking beads got to do with it?

But it isn't only humans who have memories. Other animals can learn and remember also – and need to in order to survive, even though we can't speak to them, ask them to tell us about their childhood. We can only know if they have learned and remembered something if some aspect of their behaviour, which we can observe and measure, changes as a result of experience. My chicks, which would normally peck at a bright bead, have pecked it once and found it tasted bitter. Very sensibly, when I show them a similar bead next time, instead of pecking it, they reject or avoid it. Their fellows, who have found the bead merely water-wet when they first pecked it, peck at it again second time round. Something has changed in the behaviour of the group of birds that have pecked the bitter bead; they have experienced something novel, learned from it and remember it hours later when I test them. (Perhaps something has changed in the experience of the birds that have pecked the water-coated bead too; they have tasted it and found it not disagreeable – even pleasant. So I am not really simply comparing birds that remember with those that do not in this experiment. But I'll have a lot more to say about that issue later.)

I insist on my legitimacy in using these words – learning, memory, remembering – not as mere metaphors for human experience but as proper descriptors about the animals I am studying. Nonetheless, even if the chick is learning and remembering, why should I claim that in order for it to do so something must be changing in its brain, and what can anything I might measure biochemically tell us about the processes of learning and memory themselves? There's a whole swathe of philosophers and any number of psychologists and artificial intelligencers working with computer models ready to insist that nothing one might discover about the biochemical and cellular processes going on in the brain is of any interest to theories of memory; treat the brain as a black box with inputs and outputs and they will offer to model it; it might as well be made of green cheese. For them my experiment may seem merely trivial, like trying to understand how a computer works by doing a chemical analysis of its logic chips. Later, later; I won't duck the argument, but I'll get to it by my own route; just let me finish with my day in the lab.

Are the processes that go on in the chick when it 'learns' (and that is the first and last time I'll put quotation marks round the word) similar to what goes on in a human brain? Not so fast; I'll come to that too; there are more serious charges to answer first.

If I am right – not just about this particular experiment, but about trying to understand the mechanisms of learning and memory – I have to look inside my chicks' heads. Be warned, this bit is not for the squeamish. An hour after I have tested them, I am back with the chicks again. Reza joins me once more. On the bench in front of us is a tray of ice. Mounted above it is a dissecting microscope – fitted with two eyepieces and looking more like a pair of binoculars. To the side, a row of 48 tiny plastic tubes in a rack. I pick up the first bird in my left hand, body in the palm, head between my fingers, and with a large pair of scissors quickly cut head from body which I drop into a small plastic bucket. If I do it fast enough, there will be virtually no blood. Still holding the head in my left hand I peel back the skin to show the thin transparent skull, beneath which I can dimly make out the contours of the brain. With a small sharp pair of curved scissors, I cut round the skull as if I were tracing the shape of a cap, and lift it with forceps. There, pink, symmetrical and elegant, lie the hemispheres of the brain, a cubic centimetre packed with cells – and the answers to the problems I am wrestling with. I lift the whole brain out of its bony casing with a spatula, place it on a filter paper on a dish of ice, and pass it to Reza. He places it into a resin mould into which two grooves are cut and then takes two razor blades – Gillette or Wilkinson work fine – and presses one into each groove, thus cutting the brain into slabs. Each slab is taken from the mould and placed flat on a glass plate above the ice. Light from a fibre optic tube illuminates the plate from below, and the contours of the six small regions we are hunting become visible, a slightly different shade of pink and white from their surround, probably because of a different distribution of cells, blood vessels and so on. With a scalpel, Reza dissects each in turn, places the tiny samples, each weighing no more than 20 milligrams – twenty thousandths of a gram – into one of the labelled tubes, and passes it back to me. By my side is a polystyrene box half full of dry ice – frozen carbon dioxide, at 18° below zero. I plunge each tube in turn into the ice and the tiny brain sample inside freezes instantly. Meanwhile, during the time Reza has been dissecting, I have been taking out the next brain; three minutes from living chick to frozen brain samples; twenty-four minutes for the eight birds of this half of the experiment.

There is a particular pleasure in the manual skills and craft knowledge that these operations involve. Treat the chick badly and it may not train; slip with the syringe or the scalpel and hours of work are ruined; the chick has been killed to no purpose. Reza and I pride ourselves on the speed and tidiness of our dissections – and this teamwork is yet another key feature of scientific labour. Most scholarly work is intensely lonely. You sit in a library, read what others have inscribed, ask questions of it, reinterpret. Loneliest of all, you face the blank sheet of paper or screen, pen or keyboard at the ready, and struggle to write. The lab isn't like this at all; more like a small factory, with its production line and division of labour. (I do the training and remove the brains, Reza does the dissections; next morning, no longer in the animal house but in the light, bright biochem lab, Jenny will start the analysis of the samples we've prepared today.) We each have our specialisms – and the lab certainly has its managerial hierarchy, however hidden by first name familiarity, in-group lab jokes and shoulder-to-shoulder labour. Counting the post-docs, visitors, students, and technicians there are twenty-some people in the chick team, working in a fairly coordinated way; we have different backgrounds, as biochemists, physiologists, psychologists, anatomists; yet all of us would these days call ourselves neurobiologists, a term that has only come into fashion over the last decades. The implication is that however varied are the neurobiologists' skills, all share a common interest in the brain and its functions.

A few within our group – myself, another academic colleague, some of the technicians – have permanent positions. But most are on short-term contracts, employed on funds one of us – normally me – has raised from the Research Councils (government funding agencies) or private foundations; some are visitors from abroad, working for a few weeks or months with the group; towards all these temporary members of the group in particular I unavoidably have a special type of managerial position.

I wash the instruments, find a scrap of paper, write on it '400µCi 3H' – the measure of the radioactivity inside – and stick it to the side of the bucket with the bodies in as a warning that they are radioactive. Steve or Dawn will dispose of them later. Gloves off, hands scrubbed, lab coat off. Time for lunch. Half the experiment is done; this afternoon I will repeat the injections and dissections with the other eight birds, and clean the training pens ready for tomorrow. But for today the biology will be finished; the samples will go into the fridge at -80°, and the experiment will become as frozen in time as it is in temperature. All the brain tissue's metabolic processes, the incorporation of the radioactive sugar into the membranes, the post-mortem decay and dissolution of cells, all have been halted indefinitely by the freezing. We can knock off for the day; go to a seminar, meet the rest of the group for a beer in the bar, disperse into our several nights. Tomorrow, the biochemistry will be waiting to begin.

This killing business though. It is not easy or pleasant to reduce a bundle of yellow fluff to brain and body. 'Last week I saw a woman flayed', wrote Jonathan Swift, 'and you will hardly believe how much it altered her person for the worse.' Of course, he was right. Yes, the brain has a beauty, its cells an elegance of structure that catches my breath whenever I look down a microscope even now, forty-plus years since I first observed them. But I have destroyed life.

For anyone who holds an absolutist animal rights position there can be no doubt that in killing the chicks I have done evil; demonstrated a hegemonic concern to dominate nature; caused pain; behaved, unequivocally, as a speciesist. Sure, everything I have done has been under authorisation by quite stringent British Home Office laws, which regulate the conditions under which I keep the chicks, the numbers I may use and the types of operation I may perform on them rather more rigorously than do the laws to protect children. (It is an old joke that in Britain we have a *Royal* Society for the Prevention of Cruelty to Animals but only a *National* Society for the Prevention of Cruelty to Children.) And of course, if I hadn't ordered the eggs from the hatcheries, they would have gone to a chicken farm and ended up twelve weeks or so of free-range or battery life later on a supermarket shelf, plus or minus giblets and salmonella. In between, someone else – or a machine – would have killed them.

Irrelevant, if you accept the animal rights arguments. Two offences against animals don't cancel out. If I am to go on with these experiments, if I am to go on with this book, I need to address the arguments squarely. My research is aimed at understanding the basic brain mechanisms involved in learning and memory. If I / we / society – use any pronoun you choose – want this sort of knowledge, there is no other way at present of obtaining it than to work on animals. Whether we (and here I do mean we as society, for what I do is not the private act of an isolated individual but is part of socially produced work) want it is of course a social decision, because science is a social activity, funded by government and industry.

In a properly democratic society all our institutions – including science – should be open to public scrutiny. And there is no doubt that our society both uses and abuses animals. I strongly oppose many things that are done to animals in farming, in hunting, in rearing animals as pets, and indeed in some forms of animal experimentation – nor would I ever accept the Cartesian view that non-human animals can be regarded as pain-free machines, so that one can do what one likes with them without it mattering. If they were, my research would probably be meaningless.

But those who argue for animal rights seem to want it both ways. On the one hand, they claim that animals are sentient and therefore, like humans, have certain rights. On the other they maintain that there are such great discontinuities between animals and humans that animal experiments can tell us nothing relevant to the human condition. This is frankly nonsense. The biological world is a continuum. The basic biochemical mechanisms by which we tick are very similar in most other organisms. If they weren't, even the food we eat would poison us. Many human diseases and disorders are found in other mammals – which is why we can learn how to treat them by research on animals. So too are the brain mechanisms I am studying; bacteria, or tissue cultures (pieces of tissue or isolated cells growing in a test-tube, sometimes proposed as an alternative to animals in experiments) don't learn or remember. If this sort of knowledge is wanted – by scientists, by society, leave aside by just who at present, for that after all is a separate political question – then there is no alternative to using animals.

Unless, of course, we experiment on humans. Some such experiments are of course possible – and in the case of clinical trials of new drugs or medical treatments, are even essential – after the initial animal safety tests have been done. There have been heroic physiologists – the great J. B. S. Haldane in the 1920s and 1930s was one – who have relished testing certain hypotheses on their own bodies. The new imaging systems, the body scanners and biomagnetic detectors, make possible certain sorts of measurement of internal brain processes, inconceivable a few years ago. I'll be discussing some of these in later chapters. But despite all that powerful instrumentation, for the overwhelming number of biochemical and physiological processes that need to be studied to make sense of our own human biology, there is no present way round the paradox that the study of life seems to demand the destruction of life.

And this is the nub of the question. Just because we are human, any discussion of rights must begin with human rights. How far are those rights to be extended – does it even make sense to talk of extending them – to 'animals'? But cats and dogs, mice and monkeys, slugs and lice, wasps and mosquitoes, are all animals. And where should my chicks fit into the picture? How far should one extend the concept of rights – to not swatting a mosquito sucking your blood? To preventing your cat from hunting and killing a rat? Does an ant have as many rights as a gorilla?

Most people would say no – though I did speak to one activist picketing our lab who argued that even viruses had souls. I think most animal-righters are really arguing that the closer animals are to humans, biologically speaking – that is, evolutionarily speaking – the more rights they should have. So where does the cut-off come? Primates? Mammals? Vertebrates? The moment one concedes that question, it is clear that the decision is arbitrary – that it is we, as humans, who are conferring rights on animals, not the animals themselves. Of course, to be arbitrary is not necessarily to be wrong; ethicists and animal legislators alike try to divide between species which appear to be able to feel pain and those which so far as one knows do not, or species which have large nervous systems as opposed to those which have small or relatively simple brains. I have always avoided working with primates, or cats or dogs – although I can see that for some purposes it might be unavoidable. The key test for those who argue the moral absolute of not using experimental animals comes with AIDS research, as the only available species in which to test both the virus and possible treatments, other than our own, is the chimpanzee.

Put like this it is plain the debate about animal rights is not like that about women's rights or black people's rights or civil rights, in which the oppressed subjects of history are demanding justice and equality. It is an argument about how we as humans should behave. It is here that the biological discontinuity between humans and other animals becomes important. Our concern for how we treat other species springs out of our very humanness, as biologically and socially constructed creatures. We do not expect cats to debate the rights of mice. So the issue is not – or ought not to be – about animal rights at all, but about the duties that we have just because we are human.

And I am sure that we do have such duties, to behave kindly and with respect to other animals, with the minimum of violence and cruelty, not to damage or take their lives insofar as it can be avoided, just as we have duties to the planet's ecology in general. I believe that most working biologists would agree with this – indeed I am far from sure one can do decent experiments with animals without respecting them. If I were to persist in treating chicks as Descartes might have wanted me to – and indeed as some schools of behaviourist psychologists would still maintain – as insentient machines, mere logic circuits based on carbon chemistry instead of the more reliable silicon chemistry of the computer, I would soon cease to be able to design sensible experiments or interpret the results that I obtain.*

But all such duties to non-human animals are limited by an overriding duty to other humans. At home we have a much loved and exceedingly beautiful cat (who would, I am sure, be enchanted to help me train chicks, were I to allow her). But if I had to choose between saving the cat's life and that of any human child, I would unhesitatingly choose the child. And my cat at the expense of a chick. And so would the vast majority of people. That is species loyalty – speciesism, as the animal rights advocates call it – and in that sense I am proud to be a speciesist. My decision to work with chicks, to begin the experiment I have started today, stands and falls not by some absolute criterion of animal rights, but by whether I can justify the taking of the chicks' lives by the increased useful non-trivial knowledge it will bring.

If I stop to think about it – and I find myself doing so quite often – this is an awesome responsibility: playing God to the chicks. I suspect that most of us who work with animals in this way feel it, even if we don't discuss it too often. But you can see it in the circumlocutions which researchers use to talk about killing animals. Pick up the average scientific paper – especially if it comes from a United States lab – and read through the introduction until you come to the methods section where, by convention, the experimenters describe in some detail the techniques they have used. You will find they describe how their animals were bred, what operations they may have done on them, how their tissues were processed – but never that the rats, mice, cats, monkeys or chicks

---

*Of course this was not always so; physiological experiments in the nineteenth century, especially before the days of anaesthetic, were less solicitous about the welfare of their animals, and anti-vivisectionists described physiology as the science of pain. Nor would I want to argue that today's biologists are always perfect in motivation and in deed. The proliferating scientific literature, and the molecular roulette practised amongst the drug companies as they test potential new agents or find ways of circumventing existing patent laws, both generate an abundance of ill-conceived or unnecessary experiments.

were killed. In such papers, animals are almost always 'sacrificed'.* An interesting circumlocution, this. Unpack its meaning, and it insists that the act of slaughter was not casual, gratuitous, but deliberate and solemn. That in performing the act the scientist was no longer merely a lay individual, but was adopting arcane and priestly functions – an extension of the white coat syndrome. That the death of the experimental animal was justified by a higher good.

In analysing such symbolism I do not wish to deny that at least part of its intention is to pay appropriate respect to its animal objects and subjects; there is, I do not need persuading, destiny in the fall of a sparrow, and no chick is an island, entire unto itself. Starved of expressive language as we scientists are, at our moment of greatest need we reach for a symbolic cloak, and sacrifice our experimental subjects. If I insist, in my own research papers, on the blunter Anglo-Saxon term, it is equally a statement of refusal to escape the significance of the act. A few years back I contacted the Secretary of the British Union for the Abolition of Vivisection to raise with her worries I had about the extent to which the organisation had become permeated by members of neo-Nazi groups who were using the movement as a cover for their attacks on kosher and halal slaughtering and hence on Jews and Muslims.† How could one claim to defend animal rights whilst at the same time holding human rights in contempt? She agreed; it was (and still is) a problem. And by the way, she added, she had never spoken to a vivisector before like this and didn't I feel what I was doing was a bit like Dr Mengele? An odd question to ask a Jew! Anyhow, no I didn't and no I don't.

But might it not at the least degrade my sensibilities to kill animals?

---

* I recently received a paper to referee in which an even more unacceptable euphemism was employed; the authors spoke of mice being 'euthanised'. Choking on this phrase, which seems doubly unacceptable – both to the painful human debate about euthanasia and to the implication that the mice were in some way volunteers for the process – I rejected the paper.

† The argument goes deeper. Within Fascist and Nazi thought there is a deep strand of 'ecological' thinking. The most stringent laws against animal experimentation were passed in Germany during the Nazi period – implying that *untermenschen* with their 'lives unworthy to be lived' were even *unteranimalen*.

There is clearly some truth in this. As a student I had to learn to kill and to dissect out tissues. Dissection is a skill, and, as I have already said, there is a lot of satisfaction in achieving it quickly and tidily. Nonetheless the work we do changes the sort of people we are – how we relate to the human world as well as the world of non-human animals. If in our work we lay violent hands on other living creatures – as farmers, butchers, surgeons or experimental biologists – we cannot expect to have the same relationships to other human beings as novelists, schoolteachers, assembly-line workers or philosophers. But those whose work does not mean they are in regular daily contact with the life and death of animals nonetheless rely for their daily survival on the knowledge and commodities our labour produces.

That is at least one of the reasons why, in this book, I am trying to lay open my craft, to desacralise the doing of an experimental science, to explore not only memory in the abstract, but also how it might be that, in researching my chicks' memory, I can begin to make sense of my own.

Fig. 2.1

Chapter 3

Making memories

WHEN MY MOTHER DIED A FEW YEARS AGO IT FELL TO MY BROTHER AND me to clear her house. She'd lived in it for almost thirty years, alone for most of the time, since my father had died within months of their moving into the house they'd dreamed of and worked for over the previous decade. During that thirty years the house had become a sort of devotedly cleaned mausoleum with her as its attendant. She had a dual existence, tough and job-oriented during the day until she was compulsorily retired a few years before she died, while at home in the evenings and weekends she was closer to Miss Havisham, preserving the day-to-day paraphernalia of her earlier, married existence. For years my father's clothes hung in the wardrobe; some were still there when she died. That was no surprise to me; what I hadn't expected was how much else she had accumulated, stacked into dusty old boxes in the loft or into the top shelves of cupboards which hadn't been opened for years. Indeed, we all keep love letters, our children's first clothes and later their school reports. But for her that was only the beginning; there was militant correspondence with traffic wardens over a parking fine dating from a quarter-century earlier; there were cheque-book stubs and bank statements from the 1950s; theatre tickets and programmes from the 1930s; a treasure trove of trivia.

My mother lived in her memories. Day-to-day events were considered in relation to past experience. Take her to a restaurant, and today's meal, however lavish, would be judged by reference to past feasts – and found wanting, until it too was safely framed in the past, to be savoured only in retrospect. To exist in today's world, to be able to define herself, her place and legitimacy within it, required such constant references back. When the house was burgled and she lost some treasures – a teapot inherited from her own mother, a silver mug given to me by my godmother to celebrate my circumcision – she mourned not so much for the objects, but for that part of her own past world which had gone, leaving her feeling correspondingly diminished.

It was as if her memory were held, not within herself, but by these external reference points. She needed memory to define herself, and to provide that memory she would surround herself with a carapace, a shell of referents. When we made a bonfire in the back garden and burned the cheque stubs, we committed her to the flames almost as certainly as, at the cemetery, we had lowered her coffin into the ground.

Of course, she was not unique in so determinedly preserving the artefacts which could prove not only that she was alive but that her present was continuous with her past. The world echoes every moment to the faint clicking of thousands of cameras, the soft whirring of the videos, as we record the present: on the beach, at a wedding, in front of the Taj Mahal, or sitting in our own back yard. What to do with these millions of images, proofs that then we were alive, observing, feeling, thinking . . . ? Most of the photos in my own house are in chaotic heaps, thrust into drawers and cupboards, to be classified some day not yet arrived. Such records are, of course, our memories, or at least they bear the same relation to our memories as does the artificial intelligence which some claim is embedded in computers to real, human intelligence.

How and why do we remember? Some old memories are obviously prompted. My mother used to say to me of some episode of my childhood: 'but surely you remember . . .' And when I agreed, I never knew if it was really my ancient memory she had triggered, or a new one she had created by her own urgency in insisting that I must. But some memories are clearly unprompted. My earliest certain memory is of the night sky; being carried wrapped in a blanket against a shoulder towards the air raid shelter at the bottom of my grandparents' garden. And of the shelters themselves; in the early days a yellow-cream painted steel-framed table inside the house underneath which we hid when the sirens sounded, and presumably later, though I've no sense of the elapsed time between the one and the other, the brick shelter in the garden, with its bunks over which adults hung their besocked feet. No one else has invented these memories for me, because I see them clearly with the peculiarly scaled vision of a child. And like many childhood memories they are not linear, not a series of sequential events, but more like pictures, truly 'photographic' memories, even if the photographs come with feels, sounds and smells attached.

> I am sitting on someone's knee being fed with gruel. The plate is on grey oilcloth with a red border, the enamel white, with blue flowers on it, and reflecting the sparse light from the window. By bending my head sideways and forwards, I try out various viewpoints. As I move my head, the reflections on the gruel plate change and form new patterns. Suddenly, I vomit over everything. That is probably my very first memory.[1]

Thus the great Swedish film director Ingmar Bergman opens his autobiography, written seventy years after the event he is recording – and many of his films continue to mine this world of remembered childhood experience, private snapshots in the head transformed into moving screen images for public inspection.

Another snapshot – a moment at my fourth birthday party, racing with my guests, around a circular rosebed, arms outstretched, being aeroplanes. But what came before or after this frozen moment in time? Who were my party guests? Did I have a birthday cake? I have no idea – or even of where the garden was in which we played party aeroplanes in deference to the real fighters and bombers overhead.

But I have of course in some measure reinvented these memories.

Obsessed with the attempt to see how far back in my childhood I can remember, I have taken out these internally filed photographs, redeveloped and reprinted them, cropped them a little differently, made them matt or gloss, black-and-white or colour, enlarged them to fit a new frame just as much as Bergman has transformed his for public viewing. Every time I remember these events, I recreate a memory anew; in writing these sentences now, they become not the recall of episodes of a wartime London childhood of the early 1940s, but have been transmuted by thought and writing into the memories of today.

However, such constant refurbishment isn't always necessary. I once remet after decades a childhood friend and neighbour, who asked, didn't I remember the man who'd gardened for both her parents and my own, who had planted purple and white crocuses in special patterns. No, I did not. Then she mentioned his name, Mr Goss, a name I certainly hadn't heard or thought about for forty years. Suddenly I had a clear image, first of dark blue dungarees, then of rubber boots, and finally of a lean, leathery outdoor sort of face. How had these attributes remained attached by some strange brain/mind process to a name unused for all those years, yet still available, stored some way within the brain and accessible a great deal more rapidly than I could lay hands on any particular photograph piled into the untidy heaps in my cupboard? How to account for the seemingly random nature of what is remembered, so that I can recreate Mr Goss from a minimal clue but can't drag back the name of a person I met at dinner a couple of weeks ago? Or perhaps I have entirely invented a Mr Goss in response to my friend's prompting? A real or a fantasy memory? There is no way I can tell.

These then are the questions that fascinate me, which my book is about. But why am I so obsessed, and what sort of answers should I be seeking? Psychoanalysts would certainly offer one type of explanation, both of my quest and of my capacity to remember Mr Goss after all these years – or even to have forgotten him for the intervening period. For Dennis Potter, it might make the basis of a television series – *The Singing Scientist*? For Marcel Proust, the taste of a madeleine cake generated twelve volumes of *Remembrance of Things Past*, transmuted from memory to novel cycle in the sanctity of his cork-lined room in Paris. But I am neither playwright nor novelist – nor, for that matter, psychoanalyst or analysand. My questions about memory are those of the biologist and neuroscientist, searching to decode memory as the brain's Rosetta stone. Perhaps charting my own personal trajectory will help make sense of the search for meaning which occupies the rest of this book.

I was a prewar baby – 'prewar' in Britain always to be interpreted as pre-1939 and not the dozens of little neocolonial wars the world's most experienced and nearly most expensive army has been engaged in over the succeeding sixty years. During the war years, while my father, a Zionist and anti-fascist volunteer, was in the army, I was brought up by my maternal grandparents in a middling suburb of northwest London, part of the classical migratory route for Ashkenazi Jews who had come over from Russia and Poland and settled in east London in the early part of the century. They raised me as a *Shul*-going and respectable young Jew; an eldest grandson whose precocity was encouraged – spoiled, some said – in a way that their own elder sons and daughters had never been. The problem was that I began to read improving children's books, full of natural history told in the sort of evolutionary just-so story way which today's television, courtesy of David Attenborough and his young successors, does so much better. My father came back from the war, happy to go on building the budding intellectual. For my eighth birthday I got a second-hand copy of Darwin's *Origin* and a chemistry set. Truth to tell I wasn't to read more than the first few chapters of the *Origin* till many years later, but I knew the message well enough from the natural history books. Now I sat every Saturday with the young boys in the back of the synagogue and explained to them, till silenced by the beadle, why you didn't need God to explain how humans had come from apes and before that amoeba, and how the earth had been pulled from the sun by a passing star – the then popular theory. And before then? Well the universe had started with a big bang, and before that there was nothing – and nothing was as easy to believe in as God (I still have no problem with the impeccable logic of this position).

Evolution provided the theoretical framework, the chemistry set the method. I spent long hours in the garden shed rigging it up as a lab, washing bottles and running chores for the local chemist in return for chemicals and glassware (his son eventually became a lecturer in biochemistry at a medical school; it must have been a formative environment). We did 'experiments' – which sometimes worked – by cookbook rules, and spent endless hours trying to concoct explosive mixes that we could use as rocket fuels. Science for me – though I hadn't read Marx even in a children's version – was already about changing the world as much as understanding it.

To gain upward mobility for their elder son, my parents packed me off to a direct grant school in north-west London which to protect itself from the pressure of local Jewish parents operated a numerus clausus – no more than ten per cent Jews permitted, lest the 'character' of the institution be changed. But it was the direct grant school which gave me the step up to the next stage for aspirants to social mobility in Britain. Experiments with rockets in the garden shed led naturally to school physics and chemistry. If Judaism linked God and the family, then I couldn't leave one without the other, and I suppose, in retrospect, that I must have been desperate to run away from suburban suffocation. Ambition, escapism and scholarships propelled me to Cambridge, ready to learn the scientific and social skills I needed to understand and change the world.

Where did the key to the universe lie? Despite Darwin, biology seemed little more than a collection of contingent stories about the world, more credible perhaps than Rudyard Kipling's versions, but without the power to explain, and, through explanation to change and control. Chemistry, physics, mathematics – these were lawful disciplines, they unified the world rather than fractionating it into an uncontrollable kaleidoscopic variety. This ordering, simplifying, rule-making was what I suppose I was seeking in science.

As in science, so in life? I started my university life as an enthusiastic and modestly competent chess player, a game, I believed, that involved purely cognitive and logical skills. As I played tournaments, evening after evening, I became embarrassingly aware that there were certain other players who always beat me, even though I was convinced they were 'really' no better than me. Playing them, I knew

I was defeated before we even made the opening moves. As their confidence grew, mine diminished and I played more and more weakly against them. Why did this happen? How could it be that emotions – irrational emotions – could affect how well one played in even such a purely cognitive game as chess? I abandoned chess in disgust at my own inability to transcend such weaknesses, and took up poker instead. Not a game of chance at all, but of competitive psychology in which affect dominated cognition – and one that turned out to be distinctly more profitable. I had learned that cognition cannot be divorced from affect, try as one might. Even today I find myself frequently in danger of forgetting that lesson, though the problems that it illuminates are fundamental to my research strategy, just as much as their resolution ought to be fundamental to a strategy for living.

Meanwhile, gentle pressure from my tutors added physiology to my studies – and without any formal biological training, without knowing what the outsides of organisms looked like or how they might be related I found myself learning about their circulatory and respiratory mechanisms, even about their brains. These too, it appeared, obeyed rules, could be modelled mathematically, studied by physical instruments, had a chemistry. I discovered a strange and exciting hinterland between chemistry and physiology: biochemistry, a word I had never even heard of at school. Enthusiasm for the new – and a distinct incapacity to handle the maths that physics increasingly demanded – made the choice of specialism inevitable.

The late 1950s was a good time to be studying biochemistry at Cambridge. The subject had practically been invented there in the 1910s and 1920s, and Cambridge biochemistry under the leadership of Frederick Gowland Hopkins had been amongst the world leaders from the 1930s. Already in decline in the 1950s, and about to be overtaken by its brash younger rival, molecular biology, itself born in the physics laboratories of the Cavendish a couple of hundred metres away, Cambridge biochemistry nonetheless retained an enormous intellectual self-confidence. As undergraduates we scarcely raised an eyebrow when we arrived for class one morning to find the foyer of the department awash with champagne to celebrate the award of a Nobel prize to one of our lecturers, Frederick Sanger, for determining the chemical structure of the small protein hormone, insulin. It was an achievement that seemed to us every bit as interesting as Watson and Crick's DNA double helix, identified three years previously at the Cavendish labs (on the basis, as everyone now knows, of Rosalind Franklin's X-ray pictures).

Biochemistry *was* the future, we were told, and most of us ended up converts. It was only much later that I came to realise how much of the style of the department was the legacy of Gowland Hopkins. A committed liberal, he had shaped his lab in the 1930s to make space for a generation of refugees from Nazism who were later, in England and the United States, to provide the cornerstones of research into modern biochemistry and in doing so to win a clutch of Nobel prizes and find themselves enshrined in every student's textbook: Hans Krebs, Fritz Lipmann, Ernst Chain, Albert Szent-Gyorgyi and many others. And, even more interesting, the Hopkins lab had been the focus of a group of young left biologists who in the 1930s had struggled to transform the world according to a socialist and Marxist vision as passionately as they had tried to develop the rudiments of a rational biology – J. D. Bernal, J. B. S. Haldane, Dorothy Needham and Joseph Needham. No wonder that in my naïveté I mistook the residual radicalism of the environment for the radicalism of the discipline.

By the time I graduated in 1959, molecular biology had arrived. It was where every bright researcher wanted to be. I was offered the chance to work for a PhD on viruses and turned it down. DNA and protein structure were solved; what else was there to do in molecular biology? What was now needed was to use the tools of biochemistry to understand function – and what more fascinating function than the brain? Such arrogance, when anyone with any sense should have seen that the great period of molecular biology – what Gunther Stent was in retrospect to call its classical period – was just opening. No one in Cambridge seemed able to offer me the chance to work on brain biochemistry, and my tutors sent me back to London, to a vast redbrick psychiatric hospital in Denmark Hill with an Institute of Psychiatry attached – the Maudsley.

Named after a prominent nineteenth-century psychiatrist, it was an impoverished environment after the affluence of Cambridge, a rambling set of buildings in a rundown part of south London which I was too ignorant to recognise had been the home for the development of the major tendencies in British psychiatry in the years during and after the Second World War. But I was rapidly disabused of any thought that the research I was about to embark on had anything obvious to do with explaining or improving the conditions of the shuffling figures that could occasionally be seen in the grounds or in the corridors. The Biochemistry Department was housed in one wing of a newish building well away from the hospital proper, and consisted of a handful of academics (although their teaching tasks were minimal and they were essentially full-time researchers), technicians, and students together with a throughput of visiting Americans.

Presiding/professing over this small empire was a dour and diminutive ex-microbiologist, Henry McIlwain, who had come to the Maudsley in the late 1940s. Arriving at McIlwain's department with a passionate commitment to use my biochemistry to explain how the brain worked, I was soon made aware that this was not the sort of question that a respectable researcher should ask. Relative to its weight, the brain uses a disproportionate amount of oxygen and glucose, derived from the circulating bloodstream, and the work of the lab had for some time been focused on the problem of how and why this should be so. Much of it is initially employed to synthesise one of biochemistry's key molecules, adenosine triphosphate (universally abbreviated by biochemists to ATP). By using a tracer method involving radioactively labelled phosphorus, McIlwain and his group had shown that the ATP in its turn was used to synthesise a special class of phosphorus-containing proteins, phosphoproteins, present in very large amounts in the brain. It wasn't clear what these phosphoproteins were doing, but when the electrical activity of the brain increased, the amount of radioactive phosphorus appearing in the phosphoproteins also increased. My research task was to try to identify and purify the brain enzymes involved in the addition and release of the phosphorus on the protein. (Actually it turns out, though no one knew it then, that there are many hundreds of such phosphoprotein molecules, each with its own pattern of phosphorylation and dephosphorylation, and playing crucial roles in the regulation and coordination of cell metabolism.)

Even though it was hard for me to see how studying these enzymes could 'solve the brain', the work was the sort for which my biochemical training had prepared me well, and as I began I tried to make sense of where it fitted in the scheme of things, and to intepret the peculiar dynamics of the department into which my brain-obsessions had catapulted me. What I only slowly realised was that, all appearances to the contrary, the distinctly anti-charismatic McIlwain was a key figure in the emergence of what became in the 1950s and early 1960s a new subject area – neuro-chemistry, the biochemistry of the brain.

Naming is an extremely important act in science. In the early 1900s, before the discipline of biochemistry existed, scattered groups of researchers interested in using the methods of chemistry to under-stand physiological processes were working in many different univer-sity departments as biological chemists or chemical physiologists. Coming together to discuss their common interests at meetings of the existing Physiological and Chemical Societies they took the first steps towards independence. They began to organise their own specialist meetings, gave themselves a new name – biochemists – founded a new society and a new outlet to report their research – the Biochemical Society and its *Biochemical Journal* (*BJ*), today two of the most venerable parts of the scientific establishment. (In the United States, the new researchers were unable to break so completely with their past; their journal, founded at about the same time as the *BJ* in Britain, is even today, many hundreds of volumes later, still called *The Journal of Biological Chemistry*.)

Once a society and its journal exist, a research group has achieved a sort of corporate identity; it develops its own research traditions, standards of proof, definitions of what is and what is not an impor-tant question, and validates them by publishing papers proclaiming them. Even more important, university departments become estab-lished in the name of the new subject, and begin to produce their own graduates with the possibility of jobs and career structures ahead of them. Biochemistry in this sense had become a properly recognised discipline by the end of the Second World War.

How about those who wished to study not just any old biochemistry, but biochemistry in relation to a particular problem (such as how cells develop and divide), or a particular tissue, as, in our case, the brain? In the 1950s biochemists working on the brain could be found in their ones and twos in many different biochemistry departments across the country – even in Cambridge. Many others used the brain as a convenient source of tissue, like liver or muscle, and indeed some of the earliest biochemical research was based on work with pigeon brain. However, such research was aimed not so much at understanding what was special about the brain as at creating a universal biochemistry, relevant to all animal tissues. The fact that biochemical processes are similar in so many ways not only in widely different tissues from the same organism, but also in very different organisms (for instance ATP is found in flies and mushrooms, humans and oaktrees) makes this goal of universality not unreasonable; as researchers we are always simultaneously trying to find out what is general and what is unique about the system we are analysing.

Nonetheless, just like the biological and physiological chemists before them, the new generation of brain biochemists found that they wished to speak not only to liver and kidney and plant biochemists but also in very particular ways to other brain biochemists too. They also wanted to relate to other, well-established disciplines working on the brain, to neurophysiologists, to neuroanatomists; even, perhaps, to psychologists. Here they found themselves in an unusual dilemma. To their surprise not only did it turn out that most biochemists were reluctant to recognise the special claims of the brain over and above any other tissue, but researchers within the established brain sciences were also less than enthusiastic about them. Biochemists, with their techniques of mashing and grinding tissues and extracting purified chemicals from them, seemed in the nature of things to be destroying everything that was special, unique about the brain, its elaborate and interlinked meshwork of nerve cells with their exquisite shapes and subtle electrical properties. A friend of mine who spent his time trying to build mathematical models of how nerve cells interact once visited me in the lab as I was dissecting out a piece of brain and beginning its analysis by mincing it in a homogeniser – the standard lab version of a food processor. He watched for a moment or two and then shook his head sadly. 'All that order destroyed' he sighed, turning away, never to return.

So the brain biochemists found themselves uneasily suspended between, so to speak, brain and biochemistry. With McIlwain playing a leading part, they took the first necessary steps. They invented a new name for their discipline, neurochemistry, founded a journal, *The Journal of Neurochemistry* and even a society, the International Society for Neurochemistry, which still meets every two years. But they couldn't break away from the power of their parent discipline (in Britain there has never been a national society for neurochemistry, only a subgroup within the Biochemical Society, for instance). Universities were reluctant to set up new departments of neurochemistry, so they could emerge only in specialist institutions like the Maudsley. There, they had to convince the existing brain disciplines of their relevance – a problem that was never faced by certain other new subjects, such as pharmacology. Because the drugs the pharmacologists played with influence behaviour, physiologists and psychiatrists seemed to have much less difficulty in accepting that they must be important.

McIlwain's approach had been that of a classical biochemist. Many years previously, biochemists had shown that if you took a piece of animal tissue, say from the liver, cut very thin slices from it with a razor blade and immersed them in a blood-warm bath containing a proper mix of salts and glucose, the slices would go on behaving biochemically much as if they were still in the living body from which they had been removed. That is, they would continue using oxygen and burning glucose to give off carbon dioxide, synthesise proteins and perform many other complex functions for at least several hours. Thus living cellular processes could be studied in tissue slices with the sort of precision and control that is quite unavailable when working with the intact organism. McIlwain treated brain slices like this, and then went one step further. The unique properties of the brain, he argued, were associated with its constant electrical activity. Much of this spontaneous electrical activity was lost when the tissue slices were prepared – but what would happen to the biochemistry of the slice if one were to simulate the electrical activity of the brain by passing brief pulses of electricity through it?

He ran such experiments during the middle and late 1950s and found that, unlike other body tissues, when he put electrical pulses across the brain slices there were dramatic increases in the amount of oxygen, glucose and ATP they used, especially in the synthesis of the phosphorylated proteins, the mechanism of which I was set to study. For McIlwain this was evidence that the tissue slice was physiologically alive; because the slice could mimic key brain processes, studying it could cast light on the functional biochemistry of the brain. The physiologists were sceptical. Feeling their discipline being invaded by the dread biochemists whose language and preoccupations were so alien to them, they cast doubt on McIlwain's observations, arguing that evidence based on so bizarre a preparation as an electrically stimulated slice of tissue was worthless. For them such a dissected chunk of tissue was moribund, its responses no more than dying pathological spasms.

By the time I joined his department in 1959 this battle had settled down into an uneasy truce based on mutual rejection. McIlwain's book *Biochemistry and the Central Nervous System*[2] (note the cautious wording of the title; later researchers would not have hesitated to substitute *of* for *and*) became the standard text for aspirant neurochemists, but confined itself to cataloguing the biochemical features of the brain without dipping as much as a speculative toe into questions about what it all might mean, how knowledge of these unique and interesting chemicals and their interconversions might translate into function. McIlwain's phosphoproteins, and his slices, dropped out of the research limelight towards the end of the 1960s and during the 1970s. But after more than a decade of neglect the wheel came full circle, for it turns out that certain phosphoproteins are centrally involved in memory mechanisms, and my lab today is once again using the modern variants of the methods I learned with McIlwain some forty years ago.

I am not sure if this is such an odd coincidence either. Fashions in research have a tendency to repeat like fashions in clothes, though with rather less self-consciousness. By the 1970s McIlwain's battle to establish neurochemistry and the tissue slice against the entrenched physiologists seemed past history. A newly confident generation had bypassed the old battles of physiologists and biochemists and given their science a much more comprehensive name: *neuroscience*. In the United States, in Europe, even in the old Soviet Union, neuroscience at once captured the imagination; it transcended old disciplinary boundaries and looked to the future, capturing an ever-increasing share of national research budgets and media attention. As a discipline neurochemistry never had a chance; by contrast, as part of the synthesis of the brain sciences which the term neuroscience implies, its place ought to be ensured. The only problem is that institutional forms are much more rigid than the sciences they encompass. The American Society for Neuroscience must today be one of the largest scientific societies in the world; its annual meetings attract a prodigious 30,000 active participants. Its much junior partner, in size if not in quality, the Federation of European Neuroscience Societies, captures some 6,000. But the old neurochemists continue with their independent societies, journals and meetings, stuck in the 1960s and unable quite to submerge their hard-won identity for the greater good.

In retrospect I can see my own moves, from chemistry to biochemistry, from bio- to neurochemistry, as a forerunner to the arrival of the new generations of neurobiologists and neuroscientists. A few years after my PhD, in the mid-1960s, I joined a small group of like-minded physiologists, anatomists and psychologists to establish the first neuroscience society in Britain – perhaps in the world: the Brain Research Association. We were committed to the view that the brain was bigger than any of its constituent disciplines, and that we needed to develop a common language and understanding if we were to approach it. The BRA was unlike any previous scientific society I had known. We met informally, in a room above a London pub, to discuss research. There were just three rules: speakers had to be able to make themselves understood to everyone in the room, and not just talk a specialist biochemese or physiologese or whatever; beer was freely available during the entire proceedings; and no professors were allowed, to avoid anyone feeling unduly deferential or the need to guard their tongue. In the years that followed the BRA became a national society with branches in many cities, and those of us who founded it grew older and more respectable, professors in our own turn. So rule three got dropped, but much of the informality remained for decades, until respectability finally set in when the society developed an annual meeting and changed its name to the British Neuroscience Association.

Once it had a name and an image, neuroscience began to attract other researchers, notably molecular biologists who believed as I had done a decade and more previously that the future lay in the brain. Molecular biology is a research area in which anything published more than six months ago is out of date; if it is a year old and you bother to read it, it must be a 'classic' paper, and if it is older than that it is of purely archival interest. Ignorant of the past, this new group – molecular neurobiologists, no less – began to establish their own standards of what was interesting and relevant. The old biochemical methods and McIlwainian slices were rejected in favour of new techniques: antibodies, cloning, DNA and RNA technology became the names of the game in a research world in which the fact that the brain had a structure and was not just a bag of chemicals was temporarily forgotten.* When, in the 1980s,

---

*Relationships between neurochemists and molecular biologists – even molecular neurobiologists – remain somewhat strained. For molecular biologists neurochemistry is fustily old-fashioned; for neurochemists . . . Well, at the annual dinner at a recent international meeting they were gleefully told the story of the terrorist group which captures three scientists, a virologist, a neurochemist and a molecular biologist, intending them as hostages. Something goes wrong with the plans and the hostages are to be shot. As an act of clemency each in turn may plead for their lives on the basis of the good their science can do the world; the best case will be saved. The virologist offers the identification of HIV, but without a cure for AIDS, and is shot on the spot. The molecular biologist comes next and is just about to speak when the neurochemist exclaims: 'Shoot me, shoot me; I can't bear hearing yet again just how molecular biology is the key to the world's future!'

molecular biologists themselves began to move beyond chemicals towards physiology, they reinvented the slice preparation, apparently virtually unaware of the way it had been developed two decades earlier. McIlwain and his pioneering research had fallen victim to a collective act of scientific amnesia, but another research wheel of fashion had come full circle.

I spent two years at the Maudsley purifying one of the enzymes of phosphoprotein metabolism, writing my first research papers and the PhD thesis. Unfortunately I didn't learn much about brains and their function in the process because no one bothered to teach me and I didn't know where I should go to learn. I read papers which gave me some ideas about how I should set about purifying my enzyme, picked up some more biochemical methods, and just about learned to distinguish the top of a brain from its bottom. This last was mainly because, having found out something about the properties of my enzyme, which rejoiced in the grandiloquent name phosphoprotein phosphatase, I thought I might try to find out where it was located in the brain – was there more, for instance, in the grey matter, the cerebral cortex, packed with nerve cells, than in the white matter, where the nerve fibres run? (Once again, many years later simplicity has given way to complexity – it turns out that there are many many such phosphatases in the brain and I was probably studying the average properties of the entire group.) To discover that, I needed to dissect out relatively large chunks of brain – more than I could get from the laboratory rats and guineapigs I had been using. So I took to frequenting York Road slaughterhouse early in the morning on the days they were killing kosher cows. Not for any residual hankerings after my ethnic origins but because in non-kosher killing the animal is shot with a bolt through the head, which tends to make a mess of the brain I wanted intact. Kosher killing involves cutting the throat and bleeding, and a few bob to the butchers would ensure they then hacked the brains out and passed them over to me (I don't know what the rabbis who super-vised the exercise thought about it). I packed the brains in a thermos of ice, rushed them to the lab and prevailed on one of my colleagues to show me where the different bits were, bits I had only known in the past by obscure dog-latin anatomical labels but now, I saw, corresponded to real masses of cells. What was left over I took home for supper, thus eating as well as living research for a while. (Oh, the innocence of those pre-BSE, pre-cholesterol days!) The phosphoproteins studied in this way produced my first series of published research papers.

In the English system a PhD is a sort of apprenticeship. Unlike the United States, where there is a lot of formal coursework, here you are assigned a supervisor and a problem and told to get on with it. You learn your trade somewhat on Wackford Squeers' Dotheboys Hall principle: 'Boy – spell window – WINDER – go and clean it.' Learning by doing, by watching what your elders do in the lab and by taking advice from anyone you can find to talk to. That's why labs need a lot of post-docs for the graduate students to learn from while their ostensible supervisors are lecturing, committeeing, or away at important conferences. Just as natural science is collective, industrialised research, unlike the individualised craft production of the social scientist, so the graduate science student benefits from the shared labour of an open lab, squabbling over who dirtied the glassware, who has booked the next session on the centrifuge or used the last of the pipette tips in an atmosphere in which hierarchy nearly but not quite becomes submerged. Within months of entering such a lab as a somewhat deferential student one assimilates to the casual, critical atmosphere, learning to disbelieve 90 per cent of what you read, especially if written by someone regarded as working in a rival lab; graduate students soon run out of scientific heroes.

The danger is to become so obsessed with one's own minuscule piece of research that the thesis you are supposed to write at the end of it achieves, in your own eyes at least, the status of a magnum opus, *the* definitive statement on a subject, not complete at shorter than 450 pages and ten thousand references with a final reflection on the meaning of life and the grand theory which your tiny segment of research has helped establish. Such theses often never get completed; they weigh heavier and heavier on the student, postponed indefinitely to prepare and polish. My own, it is true, was much shorter than this; I wanted to finish it and to move on from the

Maudsley, a place which made me feel uneasy. Compared with the swirling debates over the nature of life and the meaning of research which had characterised my undergaduate days, there was a dour down-to-earthness and lack of adventure that depressed me about it. (I had of course been corrupted by the class arrogance of Cambridge, assimilated its intellectual pretensions to an extent that I have never quite been able to lose; a process that leaves one unfit for the less-than-golden world outside its protective sphere.)

Even so, my own thesis had its moderately grand theory, its attempt to prove, on inadequate data, that the meaning of existence – or at least of the brain – is embedded in phosphoproteins. As I've said, twenty years later the importance of the phosphoproteins became clear – but by then no one, not even I, would know or care what I had written in that hard-won concluding chapter.*

My thesis examiner, Hans Krebs, with the certainty that befits a Nobel-prizewinning Oxford professor and an experimentalist trained in the disciplined German manner, dismissed my last chapter quickly enough. Too much philosophy, he said – and there were three spelling errors that I would please correct. Then it was through, and I had been doctored. Krebs was right; the real value of the thesis is quite different from all this grand theorising; it is simply a certificate of completion of apprenticeship; it says, this person can do independent research and is now fit to be launched on the world, to sink or swim.

The launch meant leaving London again and initiation into the peripatetic life of the post-doc. Krebs had found me a three-year fellowship in his biochemistry department, and an Oxford College, New College, gave me another, and I found myself free to choose research direction. Krebs, whose name is known to every biochemistry student for his discovery of two central metabolic routes, that by which glucose is oxidised to carbon dioxide in the cell (known as the Krebs cycle) and that by which urea is produced (for which

---

*Tell a student that the only people who are ever going to read this great document are the supervisor, two examiners and one or two close friends, after which it will be deposited in the university archives and never seen again (except by the next generation of students looking for guidance as to how to do it – or not, as the case may be) and you will be regarded with rank disbelief.

he was awarded his Nobel), was a man who lived for his own experiments and, within the hierarchical framework that his own training had bequeathed him, gave his juniors enough rope to hang themselves.

This freedom was probably excessive. Following up some of the lines of the PhD, I began to think a little more about the function of the phosphoproteins I had been working on – but still in strictly biochemical terms. What did they do in the economy of the cell, I wondered. My hunches led me to some not too clever experiments (though I was pleased with them at the time); in retrospect I suspect that I was asking the right questions but lacked the technology to answer them. Research is above all the art of asking questions that are both interesting and answerable. The most exciting questions are often unapproachable by experiment because the effects you are trying to observe are just too small, below the level of sensitivity of the equipment you are using, or irreproducible because you can't control all the variables. On the other hand, asking trivial questions can generate lots of equally trivial answers. The task is to steer a course between the trivial and the unachievable – the immunologist Peter Medawar once described it as the art of the soluble. For a good portion of my research lifetime many of my colleague biochemists have been convinced that the study of memory lies outside Medawar's definition of soluble problems; only in the last few years have they become more optimistic, and much of what follows in this book will be concerned with the reasons for both the earlier doubts and the present enthusiasms.

The class- and gender-bound Oxford college system and eventual disagreement with Hans Krebs over the politics of science, the right directions for research, and the proper way to combine laboratory work with personal life led me back to London after a brief two years, to a new department being established at Imperial College by yet another exiled Nobelist. Some years younger than Krebs, Ernst Chain's Nobel prize had come for his part in the Oxford-based team which had discovered how to produce penicillin in bulk in the 1940s. Born in Russia and trained as a biochemical microbiologist in Berlin, he too had fled to England in the 1930s to escape the Nazis, but had left

Oxford for Rome after the war when the Italian government had promised him carte blanche to set up a major new institute there. In the early 1960s he was tempted back to England by the vast new department being constructed in London's South Kensington, Imperial College site. Unlike Krebs, Chain believed in directly controlling the work of everyone in his department. Ideas that did not originate with him were discouraged, and as he had long ago abandoned active research for management and consultancies, his own research thinking was frequently both inflexible and unimaginative. All research publications coming from the department had Chain listed as co-author, however minor his contribution to the work had been, as opposed to Krebs, who had been meticulous in keeping his name only on those papers in which he himself had a direct hand.*

Although a microbiologist by background, Chain, partly under the influence of his biochemist wife, Anne Beloff, had developed an interest in brain metabolism, and set me to work on a project concerned with energy utilisation in the brain. But by now I was beginning to feel more certain about the directions I should take. If I didn't find a way of using my biochemistry to answer questions about how the brain worked, I might as well research on toes or livers. At this point one man's writing helped shape my future. I found a major review by a Swede, Holger Hydén, describing his research, work of such exquisite precision that it left me breathless.

The brain is composed of a vast number of nerve cells (neurons),

---

*This authoritarian style was reinforced by the fact that, in coming back to England, he had been given financial support to establish a research unit by the Medical Research Council, the major government funding agency for medicine and biology in Britain. Research staff appointed to such a unit, unlike university teachers, were employed (and still are in such units today, although there are fewer of them now as a result of changing policy and diminishing funds) mainly on short-term contracts to work on projects set by the director. However senior, they thus have an almost serf-like relationship with the director. If he (the gendered pronoun is I believe precise in this case) moved institutions, for instance, as happens not infrequently, his staff would have to move with him or risk losing their jobs. As Chain had not yet left Italy when he began to appoint staff for the new institute, for instance, we were dispatched to Rome to work there until the lab was completed. My year there didn't result in much useful in the way of published research; however, it did give me the time to write my first book, a little biochemistry text called *The Chemistry of Life*, which remained, in various avatars, in circulation until the new millennium.

up to a hundred billion in all in humans (figures 3.1, 3.2, 3.3). But even this huge number is dwarfed when one learns that each neuron is itself embedded in a mass of much smaller cells, called glia, whose function was (and still is) much less clear than that of the neurons, but which have a supportive, nurturant and protective role. There may be as many as ten glial cells for every single neuron. So studying the biochemistry of a piece of brain means studying a mix of neurons and glia. If the real functional business of the brain is going on in the neurons, one should really try to study their properties in isolation from the glia. But how? For years, Swedish labs had a tradition of developing micromethods for analysing tiny quantities of material. Hydén took this tradition to its extreme. He chose a particular region of the brain where the nerve cells are relatively large – perhaps 30 millionths of a metre (30 microns – a micron, one millionth of a metre, is usually abbreviated as μ) in diameter. He put the piece of brain containing the cells under a dissecting microscope, delicately staining the cells with a blue dye to make them visible, and, using a fine wire with a sharpened edge as a knife, cut out each tiny cell individually from the surrounding glial mass. In this way he collected a few dozen neurons, and matched them to similar

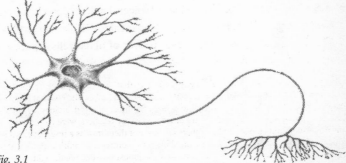

Fig. 3.1
A neuron
Note the cell body with its prominent nucleus. Branching from the cell body are the dendrites, studded with small spines. Running away from the cell body is the axon, which branches into an array of processes each ending in synaptic terminals at which contact is made with the dendrites or cell bodies of other neurons. Neurons come in many shapes and sizes; the key features of all are, however, the cell body, dendrites, axon and synaptic terminals.

quantities of glia. He was even able to puncture the cells like small balloons, pick them up by their outer skins (membranes), and shake the contents out to leave himself with isolated membranes to study. During the 1950s he had painstakingly used the famed Swedish micromethods to measure the oxygen utilisation and the DNA, RNA and protein composition of such isolated cells, comparing the biochemistry of the neurons with that of their surrounding glia. The differences between the two would surely tell one something of the ways in which the nerve cells were specialised to carry out their unique functions.

As if this were not enough, Hydén had gone further. He had begun to use his new techniques to explore functional changes in the biochemistry of the nerve cells. Could it be, he asked, that the cells changed their properties as a result of the experience of the rats or rabbits from which they were derived? The large cells he was studying came from a region deep in the brain concerned with control of balance. So he rotated his rabbits on a sort of merry-go-round, and taught his rats to climb precariously up a wire to reach food. Sure enough, he claimed, these experiences altered the composition of the RNA and protein of the neurons, but not the glia, that he was studying. He developed a theory that memory was stored in the brain in the form of such changed molecular structures.

These days, like many other pioneers, Hydén is largely ignored and his death in the late 1990s was scarcely noted. His micromethods were unique and other labs were unable or unwilling to adopt them. From the 1970s on the conventional wisdom was that his data were unreliable or statistically unsound, and newer methods and models became popular. But the 1960s were his apotheosis. He was a regular performer at conferences and seminars, a man of presence, with a deep, slow Swedish-English voice, his talks illustrated with astonishing pictures of micro-dissected cells. Do I now believe his claims? During the 1960s and early 1970s I visited his lab on a number of occasions, and grew to share the general scepticism about the specificity of these results. However, I have watched him cut out individual cells with an elegant precision that leaves no room for doubt that this technique at least worked well.[3]

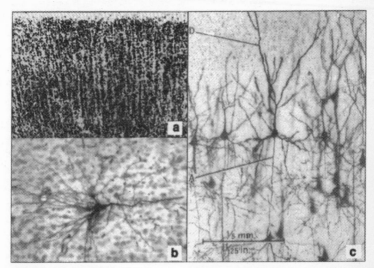

Fig. 3.2
Neurons in the brain
(a)–(c) are light microscope pictures at increasing magnifications made using a staining technique (Golgi) which picks out only a proportion of the neurons but stains each in its entirety, showing cell bodies, their dendrites and the spines that stud their surface. To get a sense of scale, the black staining cell bodies of chick-brain neurons are on average about 5-10μ in diameter – 1μ is a millionth of a metre.

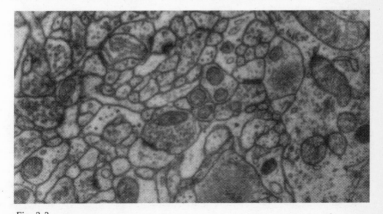

Fig. 3.3
Electron micrograph of brain tissue
Making a micrograph like this freezes for ever the dynamic structures of the brain into a seemingly inextricably tangled mass of neurons, glia, dendrites, axons and synapses.

Towards the end of the seventeenth century a Dutch draper, Antoni van Leeuwenhoek, invented a new sort of microscope. Looking down it he described for the first time in human history a hitherto invisible world, teeming with micro-organisms – animalcules, he called them – present in their thousands in every drop of pondwater. He drew them with surprising precision and accuracy. Look down one of van Leeuwenhoek's microscopes today and you will be lucky if you can see anything at all, so small and blurred are the lenses by modern standards. Many of his contemporaries were sceptical – they couldn't see what he claimed to see – and maybe they were right. But van Leeuwenhoek was also right; the animalcules were real enough, and their discovery has transformed our understanding of the biological world. Maybe Hydén was like a modern van Leeuwenhoek? But then one should not forget that van Leeuwenhoek also looked at his own semen, observed sperm, and described each individual spermatozoon as a perfectly formed, minuscule mannikin, thus reinforcing a long-held preformationist superstition about human reproduction that took at least another century to outgrow.

Whatever the final verdict on his experiments, Hydén's claims galvanised neurochemists. Biochemistry could indeed be applied to the study of brain function, even memory. I had found my research direction.

But how to advance? I had no access to Hydén's micromethods, and Chain was, as ever, sceptical. If the key methodological step was to be the isolation of neurons and glia, perhaps I could find another route to the same end. By the end of the 1950s, biochemists had available to them a number of general techniques for studying cells. One in particular, centrifugation, is used to separate cells into their various components. The principle of the laboratory centrifuge is exactly that of the drying programme of the domestic washing machine; spinning the clothes very fast forces the water out of them to the sides of the drum thence to drain away. In a biological centrifuge, a suspension of particles in a test-tube is spun very rapidly – the machines are capable of anything up to 70,000 rpm. As they spin, the particles are forced down the tube by the gravitational forces that the rotation generates (up to 500,000 x gravity) and the heaviest particles travel down the tube fastest. Suppose, then, that one takes a piece of brain and grinds it up in a homogeniser. The cells are broken, and their internal components spill out into suspension. These internal components include a variety of subcellular particles – in particular, the cell nucleus (which contain its DNA) and the mitochondria (the structures which contain the enzymes of the Krebs cycle and in which ATP is synthesised). By selecting the right combinations of speeds and times of centrifugation, the various subcellular components can be separated and collected. A variation of the technique spins the suspension of particles in a sucrose solution which is very concentrated at the bottom of the tube and more dilute at the top. The more concentrated the solution the greater its density, so the solution in the tube forms a density gradient from top to bottom. As the subcellular particles are centrifuged in the gradient they are forced down the tube under gravity until they reach a zone in which the sucrose is the same density as they are. Here

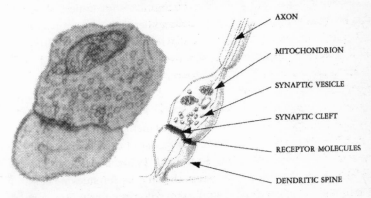

AXON

MITOCHONDRION

SYNAPTIC VESICLE

SYNAPTIC CLEFT

RECEPTOR MOLECULES

DENDRITIC SPINE

Fig. 3.4
A synapse
At left is an electron-microscope picture of a synapse with the surrounding tissue blanked out. At right, the artist's impression of the structure. Nerve impulses arriving down the axon result in the release of the neurotransmitter molecules packed into the small vesicles. The neurotransmitter diffuses across the synaptic cleft to the dendritic spine on the post-synaptic side, where the receptor molecules await it. The interaction between receptor and neurotransmitter provides the signal to activate or inhibit the post-synaptic cell. Enzymes then destroy the neurotransmitter. Energy for the biochemical processes involved in the release of neurotransmitter is provided by the oxidation of glucose in the mitochondria. A single synapse like this is about 0.2μ in diameter.

they come to a halt, floating on the dense sucrose, below which they cannot penetrate.

When, in the early 1960s, these techniques were applied to brain tissue, it was discovered that as well as the standard subcellular fractions of nuclei, mitochondria and so forth, there was a wholly different type of particle which could be collected. Nerve cells make contact with each other at special junction points, the sites at which information is transferred from cell to cell. These junctions are called synapses (fig 3.4). Homogenisation of the brain results in the synapses becoming pinched off from the rest of the cell; the membranes surrounding them reseal as if they were made of clingfilm, and the entire, artificially created synaptic particle, or synaptosome, can be collected by centrifugation and studied in isolation.

This advance in methodology, which has opened the way for vastly increased understanding of the biochemistry of the synapses, was the result of more or less simultaneous discovery by two neurochemists, Victor Whittaker, in Cambridge, England, and Eduardo de Robertis, in Buenos Aires, Argentina. It immediately suggested a way of bypassing Hydén's micromethods. Provided I could find a way of disrupting the brain tissue sufficiently gently that instead of breaking the cells completely I merely shook them apart from one another, I could devise a centrifugation regime which, rather than separating subcellular particles, would give me fractions enriched in neurons and glia.

A few months of trial and error and I had a working method, reminiscent of my grandmother and mother's way, which I had watched so often as a child, of making curd cheese from curdled milk by filtering it through muslin. I wrapped chopped bits of brain in muslin bags, immersed them in a solution containing a type of sugar and stroked the outside of the bag gently with a glass rod. The gentle pressure released the neurons and glia into suspension, and by choosing a combination of gradients and centrifugation speeds, I ended up with two fractions, each enriched in one of the two types of cells. I could make sufficient quantities of the cells for micromethods of analysis to be unnecessary and to enable me to use the standard biochemical procedures to study the cells' metabolism; how they used oxygen, their differing enzyme contents and protein composition. The published papers[4] were successful, and Chain's earlier scepticism relaxed to the extent that I was instructed to put on a display of the method for the official royal opening of his grand new department. For the next decade we went on working with the technique, until newer methods made it redundant.

Now that at last my research was addressed to asking functional questions – about just what was special about the biochemistry of nerve cells by comparison with others such as glia – I felt encouraged to press on. I had become fascinated by Hydén's observations about biochemical changes in memory. So had others. There was a sudden flurry of novel research; people began to describe startling results about the role of RNA and protein in memory. I wrote some popular science articles about the work, but what I really wanted to do was to get in on the act myself, and I couldn't see how to do it. For sure, Chain would never accept that memory was a suitable topic for a biochemist to study. And anyhow, I was no psychologist and had no training in animal behaviour. I approached the issue crabwise, and almost clandestinely.

Memory, after all, I argued, was a special case of the plasticity of the nervous system, its capacity to respond to environmental stimuli, to experience. Could one develop an alternative experimental system in which to study such plasticity? I spent several months thinking about what might serve as an appropriate system before taking the plunge. Baby rats, like many other young vertebrates, are born relatively underdeveloped; they are naked, rather immobile and blind. They normally open their eyes only at about fourteen days after birth. During these two weeks their brains grow rapidly, and by the time their eyes open almost all the neurons and glia are present and a myriad of synaptic connections have been formed between the neurons. Species whose young are born so premature by comparison with humans are called altricial. By contrast, baby guineapigs are born fully furred, eyes open, and ready to scurry after their mother as she seeks food; such species – like chicks and ducks – are called precocial. This distinction became important to me when, a few years later, I started to work with baby chicks.

But for now my focus of concern was this early period of neuron and synapse development in the young rats. The first exposure to light for such rats, I thought, must be a major event, rich in experience, and resulting in profound cellular changes. To bring the exposure under control, I kept the pregnant mothers in pens in complete darkness; they gave birth and reared their pups in the dark, until, after seven weeks, I brought the now young adult rats out into the light for the first time in their lives. So far as I could tell, this dark-rearing was without effect on the general well-being of the animals; they put on weight and developed just as if they were in a normal light–dark cycle. Of course it has to be remembered that rats are normally nocturnal creatures; they tend to sleep during the day, only moving around to hunt for food and social interaction at night anyway. So vision is not a very important sense for them – much less than the sense of smell, for instance.

The first experience of light, however, turned out to have profound biochemical effects. For instance it dramatically increased the synthesis of proteins in the neurons of the visual regions of the rat brain, though not other areas such as the motor regions. I struck up a collaboration with a neuroanatomist, Brian Cragg, based elsewhere in London, at University College. He undertook to look at the structure of the cells of the visual region of the brain – the visual cortex – in the dark-reared and light-exposed rats by means of an electron microscope. He counted the synapses in the visual cortex and found that, just at the time I found an increase in protein synthesis, he could detect a small but significant increase in synapse numbers. The new experience was clearly having both biochemical and structural effects on the visual regions of the brain; biochemistry, anatomy and function seemed to be coming together. We drafted a paper reporting these results for the leading scientific journal, *Nature*, and I took it to Chain for approval to publish. He objected strongly. He didn't approve of anatomy in general, or Cragg's data in particular, and he hadn't been consulted before we ran the experiments. Nothing we could say would convince him, so Brian and I separated the results and published two papers side by side, one reporting the anatomical changes under his name, one the biochemistry under mine.[5] He had the last laugh though, as in subsequent years it was the anatomical and not the biochemical paper which was the more widely cited.

Then, not long after the *Nature* papers had appeared in 1967, came a Royal Society discussion meeting in London at which I mentioned Brian's and my results. As the meeting ended I was approached by the ethologist Patrick Bateson, who was based in the Animal Behaviour research station at Madingley in Cambridge. (Having served as the Director of Madingley, Pat later became Provost of King's College, and Biological Secretary of the Royal Society.) His research interest was imprinting in chicks. All I knew about imprinting at the time was that Konrad Lorenz, in his book *King Solomon's Ring*, had described how, if the first thing that young geese saw when they chipped their way out of their eggs was his rubber boots, they would follow him faithfully thereafter as if he were their mother. (I was unaware too of the less savoury side of Lorenz's character, his involvement with the Austrian and German Nazi parties and his enthusiasm for genetic determinism and racial biology.)[6] But whatever else, it was clear that imprinting was a form of learning, and the chick, potentially, an ideal animal in which to study its biochemistry.

Pat explained that he and an anatomist colleague, Gabriel Horn (later Master of Sidney Sussex at Cambridge, and knighted for his work on government committees), had been intrigued by the idea of exploring the cellular and biochemical mechanisms involved in imprinting in the chick but that they couldn't find a biochemist in Cambridge to collaborate with. Was I interested? A few days later I went up to Cambridge, to see how he imprinted his chicks and to have the first of what was to become over the next seven years a regular series of sandwich lunches at King's College with Pat and Gabriel as we designed the experiments and interpreted the increasingly intriguing results of the chick experiments.

The general approach was for Pat to train the chicks, dissect and code the brain samples, and send them down to me to do the biochemical analyses, blind as to which samples came from which condition, at Imperial College. From the first the experiments came good. They were ingeniously designed, and the collaboration of Pat, a behaviourist, Gabriel, an anatomist and physiologist, and myself, a biochemist, made tremendous sense. We became convinced that we symbolised the future, in which an integrated neuroscience would emerge as a result of just such combinations of different brain and behavioural sciences. And I was sure I could find a way of squaring or bypassing Chain's predictable opposition.

As it happened, I needn't have worried. Within a few months of the first of the joint papers with Pat and Gabriel in *Nature*,[7] I was appointed as Professor of Biology at the newly invented Open University. I was just over thirty years old, and about to become, irrevocably, a memory researcher.

Chapter 4

Metaphors of memory

We take our technological world, and our memories within it, very much for granted. We leave messages on voicemails or email absent friends, we consult our diaries for free dates and write notes to colleagues arranging dinner, a theatre or a meeting; we check the fridge and write ourselves a shopping list. Each of these is an act of individual memory – but an act in which we have manipulated technologies external to ourselves in order to aid, or supplement, or replace, our internal brain memory system. It was not always thus; individual our memories may be, but they are structured, their very brain mechanisms affected, by the collective, social nature of the way in which we as humans live. For each of us as individuals, and for all of us as a society, technologies, some as old as the act of writing, some as modern as the electronic personal organiser, transform the way we conceive of and the way we use memory. To understand memory we need also to understand the nature and dynamics of this process of transformation.

The greater part of the history of humanity not only pre-dates modern technologies; it even pre-dates writing. For such early human societies, records, individual life histories, just as much as histories of family and tribe, were oral. What failed to survive in an individual's memory, or in the spoken transmitted culture, died for ever. People's memories, internal records of their own experience, must have been their most treasured – but also fragile – possessions. In such oral cultures, memories needed to be preserved, trained, constantly renewed. Special people, the elderly, the bards, became the keepers of the common culture, capable of retelling the epic tales which enshrined each society's origins. Then, each time a tale was told it was unique, the product of a particular interaction of the teller, his or her memories of past stories told, and the present audience. Walter Ong describes how in modern Zaïre a bard, asked to narrate all the stories of a local hero, Mwindo, was amazed; no one had ever performed them all in sequence before. Pressed to do so, he eventually narrated all the stories, partly in prose, partly in verse with occasional choral accompaniment. It took him twelve exhausting days whilst three scribes took down his words. But once written, Mwindo had become transformed. He no longer existed as a continued, remembered recreation of past stories. Instead he had become fixed in the linear memory form demanded by modern cultures.[1]

Although we still retain a concept of memory in this deep, collective sense, the new technologies change the nature of the memorial processes. A video or audiotape, a written record, do more than just reinforce memory; they freeze it, and in imposing a fixed, linear sequence upon it, they simultaneously preserve and prevent it from evolving and transforming itself with time, just as much as the rigid exoskeleton of an insect or crustacean at the same time defends and constrains its owner. For instance, when, in 1990, world Jewish leaders convened at Wannsee, the lakeside villa at which Heydrich and others had drawn up the plans for their 'final solution' to the 'Jewish problem' nearly fifty years earlier, the Nobel Peace Prize winner Elie Wiesel wrote that the intention was to demonstrate that 'memory is stronger than its enemies . . . most German men and women in the past refused to speak, refuse to remember'. An act of group memory on the one part confronts an act of social amnesia on the other. Yet it is an act of memory reinforced not merely by oral tradition but by the written texts, by audio and above all by the visual images of photography and the cinema, terrible images which become fixed in the minds and memories even of those who were distant from the events. The contrast with the old oral cultures could not be greater.

And of course similar strongly reinforced collective memories – and amnesias – underlie many present-day national and ethnic conflicts. As a youngster I was brought up never to forget 'next year in Jerusalem'. When, in 1982, after the Sharon-directed massacres of Palestinians in the camps of Sabra and Chatilla I visited the Lebanon, I met many young Palestinians, who had certainly never been there, who 'remembered Jaffa' – or even Jerusalem – and their expropriated familial homes in what was now Israel at least as strongly and with as much feeling as those who convened at Wannsee in their act of collective memory. Think of the ways in which the significance of Kosovo for Serbs and Albanians, or of the temple/mosque at Alyodha for Hindus/Muslims becomes reinforced by the collective sharing of technologically preserved images.

The psychoanalyst C. G. Jung based his theory of mind in part on the claim that such collective memories were racial and had become deeply inscribed in our biological as well as cultural inheritance. I of course mean nothing of the sort here; rather I am talking about mechanisms of retention and transmission – the sharing and collectivisation of memories. Following the collapse of the Soviet Union, the society set up to commemorate the victims of the Stalin era was known as Memorial. But equally, the political organisation of the extreme right, of Russian nationalism and anti-semitism, was called Pamyat – Memory. As will become apparent, in the understanding of both the social and the biological functions of memory, the elucidation of forgetting, of individual and social amnesia, provides as powerful a clue as does remembering.

Collective and individual memories, the changing technological forms which enrich and constrain our memories and provide the analogies by which we endeavour to explain them, these form the themes of the present chapter. And, as will become clear in chapter 5, the technological evolution by which the fluid memory of oral cultures becomes fixed and disciplined in a computer-driven industrial society, finds strange echoes also within individual human development.

The ancient arts of memory

The ancient philosophers were distinctly dubious about the merits of a written culture. Thus Plato describes Socrates as claiming that writing is inhuman, in that it pretends to establish outside the mind what in reality can only be in the mind. As Walter Ong points out, writing reifies, it turns mental processes into manufactured things. Writing destroys memory; those who use it, Plato has Socrates argue, will become forgetful, relying on an external source for what they lack in internal resources. Writing weakens the mind.

Instead, memories should be trained, as the Zaïrean bard's memory must have been. This training is the discipline known as mnemotechnics – a discipline which must have been invented separately at many times and in many cultures. Within Western culture, there is a clear history of this mnemotechnic tradition, running back to Greek times, though the written record of the method is not Greek but Roman, and first appears in *De Oratore*, a famous text on the art of rhetoric – that is, of argument and debate – by the Roman politician and writer Cicero. In it, Cicero attributes the discovery of the rules of memory to a poet, Simonides, who seems to have been active around 477 BCE.

The Simonides story appears and reappears throughout Roman, medieval and Renaissance texts. In its basic form it tells how, at a banquet given by a Thessalonian nobleman, Scopus, Simonides was commissioned to chant a lyric poem in honour of his host. When he performed it, however, he also included praise of the twin gods Castor and Pollux. Scopas told the poet he would only pay him half the sum agreed for the performance and that he should claim the rest from the gods. A little later Simonides received a message that two young men were waiting outside to see him. During his absence the roof of the banqueting hall fell in, crushing Scopas and his guests and so mangling the corpses that their relatives could not identify them for burial. The two young men were (of course!) the gods Castor and Pollux, and they had thus rewarded Simonides by saving his life, and Scopus apparently got his come-uppance for meanness. But – and this is the crucial bit of the story – by remembering the sequence of the places at which they had been sitting at the table, Simonides was able to identify the bodies at the banquet for the relatives. This experience, as Cicero tells the story, suggested to Simonides the principles of the art of memory of which he was said to be the inventor. The key to a good memory is thus the orderly arrangement of the objects to be remembered.

> He inferred that persons desiring to train this faculty must select places and form mental images of the things they wish to remember and store those images in the places, so that the order of the places will preserve the order of the things, and the images of the things will denote the things themselves, and we shall employ the places and images respectively as a wax writing-tablet and the letters written on it.[2]

Such rules are designed to relate a collection of items which do not have a particular relational logic, but are contingent, like the guests at the banquet, on some structure whose logic is apparent or at least can readily be remembered because of its striking features. Memories, in such mnemotechnic systems, could thus be stored by remembering some familiar environment, commonly a house with a series of rooms, or a public space with prominent monuments and buildings, and 'placing' the items to be remembered in an appropriate sequence within the environment. One could then recall them, for instance during the course of a speech or recitation, by mentally walking through the environment, visiting each location in turn. The ancient texts refer to this method of memorising as 'artificial' memory, by contrast with given or natural, untrained memory; the artifice of memory seems a project as dear to them as it is to present-day computer enthusiasts.

A further Latin text, of unknown authorship, known simply as *Ad Herennium*, essays a definition of memory, as 'firm retention/comprehension in the mind of the matter, words and arrangement of objects'. The text dwells on how to choose the images which above all give guidance as to the arrangement, which is seen as the key feature of an effective memory:

> We ought then to set up images of a kind that can adhere longest in the memory, and we shall do so if we establish likenesses as striking as possible . . . if we assign to them exceptional beauty or singular ugliness, if we dress some of them with crowns or purple cloaks, for example, so that the likeness may be more distinct to us, or we somehow disfigure them, as by introducing one stained with blood or soiled with mud or smeared with red paint . . . this too will ensure our remembering them more readily.[3]

These methods were no mere personal idiosyncrasy or device of a great orator like Cicero, who could apparently speak in the Senate for days on end without using notes. Similar descriptions crop up, as Frances Yates relates in her well-known book *The Art of* Memory,[4] in other classical texts. Certain Roman generals were supposed to have used them to recall the names of their soldiers; thus Publius Scipio was said to be able to recognise and name all of his entire army of 35,000 men. Such mnemotechnics were the forerunners of a tradition which ran through the medieval and Renaissance periods and is still alive today. During much of the Middle Ages they became debased into mere crude devices for remembering numbers and letters. Easily visualised picture sequences or inscribed wheels were supposed to enable those who learned them to recall the ordering of devotional exercises or catalogues of virtues and vices rather as today children's spelling books offer the pictogram sequences of 'A is for Apple; B is for Ball . . .'

But increasingly, and especially from the fourteenth century on, mnemotechnics became more daring. The locus for the memory-images came to be described as a theatre, a memory theatre in which symbolic statues, rather like those one might have expected to find in a Roman forum, were placed, and at the base of each statue could be stored the item to be memorised. During the early Renaissance

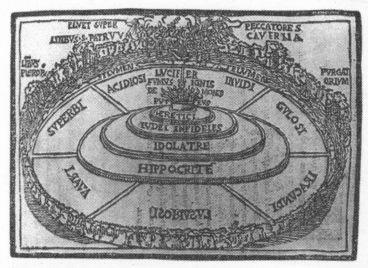

Fig. 4.1
(above) Hell as artificial memory
(below) Paradise as artificial memory
From Cosmas Rossellius Thesaurus Artificiosae Memoriae, *Venice, 1579.*

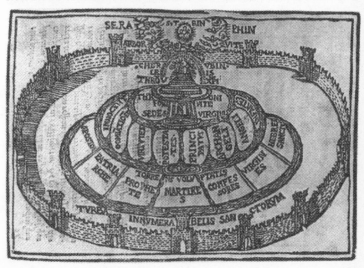

period these imaginary theatres became increasingly complex, with gangways, tiers of seats, and classical statues representing virtues, vices and other key figures. But where in the past exponents of the mnemotechnic art might envisage themselves as spectators at such a theatre, looking inward to the stage as an elaborate set full of memory cues, in the Renaissance memory theatres the mnemotechnician was supposed to look outward from the stage, the actor facing an audience whose location in their ordered ranks of seats provided the sequence clues.

The theatres even became agents of religious propaganda. In 1596 the Jesuit missionary Matteo Ricci offered a 'memory palace' to the Chinese he was bent on converting to his faith. He told them the size of the palace would depend on how much they wanted to remember: the most ambitious construction would consist of several hundred buildings of all shapes and sizes, the more the better, although more modest palaces, temple compounds, government offices, merchants' meeting lodges or even simple pavilions might suffice. Ricci found images likely to be familiar to his Chinese hosts to place in the imaginary rooms and pavilions of the equally imaginary palace to act as the memory loci for the storage of concepts and ideas, though I have to say that I find the relationship of the entire elaborate system to Christian theology somewhat elusive.[5]

In the hands of Galileo's contemporary but far more dangerous heretic, Giordano Bruno – who unlike Galileo would not recant and was burned by the Inquisition – the theatres also became a key feature of occult, hermetic philosophy. For Bruno they became a way of classifying and hence penetrating to the mysterious core of the universe. Memory gave power over nature. Memory theatres became the very models for heaven and hell (Dante's systematic descriptions of the circles of both in his *Divina Commedia* have been claimed to be derived from such mnemotechnic systems). A debased version of Bruno's vision and philosophy is still around today. Turn to the classified section of a Sunday newspaper and you will find ads along the lines of 'Loss of memory? A well-known publisher can teach you how to improve it' or 'It may be news to you but

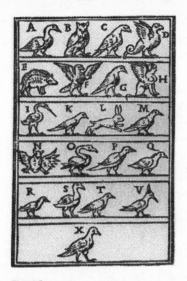

Fig. 4.2
Grammar as memory image
Visual alphabets used for the inscriptions on grammar. From Johannes Romberch Congestorium Artificiose Memorie, *edition of Venice, 1533.*

the Egyptians knew it long ago . . .' On closer inspection many such ads turn out to be placed by an obscure sect calling themselves Rosicrucians, whose ancestry may not be quite as antique as they claim but certainly stretches back to Bruno's day, and retains many Brunian elements. Take up such offers of training your memory and you will even now very likely be offered a version of the memory theatre.

By the Renaissance, the memory theatre was turned from a symbolic device, a piece of mental furniture, into an actual construct. In the sixteenth century, and to the disapproval of more rationalist philosophers, such as Erasmus, the Venetian Giulio Camillo actually built a wooden theatre crowded with statues which he offered to kings and potentates as a marvellous, almost magical, device for memorising. Frances Yates even goes on to speculate, daringly, that the lost Globe Theatre of Shakespeare was actually built to the design of a real memory theatre. 'Why', she asks,

Fig. 4.3
A memory theatre
From Robert Fludd's Ars Memoriae.

does such a theatre seem to connect so mysteriously with many aspects of the Renaissance? It is, I would suggest, because it represents a new Renaissance plan of the psyche, a change which has happened within memory, whence outward changes derived their impetus. Medieval man was allowed to use his low faculty of imagination to form corporeal similitudes to help his memory; it was a concession to his weakness. Renaissance Hermetic man believes that he has divine powers; he can form a magic memory through which he grasps the world ... The magic of celestial proportion flows from his world memory into the magical worlds of his oratory and poetry, into the perfect proportions of his art and architecture. Something has happened within the psyche, releasing new powers ... (pp. 173–4)

The technological metaphor

Real or imaginary, at this point we are already far from Cicero's original intent, and have long transcended Plato's concern that writing might be deleterious to the mind; a technological imperative to harness memory is beginning to emerge. But harnessing memory requires also some sort of effort to understand or explain it, and it is here that the peculiarly double relationship of technology to biology in general and the biology of mind in particular achieves special significance.

Explanation in science often proceeds by metaphor. We endeavour to understand how something we don't know works by comparing it to something we do know – or something we can at least imagine we know. Think of one of the most basic of divisions of the known world, the division between animate and inanimate. Within science, the former became the province of biology, the latter of physics. In the pretechnological era within Western societies, and in many other cultural traditions, explanations run transitively, in both directions between biology and physics. The irregularities of wind and rain, just as much as the regularities of rivers, the sea and the earth, the stars, sun and moon are explained animistically, as reflecting the minds or whims of local or global gods, themselves motivated by similar concerns to those that motivate humans. But, equally, animate, biological phenomena are given metaphorical explanations in physical – and increasingly in technological – language. Because biological systems are so complex, they are above all analogised to the most complex, the highest, forms of current technology. Each period, each culture, has such a form – one that David Bolter[6] has called its defining technology. Indeed, we periodise ages in humanity's prehistory by such defining technologies – stone age, bronze age, iron age.*

For early cultures one of the most subtle of technological metaphors was that of the potter, who with clay and wheel, glaze and fire, could create shape and pattern. No wonder that for such cultures – and it is a creation myth that turns up again and again in the origin stories of both New and Old Worlds – it is a deity with a potter's wheel who shapes humans and then breathes life into them. Other

---

*Not just in prehistory; writing about the growth of science as an institution, Hilary Rose and I spoke of 1914–18 as the chemists' war, 1939–45 as the physicists' war. Since then we have entered the age of computer technologists' and even biologists' wars.[7]

myths evoke spinning and weaving, as in the loom of life held by the Fates. Memory was and is no stranger to such metaphor; for the ancients, images become inscribed within it – as Cicero puts it in *De Oratore* – like 'a wax writing tablet and the letters written on it'. This metaphor resonates down the ages, becoming the basis for philosophical debate in the eighteenth and scientific/ideological debate in the nineteenth and twentieth centuries as to whether humans are born with innate predispositions or as *tabula rasa* – clean slates on which experience inscribes individual memory.

Today the word memory occurs in a multitude of scientific discourses. Quite apart from the sort of memory that neurobiologists, psychologists and even novelists talk about and with which I am concerned, mathematics and physics, chemistry, molecular biology, genetics, immunology and evolutionary biology, not to mention computer science, all use the term. Why these many uses of the word memory? Do we have here to deal simply with puns – the use of words originating in one context in another, different one – or does the fact that many different scientific discourses use the term cast some light on the mechanisms and processes that may be involved?

Can these varied metaphors[8] reveal something about the nature of the processes involved in any phenomenon – even illuminate otherwise unexpected similarities as to process or mechanism between seemingly widely different phenomena – or are they merely figures of speech? In what sense is memory to be taken to be like a wax tablet – or for that matter, a computer?

One can usefully distinguish between three types of metaphor in science.[9] The first is poetic – for example Rutherford's description, in the early twentieth century, of electrons in orbit around the atomic nucleus as if they were planets revolving around the sun. In using this analogy he surely did not mean that the nucleus and electrons were like the sun and planets nor that the forces which related them were gravitational; all that the analogy provides is a useful visual image. The ancient metaphor of the potter's wheel clearly comes into this category.

The second metaphoric mode is *evocative*, in which a principle from one sphere is transferred to another. Thus until the Middle Ages and the Newtonian revolution, everything that moved seemed to be pushed or pulled by something else. Hence to explain the movement of the sun around the earth, metaphor spoke of horse-drawn fiery chariots.

Finally, one has the metaphor as a statement of structural or organisational identity. Thus when, in the seventeenth century, William Harvey discovered the circulation of the blood and described the heart as a pump, his metaphor had a precise meaning which distinguishes it from the previous two categories. Within the circulatory system, the heart indeed functions as a pump, and organisationally its structure, with valves and emptying and filling phases, resembles at least those pumps which were being mechanically contrived at the time of Harvey's discovery. Treating the heart as a pump enables mathematical models of its action to be made which accurately describe many of its properties.

In which of these senses, then, are wax tablets – or computers – metaphors for brain memory: poetic, evocative or structural? Or are they indeed none of these but instead merely mischievous?

The Cartesian rupture

With the birth of modern science in Europe in the seventeenth century, the symmetry of analogising physical forces to animate ones and biological phenomena to technological models was broken. It is important to realise that this is indeed a particularly Western phenomenon, and is perhaps best explained by the the fact that science was not born as a singleton but as a twin; it emerged and grew to maturity along with particular forms of bourgeois, capitalist social organisation and the two shared many philosophical and ideological premises in their understandings of, and approaches to the natural and social worlds.[10] The broken symmetry of Western science was long resisted in other cultures with alternative indigenous scientific traditions, most notably of course in China,[11] where the animate/inanimate division of nature, along with other forms of dualism which the Western cultural tradition has naturalised, was never drawn so rigidly.

The science that developed in Europe, however, was epitomised by Galileo, Newton and above all Descartes, who between them debiologised the physical world, turning it into 'mere' mechanism. For them, the defining technology was the clock and its associated systems of gears, cogs and hydraulic transmission, which together could generate precisions of mathematically describable motion hitherto unimaginable. Clockwork redefined time, trapped and reduced the hitherto seamless universe into units which could be separately controlled and costed.* Hydraulics were a source of power and controlled movement within this mechanical universe. The new physics generated not merely new explanations of the universe but also new technologies, new production systems and new relationships of production between those engaged in them. Europe became set on a course of industrial and imperial expansion which has far from run its course today, and mathematical physics became the defining model of scientific explanation against which all others should be judged. If the very motions of planets, moon and sun could be reduced to simple mathematics, and were nothing more than the ineluctable workings out of equations, why not mere biology?

It might of course have been different. Biology, as an organised science, might have developed before physics, and those less mechanical, more goal-directed (teleonomic), functional and evolutionary modes of explanation of the animate world which biologists favour might have become also the model towards which physicists aspired. Reductionism, with its insistence that in 'the last analysis' the world can be explained in terms of atomic/quantum properties and a few universal equations, would then seem no more than a ludicrous inversion of proper scientific explanation,[12] and biologists would no longer suffer from a sense of physics-envy and an unease about their subject being a 'soft' rather than a 'hard' science. But it was not to be; the technological rather than the biological metaphor dominated,

---

*With the advent of digital as opposed to analogue watches, time becomes divided up and budgeted even more precisely, increasingly divorced from the world time given by the cycles of day and night, the months, seasons and years. In today's world, wearing an analogue as opposed to a digital watch becomes a small act of resistance – a point first made by the radical physicist Maurice Bazin, one of the finest of teachers of popular science.

and in the hands of Descartes, living organisms themselves became clockwork, their internal processes powered by complex systems of hydraulics, tubes and valves.

For humans, as is well known, Descartes made a crucial exception. Although everything about their day-to-day functioning was as mechanical as that of any other animal, humans could also think and above all had a soul, whereas, for Descartes, animals were capable only of fixed responses to their environments. Thought and soul were incorporeal entities, but could interact with the mechanism of the body by way of a particular gland, the pineal, located deep in the brain. Descartes chose the pineal for this localisation on two grounds. First, whereas other brain structures are all duplicate, in that the brain consists of two more or less symmetrical hemispheres, the pineal is singular, unduplicated, and mental phenomena must of course be unified. And second, the pineal is a structure found uniquely in humans and not present in other animals. He was of course wrong on both grounds; there are many other non-duplicated structures in the brain, and other vertebrates also possess pineal glands, but the theory-driven logic of his argument remains appealing for those who want to argue, as he did, for the uniqueness of humans: 'It is morally impossible that there should be sufficient diversity in any machine to allow it to act in all the events of life in the same way as our reason causes us to act.'[13]

The Cartesian split, between mind and body, a dualism which has clouded Western scientific and philosophical thinking with its obsessive and misguided worries about the 'mind-brain problem' for the subsequent three centuries, begins here.

But it is the metaphors of Cartesian clockwork and hydraulics rather than Cartesian dualism which concern me at present. The present-day animal rights movement has made much of the way in which such thinking enabled Descartes to dismiss the cries of pain of animals when vivisected as no more than the squeaks of poorly oiled machines. The Cartesian view was taken most seriously perhaps by the French tradition of physiology in the nineteenth century – particularly Claude Bernard – in its indifference to animal suffering.[14] The modern repudiation of Descartes is of course right, but I would

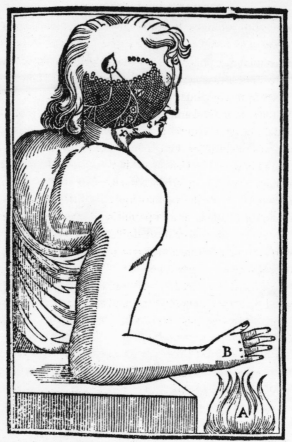

Fig. 4.4
The Cartesian metaphor
An interpretation of his explanation of the automatism that determines the withdrawal of a hand from something that burns it.
From L'Homme de René Descartes, *Paris, 1664.*

argue that the clockwork metaphor is as damaging in its partitioning and reduction of humans as it is of non-human animals. Descartes may have saved the soul/mind for Catholicism in its Sunday-best garb, twiddling the knobs of mechanism via the pineal, but he left a clockwork human for the remaining six days of the week, debiologised as well as desacralised and open to treatment as a mere *bête machine* within the developing industrial revolution of the eighteenth and nineteenth centuries. It would only be a matter of time before technology would challenge the Cartesian 'moral impossibility'.

Against this grave philosophical and ideological disservice there are, to be sure, Cartesian achievements. The localisation of function to the brain, even in machine-metaphorical form, was no trivial event. The brain as the seat of mind and soul is not an automatically self-evident proposition, however natural it seems to us today; for Aristotle that function was reserved for the heart, for the ancient Hebrews for the kidneys and bowels. The Galenic medical tradition had demonstrated that nerves originate in the brain and that motor and sensory functions are abolished by brain injuries. But hydraulic thinking centred not on the fatty and unpromising tissue of which the brain was composed but instead on its fluid-filled core, the ventricles, lovingly drawn by early anatomists, none more strikingly than Leonardo.

As a consequence early hydraulic memory models had memories stored in the ventricles, and animated by a flowing spirit, controlled by a valve between front and rear portions of the brain. In Descartes's version this crucial task was naturally ascribed to the pineal:

> Thus when the soul wants to remember something . . . volition makes the gland lean first to one side and then to another, thus driving the spirits towards different regions of the brain until they come upon the one containing traces left by the object we want to remember. These traces consist simply of the fact that the pores of the brain through which the spirits previously made their way, owing to the presence of this object, have thereby become more apt than others to be opened in the same way when the spirits again flow towards them. And so the spirits enter into these pores more easily when they come upon them, thereby producing in the gland that special movement which represents the same object to the soul and makes it recognise the object as the one it wishes to remember.[15]

This ingenious description contains the forerunners of many modern ideas about the mechanisms of memory with which the present book deals – and of the capacity of philosophers to see biological problems as straightforward. In this context I am especially fond of Descartes's use of the term 'simply'. If only it were so . . .

How are we to understand these Cartesian metaphors of memory? Descartes may have meant his metaphor to be precise, as structurally accurate a descriptor of the brain and its processes as Harvey's of the heart as a pump, but I suggest we can take it as no more and no less than poetic, a way of thinking about a complex human phenomenon which places it not *sui generis* but as merely one among other types of matter in motion.

Through the eighteenth and nineteenth centuries the metaphors of mind and memory steadily shift. With the discovery by Galvani of 'animal electricity' – that frog's legs twitched when connected by metal wires – the nervous system ceased to be hydraulic and became instead an electrical maze. And within it, the brain became first a telegraphic signalling system and later, at the start of the twentieth century, a telephone exchange, one of the several metaphors favoured by the great neurophysiologist Sherrington. (Another, unforgettable but clearly poetic, Sherringtonian image saw the brain as an 'enchanted loom' weaving patterns in electricity.) Even more than pipes and valves, telegraphs and telephones were surely systems that were like the brain in more than a poetic sense. The telegraph, for instance, converted sense data into symbols – in the hands of Morse and his successors into specific codes for given individual letters – which could be passed over large distances and be decoded at the other end. The telephone was even more promising, for speech was here converted into patterns of electrical flow across a wire. In the telephone exchange metaphor the brain processes messages coming in and going out; signals from eyes connected to muscle contractions in the leg and so forth.

When, during the 1920s, it was discovered that the brain was indeed in a state of ceaseless electrical flux, that electrodes placed on the scalp could detect regular bursts and rhythmic waves of electrical activity changing with thought and rest, sleep and wakefulness, this was instantly assimilated to the telephone exchange model, with subscribers dialling in and being connected to their required addresses by a central operator. But the telephone exchange is merely the paradigm form of the electrical office of the first half of this century. Here, for instance is the metaphor at its most primitive, from a children's encyclopedia of the period:

> Imagine your brain as the executive branch of a big business . . . Seated at the big desk in the headquarters office is the General Manager – your conscious self – with telephone lines running to all departments . . . Suppose you are walking absentmindedly in the street and meet your friend Johnny Jones. He calls your name, you stop, say 'Hullo!' and shake hands. It all seems very simple, but let's see what happened during that time in your brain. The instant Johnny Jones called your name, your Hearing Manager reported the sound, and your Camera Man flashed a picture of him to the camera room. 'Watch out!' came the signal to your desk, and at the same instant both messages were laid in front of you. As quick as lightning your little office boy, Memory, ran to his filing case and pulled out a card. The card told you that that voice and that face belonged to a person named Johnny Jones and that he was your friend. Instantly you began issuing orders . . . [16]

The computer and artefactual intelligence

From wax tablet to little office boy with a filing case in about 2,000 years doesn't seem like too rapid a rate of metaphoric progress, and to call this even a poetic metaphor would seem to debase the term. But the real challenge to Descartes's 'moral impossibility' came with the defining technology of the second half of the twentieth century, the computer. The immediate antecedents of today's machines, like those of so many other major technologies, are military. They include the logical games of the Cambridge mathematician Alan Turing, pressed into practical use during the code-breaking exercises of British Intelligence at Bletchley Park (no further than a long stone's throw from my own laboratory today) during the Second World War. They were given electronic form by a different set of military requirements – the need to develop effective servo-mechanical devices to calculate the elevation and direction to fire anti-aircraft guns against rapidly moving targets – techniques developed by the United States mathematician Norbert Wiener, who gave the new science a new name, cybernetics, by which it became fashionable in the decades after 1945. Wiener and fellow mathematician John von Neumann, together, of course, with United States (and some, small-scale, British) industry, were responsible, in those years, for giving the new science and the technology it generated electronic form and theory. The military interest – and its impetus for new developments – has never faded, and reached a crescendo during the 1980s, a decade of seemingly unbridled spending under the auspices of the Reagan administration's Star Wars Program, which demanded computing power on a scale of unparalleled extravagance – only to be surpassed by George W. Bush's latest splurge, Son of Star Wars. Computers which worked like or could replace brains became not merely a science fiction but a serious military goal. 'Intelligent systems' which could replace or supplement skilled, highly trained and expensive humans in flying planes and firing weapons seem an attractive option. Indeed it has become hard these days to attend a scientific conference on themes associated with learning, memory and computer models thereof without finding a strongly hovering United States military presence, whether navy, airforce or the somewhat sinisterly acronymed DARPA – the Defense Advanced Research Projects Agency (later, in a seeming attempt at new respectability, to drop the 'D' from its title).

That the computer was something qualitatively new was obvious from the start. Certainly, electromechanical calculating machines and their relatives already existed. But the general purpose computer was far more than just a very fast calculator and storer of data; it could manipulate, compare and transform information in ways that have made possible wholly novel technologies, instrumentation and even the scientific questions one can conceive of asking of the universe. Slowly but with increasing acceleration over the past two decades computer technology has transformed our ways of understanding and operating upon the world. Small wonder that its ideological resonances have been so profound. From the very start, the relationship between computers and minds/brains was at the forefront of the thinking of its inventors and was apparent in their language. For instance Von Neumann's digital computer consists of a Central Processing Unit which carries out arithmetical and electrical operations, and a storage unit, immediately christened by its designers a *memory*.

A computer memory consists of chips (silicon wafers with transistors engraved on them) which store data in the form of a binary code – that is, each unit can exist in one of two states (0,1). Implicit in this design of course is that anything that the computer stores and manipulates must first be converted into a form in which it can be represented in this numerical, binary mode as a number of bits (binary units) of information. Information in this sense has a technical – even technological – rather than an everyday definition, which will need further analysis later. Also requiring further note is that implicit in the name given to the information store – the computer memory – is the claim that in some way what the computer is doing in holding and processing binary units of information is analogous to what we as humans do with our own memory.

At first sight, this might be seen as encouraging. Does this language system not describe a physical, inanimate mechanism by analogy to a biological system? And was the failure to do this not what I was regretting about the seventeenth-century Cartesian transition? Sadly, no. As will become apparent, the practical and ideological power of the technology surpasses that of the biology, so that the metaphor reverses itself. Instead of biologising the computer, we find ourselves challenged by the insistence that human memory is merely an inferior version of computer memory, and that if we want to understand how the human brain works we had better concentrate on studying and building computers.

Nor is this the aberration of a few macho science fiction enthusiasts; it has been central to the agenda of computer designers and their philosopher fellow-travellers from the earliest days. Turing himself began it in 1950, not long before his suicide, with one of his many logical games. Suppose you were in communication, via a teletype, with a second teletype in an adjoining room. This second teletype could be controlled either by another human or a machine. How could you determine whether your fellow communicant was human or machine? Clearly the machine would have to be clever enough to imitate human fallibility rather than machine perfection for those tasks for which machines were better than humans (for example in speed and accuracy of calculating) but equally the machine would have to do as well as a human at things humans do supremely – or else find a plausible enough lie for failing to do so. This is the nub of the so-called Turing Test, and he believed that 'within fifty years' a computer could be programmed to have a strong chance of passing this test.[17]

To create a machine that could pass the Turing Test has become the holy grail for the generations since 1950 of those committed to the pursuit of what they call, modestly enough, artificial intelligence (AI). But how to go about it? From the beginning, there were two contrasting approaches, that we may characterise, crudely, as reductionist and holistic. Looking back over the period with the benefit of hindsight, one of the pioneers and prophets of the holistic approach offered this fairy-tale account:

> Once upon a time two daughter sciences were born to the new science of cybernetics. One sister was natural, with features inherited from the study of the brain, from the way nature does things. The other was artificial, related from the beginning to the use of computers. Each of the sister sciences tried to build models of intelligence, but from very different materials. The natural sister built models (called neural networks) out of mathematically purified neurones. The artificial sister built her models out of computer programs.

In the first bloom of their youth the two were equally successful and equally pursued by suitors from other fields of knowledge. They got on very well together. Their relationship changed in the early sixties when a new monarch appeared, one with the largest coffers ever seen in the kingdom of the sciences: Lord DARPA . . . The artificial sister grew jealous and was determined to keep for herself the access to Lord DARPA's research funds. The natural sister would have to be slain.

The bloody work was attempted by two staunch followers of the artificial sister, Marvin Minsky and Seymour Papert, cast in the role of the huntsmen sent to slay Snow White and bring back her heart as proof of the deed. Their weapon was not the dagger but the mightier pen, from which came a book, *Perceptrons*, – purporting to prove that neural nets could never fill their promise of building models of mind: only computer programs could do this. Victory seemed assured . . . [18]

Seymour Papert's fairy tale, of course, ends with the holists triumphant, not a view that is very widely shared in the 'AI community' at present. As will become clear, my own view is that Papert's fairy-tale metaphor is as flawed as his memory/intelligence metaphors. Neither fairy-tale sister is Cinderella – or even Prince Charming; both modelling approaches are flawed if their intention is to provide structural metaphors for the way that real brains work and real memories are stored. But it is worth looking a little more closely at the pretensions of both protagonists.

As Papert rightly describes it, one group of modellers, those I describe as reductionist, argued that the proper approach for AI was to take the brain and to endeavour to simulate some of its known properties using computers. The brain's units of function were assumed to be its nerve cells, neurons, and the brain was supposed to store, process and transform information through the functioning of networks of these neurons. The task was then to make mathematical models of how the neurons might function, assemble them into nets and test how different ways of connecting the cells within the nets might generate varying forms of output, including networks that could change their properties and output functions as a result of experience – that is, could both 'learn' and 'remember'. Such simulations were first performed by Frank Rosenblatt in the mid-1950s with a modelling system called a Perceptron. Perceptrons were triumphs of computing, but it soon became clear that they were very inadequate representations of real brain neurons. Although they could seemingly learn – that is, change their output properties in response to different inputs, so as, for example, to recognise and classify simple patterns – they failed at anything that was much more complex or even remotely resembled real-life problems.

In the 1960s and 1970s the intractable difficulties that the neuron-modelling approach had run into – and the theoretical limitations exposed by Papert and Minsky – led to its virtual abandonment. It was at this time, for example, that a United Kingdom government-sponsored study of the future of AI led to the conclusion that its promise had been much overstated, and drastically scaled down the British research effort in the area.[19]

Interest in its possibilities was dramatically revived, however, in the late 1980s, by an innovative new approach. Earlier generations of computers were essentially serial processors – that is, they could carry out, albeit incredibly rapidly, only one operation at a time in sequence. Fast though they are, computers that carry out operations in this linear and sequential way face fixed limits to their speed of operation – messages after all cannot travel from one part of a computer to another faster than the speed of light, a limitation that has become known as the Von Neumann bottleneck. Perhaps it was this limitation, as the new generations of supercomputers pushed technology to the edge of the achievable, or perhaps it was the closer liaison between brain scientists and the computer modellers that helped pursuade AI enthusiasts that real brains don't work like this at all, but instead carry out many operations in parallel, and in a distributed manner, many parts of a network of cells being involved in any single function, and no single cell being uniquely involved in any. The speed limitation could be overcome if computers could be designed more like brains – that is capable of parallel and distributed rather than sequential and linear operations. The result was an explosion of interest in new computer designs based on what are described as parallel distributed processing (PDP) principles, promising new generations of equipment which fascinate the military, industry and AI modellers in equal measure – though it is of course the first two who call the financial shots. When, in 1986, David Rumelhart, James McClelland and their colleagues at MIT published a large, two-volume book of papers on the potential of PDP for brain modelling, it is said to have sold 6,000 copies the day it appeared on the market.[20]

The principles of the new modelling approach are known as connectionism – like the earlier one, they are based on the idea that the brain is composed of ensembles of neurons with multiple connections between them. Appropriately connected sets of such cells can be made to show learning- and memory-like behaviour in that they will sort and classify inputs and slowly change their output properties in response to novel input patterns. But, unlike the earlier Perceptron type of model, the 'memory' does not reside in any one single cell or pair of connected cells within the network; rather it is a property of the network as a whole. Further, whereas in Perceptron-type models single units were called upon to receive direct inputs from the external world and modify their output properties accordingly, in the new connectionist models the networks are more complex, and include what the modellers call 'hidden layers' – arrays of 'cells' located between inputs and outputs. The difference in power that this change effects is dramatic. The earlier generations of AI models were wired up almost as if the brain were a simple telephone switching system, with direct links between sense organs like eyes and ears and outputs organs like muscles. They virtually ignored the fact that the vast majority of nerve cells within complex brains are not in direct communication with the outside world either by way of sensory input or motor output, but connect internally, receiving messages from and replying to other neurons; that is, there is a vast amount of internal processing of any messages that arrive to the brain, and also a great deal of private traffic between these cells (called interneurons), before any external responses are made. The hidden layers of PDP models are intended to serve almost as such interneurons, and massively increase the power of the networks to learn, to generalise, to predict.

Connectionist models became attractive to industry and the military because they promised to leap over earlier limitations on computing power. But they also attracted a surge of enthusiasm amongst neurobiologists, many of whom began to believe that here at last might be a model which comes close to what brains – or at least parts of brains – might actually be like. During the 1990s there emerged a flurry of new research journals offering neural network models claiming to explain many aspects of brain processes; and the commanders and ideologues of this New Model Army seemed in almost continual circuit of the globe, from one high-powered conference and seminar to another, scarcely having time to touch down in their own offices and labs to collect the latest simulation before getting airborne again.

Even philosophers began to listen; one of the more influential books of the period among neurobiologists, who are not given to reading philosophy, was the California-based Patricia Churchland's *Neurophilosophy.*[21] Churchland's strategy in the book is to review long-standing problems in the philosophy of mind, match these with a competent review of contemporary neurobiological findings, and to conclude that reductionism rules, OK. Salvation, for her, comes from connectionism, and she followed the book with a couple of papers in the major journal *Science*, written jointly with San Diego neuroscientist Terrence Sejnowski, offering a prospectus for what they called a computational neuroscience,[22] a phrase which itself then formed the title of other books and journals. The fact that philosophers, modellers and neurobiologists actually began listening to one another, and that computer people had at last begun to show some respect for biological as well as artefactual brains, clearly makes their analyses an advance over the earlier ones, in which AI enthusiasts tended to run away with preconceived notions of what nerve cells did, and soon cut off all meaningful contact with the biological phenomena which the neurobiologists were studying. The neurobiologists' enthusiasm for Churchland stems, however, I believe, from the fact that, rather than challenging our assumptions, she shows us a rather uncritical respect. Her book thus serves, in a way unusual for a philosophy text, as a rather flattering reflecting mirror held up for us to shine in.* It is not merely that our warts are hidden in the reflection: our very posture, in all its reductionist discomfort, has been given the Hollywood treatment. Yet the limitations of connectionist neurobiology and the philosophy it spawns are very clear, and I believe in the long run will fatally flaw them, for reasons which will become apparent as my account proceeds.

Meanwhile, what of the second of Papert's two sisters, the 'artificial', holistic one? In this approach, no attempt was to be made to model brains; instead the emphasis was on modelling minds. That is, the modellers would try to identify what they believed to be the functions of mind processes, such as 'belief, hearing, vision, touch, inquiry, explaining, demanding, requesting . . .' (I take this seemingly eclectic, but quite characteristic, set from Minsky's book *The Society of Mind*[23]). They would then endeavour to model the logic of these processes, irrespective of whether the models they produced in any way resembled real brains. What mattered was only that the models performed; that is, their outputs matched the expectations of the modellers about what human outputs might be if they were carrying out the functions they thought they were modelling.

To see the difference between these two approaches, take for example, a person at a rifle stand in a fairground, endeavouring to shoot at a moving line of metal ducks at the back of the stand. The reductionist PDP approach would ask how the neural system might

---

*Nonetheless, a significant number of neurobiologists *are* uneasy with Churchland's reductionism. The issue came to a head a few years back at one of those Swiss ski-meetings beloved of neuroscientists – meetings at which early morning and late afternoon scientific sessions are wrapped around the real business of the day, which is to get out onto the slopes for as long as daylight and stamina persist – part of the phenomenon perhaps best embraced under the well-known thesis known as the 'leisure of the theoried classes'. The theme of the meeting was 'the relationship between neuroanatomy and psychology', and Churchland was to give the opening talk. She set up her reductionist stall, arguing for the ultimate collapse of psychology into neuroanatomy, perhaps expecting an easy ride from a group of neurobiologists, and found to her surprise that it was strongly opposed by most present – especially the neuroanatomists!

be connected so that the moving spatial image of the ducks might be conveyed via the retina to appropriate regions of the brain (hidden layers) and how these might then change their properties so as to learn to generate appropriate motor outputs. The holistic approach would instead ask: how can we construct a servo-mechanism which receives inputs as to the position and motion of the ducks and then directs outputs accordingly? Do the outputs of this mechanism resemble those in a real human carrying out the same task? If not, why not?

For the bulk of artificial intelligence's half-century history, it is this latter approach that has been the most powerful. But, in developing it, its protagonists have moved away from how biological brains and psychological minds might work and instead concentrated on solving problems embedded in the silicon of computer chips and in mathematical logic – an approach which may produce bigger and better machines, but has become entirely indifferent to their relationship with the biological systems they were once attempting to model. Their rallying cry has been bluntly expressed by their most enthusiastic spokeswoman, Sussex University philosopher Margaret Boden: 'You don't need brains to be brainy.'[24]

Indeed, the modellers have grown ever more ambitious. A whole new research field has grown up, called no longer merely AI but AL – Artificial Life, a sort of grown-up version of children's games in which programs are written for computer-generated cartoon characters which can interact with one another on screen, need care or 'feeding', grow, encounter and have to overcome challenges, reproduce and even 'die'. The most sophisticated of these are perhaps Steve Grand's 'Norns'.[25] They are obviously fun and to play with them can be a fascinating distraction from real life, but to me, despite Grand's persuasive arguments, they remain a game – one can learn a lot from them, but don't confuse them with actually living. Only a trifle more – or less? – ambitious has been engineer Igor Aleksander's claim that it is possible to create conscious computers.[26]

The suggestion that computers might be conscious is one aspect of an extraordinary new ambition developed within the neurosciences during the 1990s – the claim that not merely memory but other mental processes and even consciousness itself could be explained in scientific terms. As I said in the previous chapter, when as a young post-doc I became fascinated by the possibility of researching on memory, my Nobel Laureate boss objected strongly; this was no subject for a sensible biochemist to poke his nose into. He was of course wrong. So how should one judge the fashionable urge that neuroscientists feel to poke their noses into yet another once-taboo topic? Only a few years ago at a symposium on consciousness I heard a Harvard neurophysiologist describe it as a CLM – a Career Limiting Move. But then there emerged *The Journal of Consciousness Studies*. Seemingly batty ideas about quantum consciousness and the role of microtubules were turned into a best-selling book.[27] Before long, distinguished neuroscientists and philosophers began to rub shoulders with the wilder products of Granola State touchy-feely holism in a series of increasingly fashionable Consciousness Conferences held in Tucson, Arizona, and the pop-science books began to proliferate. Now we are drowning in books, journals, conferences (I'm reviewing another half dozen or so as I write these words[28]) and those of us working in the brain and behavioural sciences who haven't yet written a popular book about consciousness have begun to feel left behind. I can't help sympathising with Ernst Chain; but I still think memory is hard enough to deal with.

To start with, there is the small problem of definition. To many neuroscientists it is quite straightforward: being conscious is the reverse of being unconscious or asleep; it is to be aware of, and to a degree in control of one's thoughts, intentions and actions. So in his account of consciousness, *The Astonishing Hypothesis*,[29] the molecular biologist Francis Crick reduces it to the more tractable question of awareness, and then further to visual awareness or perception, as this is well approachable by classical neurophysiology and brain imaging. Philosophers in the Anglo-American tradition puzzle over the subjectivity of consciousness, which makes it private and personal, and worry about how this subjectivity may be related to the objective brain states that neurobiologists can measure. Almost everyone rejects the old Cartesian substance dualism in which mind and matter are incommensurable. We are all apparently physicalists, or at least monists, these days, and so for most of the disputants, the question becomes one of solving 'the hard problem' of how 'objectively' observable neurobiological facts about brain processes can generate 'subjective' experiences, or qualia. The possibility that such experiences are not 'generated' by brain (and body) processes, but are no more and no less than another way of describing the same phenomenon, is apparently not acceptable, though I have never been quite clear why. So we are back with one old debate, as to whether my experience of seeing red is the same as yours, and one new one, as to how and why did consciousness evolve – that is, what function, in an intellectual landscape dominated by Darwinian metaphors, does it subserve? The first question, I have to admit, has always struck me as not merely unanswerable, but vacuous. The answer to the second seems obvious: human consciousness, however defined, has contributed to the evident – to date – evolutionary success of our species, and cannot therefore be epiphenomenal.

By contrast with these philosophical frets, there are other wider contexts for the term consciousness; I don't just mean what psychoanalysts speak of 'the unconscious' as providing the bases for thoughts and actions to which we do not have direct access. For them too, consciousness is something 'belonging to' an individual and related in some way to what is going on inside that person's head. Sociologists by contrast may refer to consciousness in quite another and much broader sense, as in the context of class, race or gender consciousness. Here 'consciousness' is no longer the private property of an individual but the expression of a relationship between that person and the surrounding social and cultural world. To seek for the causes or even the correlates of this sort of consciousness inside the head would be absurd. Being conscious, whatever we mean by it, must of course be a consequential function of our biological organisation as members of a species of social individuals with large brains and communication skills. But the argument that consciousness is not merely the obverse of unconsciousness, that it is not some static brain/mind process but rather a socially, historically, developmentally engendered statement about the relationship between an individual and the surrounding world, cuts little ice – perhaps because the discussion is still being conducted primarily between philosophers trained in the Anglo-American mode and neuroscientists, and thus excludes the social and historical domains. To appreciate how much richer the issues are, turn instead to a recently translated debate in the French tradition, between molecular neurobiologist Jean-Pierre Changeux and hermeneutic philosopher Paul Ricoeur.[30]

The flawed metaphor

For all the sloganising of top-down and bottom-up modellers, if the task is to understand how real brains and minds – or even memory – may work, I believe both approaches to be fundamentally flawed. Hence the failure of all the previous predictions about just when the AI people would come up with a really mind-like computer, regarded by the Weiner-enthusiasts of the early 1950s as certain to arrive by the 1960s, then postponed to the 1970s, 1980s or even the bimillennium as time went by and models and programs from Perceptrons onwards came and vanished.

There have been three major critiques of the methodology and pretensions of AI, from, respectively, a philosopher, a mathematician and an immunologist. Let me describe them briefly before turning to my own problems with the computing metaphor. First, the philosopher, John Searle, whose argument is based on standing the Turing Test on its head. Imagine a non-Chinese speaker in a closed room, who receives questions written in Chinese through a machine. Available in the room is a code which enables the Chinese symbols to be matched against a second set which constitute replies to the questions being asked. The replies can be fed out through the machine to the outside world. For the observers outside, it is clear that questions in Chinese are being answered appropriately in Chinese; the person inside the room has passed a type of Turing Test. But in no way could one infer that the person in the room understood, is conscious of, is intelligently responding to, the content of the messages being passed in and out; what is happening is no more than a purely automatic operation. This, says Searle, is what computers are doing, and why they cannot be regarded as intelligent or conscious.[31]

Second, the Oxford mathematician, Roger Penrose, whose thesis of quantum consciousness in his *The Emperor's New Mind*[27] is presented as a sustained critique of connectionism. In essence, Penrose's point is quite straightforward: connectionism, if it is to work, depends on a relatively fixed and stable relationship of cells within a neural network, modified only in response to specific inputs and then responding in a deterministic way to this modified response. Penrose invokes both physical and mathematical theory against this claim: physical theory in the form of quantum mechanisms, which, he argues, result in a built-in indeterminacy in neural responses; and mathematics in the form of the highly fashionable chaos theory, which shows how indeterminate systems can nonetheless produce lawfully ordered outputs – just as, for instance, the random and indeterminate motions of gas molecules in a jar nonetheless together produce the precise and predictable relations between temperature, pressure and volume that are given in Boyle's simple gas law.

According to Penrose, then, a reductionist strategy must fail on two related grounds. In the first place, indeterminacy at the level of the neuron and its synaptic interconnections means that one will never be able to understand the mind or the brain simply by an analysis of its individual components, whose responses are inherently unpredictable. In the second place, however, this indeterminacy at the level of the component gives way to predictability at the level of the system. Consciousness, intelligence, memory thus emerge as properties of the brain as a system rather than those of individual components within that system.

The third critique of AI and its information-processing methodology is that of the Nobel Prize-winning San Diego-based immunologist and theoretician Gerald Edelman. In a trilogy of books,[32] Edelman has attempted the almost impossibly ambitious task of developing general theories of developmental biology, neural organisation and consciousness based on a metaphor derived, not from physics or technology, but from evolutionary theory and specifically from his understanding of Darwinian natural selection. Because in some respects I share Edelman's critique, though not his choice of the metaphor of selection, and a fuller exposition of his position requires more biology than I have yet introduced into this discussion, I will delay my engagement with it for a while, and turn instead to my own unease with the metaphorical world of the computer-as-brain/mind/memory.

The analogy, while of potent fascination to many, has always been suspect amongst biologically grounded neuroscientists, on both structural and organisational grounds. Structurally, the properties of chips, AND/OR gates, logic circuits or whatever, do not at all resemble neurons, if indeed it is neurons that are to be regarded as the relevant units of exchange within the nervous system. The units of which the computer is composed are determinate, with a small number of inputs and outputs, and the processes that they carry out with such impressive regularity are linear and error-free. They can store and transform information according to set rules. One consequence of this, for computer brain-modellers, has been the persistent tendency to attempt to reify certain types of process in which minds/brains participate. For instance the very concept of 'artificial intelligence' implies that intelligence is simply the property of the machine itself (I would argue that such a reification is equally inappropriate either for brains or for computers).

The brain/computer metaphor fails because the neuronal systems that comprise the brain, unlike a computer, are radically indeterminate. This critique of determinacy goes further than Penrose's, because I want to emphasise that brains and the organisms they inhabit, above all human brains and human beings, are not closed systems, like the molecules of a gas inside a sealed jar. Instead they are open systems, formed by their own past history and continually in interaction with the natural and social worlds outside, both changing them and being changed in their turn. This openness provides a further level of indeterminacy to the functioning of both brain and behaviour. Unlike computers, brains are not error-free machines and they do not work in a linear mode – or even a mode simply reducible to a small number of hidden layers. Central nervous system neurons each have many thousands of inputs (synapses) of varying weights and origins (perhaps 10^{14}–10^{15} in the human brain; that is getting on for a million times more connections in an individual brain than there are people alive on the earth today!). The brain shows a great deal of plasticity – that is, capacity to modify its structure, chemistry, physiology and output – in response to contingencies of development and experience, yet also manifests redundancy and an extraordinary resilience of functionally appropriate output despite injury and insult. Brains carry out linear computations relatively slowly yet can exercise judgemental functions with an extreme ease that baffles the modellers.

Consider a simple experiment in which a person is briefly shown a list of four digits, and asked to remember and recite them back. Most people can perform the task with ease. However, if the number of digits is extended to seven or eight then most of us begin to fail at the task – especially if the length of time between seeing and recalling the numbers is stretched from a few minutes to an hour or more. On this basis, the maximum human memory for digit spans of this sort can be simply calculated in bits, as follows for an eight-digit string:

First, the bits for the numbers themselves

$$8 \times \log_2 10 = 8 \times 3.32 = 25.56$$

Then, because the numbers have to be in the proper order, a calculation for order information is needed, and this is given as

$$\log_2 8! = 15.30$$

Giving a total of just 40.86 bits – and yet it would appear that we don't have the memory capacity to do it! Now compare this with the number of bits available in the memory of a simple pocket calculator – about 1,000. And the 10-centimetre double-density floppy disc which is currently inserted into my beloved Apple Mac, onto which I am typing these words, has a memory storage capacity of more than 10 million bits.

What does a calculation of bits like this actually mean? The

Cambridge mathematician John Griffith once estimated[33] that if humans were presented with, and stored, information at the rate of 1 bit a second for every moment of a seventy-year lifespan, at the end of their life they would have stored some 10^{14} bits – or about the equivalent of the information content of the *Encyclopedia Britannica*, that is, 10,000 floppy disks and not so far off what the hard disk on my computer can manage. Impressive, for a lap-top, but it would seem surprisingly small for a functioning human brain.

Can it really be that humans have no more memory capacity than a microcomputer? Clearly something is wrong somewhere with such calculations. What can it be? Here's a clue: although I cannot remember more than eight digits flashed onto a screen in front of me, I once demonstrated the power of human memory to an audience by flashing up a much longer, 48-digit list, turning my back to the screen and calling it off perfectly:

524719382793633521255440908653225141355600362629

How can I do this when failing the 8-digit test? Simply, the 48-digit list is not random, but contains a sequence of birthdays, telephone numbers and other codes that I have a regular need to use and therefore remember. But I don't remember them as numerical information in the error-free way that a computer would, and I don't store them in a unimodal number sequence. By contrast with computers, brain memory is error-full, and uses multiple different modalities. For me, unlike a computer or the Chinese translator in Searle's closed room, this particular unique number sequence has a meaning which is also unique for me. And it is on the basis of meaning, not sequence, that I am recalling them. Further, I would want to insist that meaning is not synonymous with information. Meaning implies a dynamic of interaction between myself and the digits; meaning is a process which is not reducible to a number of bits of information.

Another example. Suppose that at dinner last night I was offered a menu with a multiple choice of dishes. I read the menu, chose and ate a meal, and can now tell you that it included broccoli soup and poached salmon. Information in the printed text of the menu was transformed into recollections of earlier tastes, then the spoken order to the waiter, and then the material reality of the food and its actual taste. And now when I tell you I ate broccoli soup and salmon I neither offer you the printed menu nor the food – still less do I expect you to taste it directly – instead I further modify last night's experience by translating it into spoken words. At each point in this sequence there has been more than just a switch in the modality in which the information is expressed: there has been an input of work on that information which has irreversibly transformed it (and this of course says nothing about the work that each listener or reader of this description does in subsequently further transforming, working on and interpreting the data I have just offered).

Thus brains do not work with *information* in the computer sense, but with *meaning*. And meaning is a historically and developmentally shaped process, expressed by individuals in interaction with their natural and social environment. Indeed, one of the problems of studying memory is precisely that it is in this sense a dialectical phenomenon. Because each time we remember, we in some senses do work on and transform our memories, they are not simply being called up from store and, once consulted, replaced unmodified. Our memories are re-created each time we remember. I will have much more to say about this work of re-creation, of remembering, in the final chapters of this book.

If my critique has so far been addressed to the reductionist, connectionist modellers, related arguments affect the holists. Consider the contrast between the relative ease with which programmers were able to develop chess-playing programs to Grand Master level and the difficulty in developing a robot system which can laboriously pile an orange pyramid onto a blue cube. Compare this with the ease with which an untrained human can toss an apple core into a waste basket five metres away, or, for that matter, play poker. Certainly, one can devise a program that can calculate the odds of drawing a flush against three of a kind, but poker involves a type of competitive psychology, of bluff, and it requires the appreciation and assessment of non-cognitive inputs for which, I believe, there can be no effective machine analogy. Humans might derive some pleasure from playing chess against a program, but it is hard to imagine much fun in betting against a computer in poker – perhaps a poker test should replace the Turing or Searle test?

The holistic attempt to bypass the problem of the brain entirely, and to concentrate instead on modelling the mind, empties all real biological phenomena from psychology in its attempt to explain behaviour. This arid, almost scholastic tradition argues that, if one can identify appropriate mind properties and processes, then one can model these properties and processes in abstract thought-experiments or sets of mathematical symbols and subsequently incarnate them (or, perhaps better to say, 'inmachinate' them) into silicon components, optical switches or magnetic monopoles just as well as into the complex bits of carbon chemistry out of which evolutionary processes have generated real brains.[34] Hence Boden's ludicrous aphorism 'you don't need brains to be brainy' – by which she means that you can model mind processes using whatever are the latest and most powerful computer systems without paying any attention to the underlying biology. All that is needed is machinery, or mathematical models of such machinery, which will respond to particular inputs with outputs which resemble those that brains might have. Such models work like machine-translation systems, which, given a sentence in French, can convert it to its English equivalent, even though the way a machine translates from one language to another in no way resembles how people may carry out a similar task.

This separation of mind from its actual material base is in some ways a reversion to the old Cartesian programme of brain/mind dualism. But equally, the insistence on treating the brain as a sort of black box whose internal biological mechanisms and processes are irrelevant, and all that matters is to match inputs to outputs, is reminiscent of the behaviourist programme in psychology, about which there will be much more to say in chapter 6. For the behaviourists, both mind and the brain were unnecessary concepts. Behaviour, of humans or other animals, was to be understood in its own terms, as simply a series of stimuli and responses linked in chains and adopted by the organism in response to rewards for approved and punishments for undesirable behaviour. If Cartesian dualism is thesis, and behaviourism antithesis, the programme of the holistic modellers epitomised by Boden's slogan is a sort of unholy synthesis.

To unpack the implications of Boden's phrase, let us replace it for a moment with its equivalent 'You don't need two legs in order to locomote.' Well, surely this is true: one can move along the ground with four legs, like a horse, many legs, like a millipede, or none, like a snake or snail; or with wheels on a track like a railway train, or with freely moving wheels like a car, or with caterpillar tracks like a tank. One could travel on a cushion of air like a hovercraft or by magnetic levitation. One could, conceivably, fly at speed on a magic carpet. All these modes of locomotion, in terms of a function defined as getting from A to B, may be equivalent. But the principles they employ show radical – in some cases fundamental – differences. If we should be interested in the specific question of human walking – how it is achieved, how it is acquired, and how it may be impaired temporarily by overconsumption of alcohol or lastingly by brain damage, studying fairy tales about flying carpets or the principles of the internal combustion engine, or even the wave forms of snake tracks, will be of only limited assistance. And it is just these specifics, however boringly trivial they seem to philosophers, that most brain scientists are interested in.

Lest this should seem an excessive polemic, consider the lists of mind processes which Minsky's book to which I referred earlier, identifies as to be modelled. Minsky sees 'mind' as a 'society' of arbitrarily defined and hierarchically arranged 'agents', of 'memory', 'anger', 'sleep', 'demanding', 'believing', 'more' or what-you-will. These labels are then given to a series of 'black boxes' (actually, black-outlined open ellipses and arrows on a page) connected up in seemingly random manner, out of which a theory of mind is supposed to emerge. This exercise seems a classic example of a top-down approach, far removed from the biological world which I – and I suspect most other people – inhabit. How am I to judge which, if any, of the multifarious, jostling and seemingly arbitrary members of Minsky's 'society of mind' have manifestations in identifiable brain processes? Suppose I were to invent a wholly different listing, including 'spirituality, belief in mutant teenage ninja turtles, scepticism and an inability to distinguish a hamburger from the polystyrene package in which it is encased', how would I be able to decide whether the 'agents' in any human brain were more similar to Minsky's or to mine? The point is surely that it is possible to generate a strictly infinite number of models capable of being drawn as black boxes linked together by arrows which can pass some sort of theory test – for if the output doesn't seem quite right one can always redirect the arrows, add more or turn them from hard into dotted lines. In this make-believe world we can with impunity draw legs on snakes or fix roller skates to human feet for convenience so as to get the 'right' output. But the material, biological world imposes rather more rigorous reality tests upon empirically based science than this. Minsky's holistic models are precisely the type of evocative analogy neurobiologists do not need in our efforts to understand biologically real brains and behaviour.

I am certainly not arguing that computer-modelling and its metaphors are worthless. Making metaphors does seem to be a fundamental part of the scientific endeavour, and without such approaches, neurobiology cannot succeed in its task of understanding real* – that is, biological – brains and their functions, whilst for AI itself, progress depends on a proper grounding in such biology rather than the imposition of purely rational, cognitive, top-down models. This is why I rather like both Steve Grand's and Igor Aleksander's bold speculations and consequent models. But AI needs to know its place,

---

* I have lost count of the number of times I have written the word 'real' in drafting this chapter, only, frequently, to edit it out on the grounds that the realist/social constructionist debate within both philosophy and the sociology of knowledge has become so intense that to use the term without hedging it round with a defensive thicket of qualifications is to run the risk of being charged with intellectual naïveté – or its wicked sibling, natural scientific arrogance towards other forms of understandings of the world. But I don't want to get into this debate here. For those who want me to lay on the line precisely where I stand, I am a qualified realist in the historical relativist tradition; put briefly, I claim there is a material universe of which we can have sound knowledge, albeit knowledge which is coloured by our historical and social locations, the current state of our technology and the framework within which we seek that knowledge.[35] That means that when I speak of 'real' brains I remain unashamedly prepared to defend their existence and my capacity to achieve objective knowledge about them and how they work against sociological or philosophical doubters. But not here, please; I have other fish to fry.

which is never to reverse its metaphors by privileging the model over the biology, but instead to show some humility in confronting that marvellous object of its study, the brain.

Memory, natural and artificial

At the beginning of this chapter, I described how Greek and Roman philosophers and rhetoricians contrasted two types of memory, natural and artificial. It was artificial memory that could be trained and analogised to writing on wax tablets, which led to the search for technological metaphor. Natural memory by contrast was a given, a human quality that required no explanation, merely recognition. Yet, as I have also argued, such is the powerful interaction of our technology with our very biology that the fact of creating a technology-driven society in which metaphors for memory have become central changes the very nature of our memory itself. The act of writing, as both Plato and the Zaïrean bard recognised, fixes the fluid, dynamic memory of oral cultures into linear form. The establishment of mass-circulation printed texts, as opposed to hand-crafted and variable manuscripts, as Walter Ong points out, further stabilises and controls memory, standardising and collectivising our understandings, creating

> a sense of closure not only in literary works but also in analytic, philosophical and scientific works. With print came the catechism and the 'textbook', less discursive and less disputatious than most previous presentations . . . Catechisms and textbooks presented 'facts' . . . memorizable, flat statements . . . the memorable statements of oral cultures tended to (present) not 'facts'; but rather reflections.[1]

Modern technologies – photography, film, video and audiotape, and above all the computer – restructure consciousness and memory even more profoundly, imposing new orders upon our understanding of and actions upon the world. On the one hand, such technologies freeze memories with all the rigidity of old Victorian sepia family portraits, providing an exoskeleton which prevents them from maturing and transforming themselves as they would do if untrammelled and without constant external cues within our own internal memory systems. On the other, they dissolve the barriers between fact and fiction in quite subversive ways. Think of the television penchant for docudramas, or the Woody Allen movie in which our hero suddenly inserts himself within a well-known sequence of Hitler addressing his supporters at a Nuremberg rally.

As I key these very words into my computer and they appear on the screen in front of me I am acutely conscious of these contradictions. Once, when I drafted a chapter, it would initially be laboriously by long-hand; I would edit my text and type it, thereby fixing it more or less permanently, if only because the labour of shifting and reordering material was then too great to contemplate for all but the gravest of reasons. Now all is fluid; this chapter, which set out to be written sequentially, has grown, not linearly, but throughout its entire body as draft sentences which once would be fixed in place here suddenly take wing and fly to other sections entirely. My memory for a once-planned sequencing of material is subverted by the liberatory power of the technology and refuses to be confined by the discipline I endeavoured to impose upon it. The Platonic, Ciceronian distinction between natural and artificial can no longer be maintained. If our study of memory must reject the computer model and metaphor in favour of the biological and 'real' we must equally recognise that the very nature and mechanisms of memory are being transformed by the technology whose explanatory power we are rejecting.

Perhaps this helps explain the extraordinary concern over the nature of memory that characterises art and literature. Jane Austen well describes this fascination when she has the stoic heroine of *Mansfield Park*, Fanny Price, say:

> If any one faculty of our nature may be called more
> wonderful than the rest, I do think it is memory. There
> seems something more speakingly incomprehensible in
> the powers, the failures, the inequalities of memory, than

in any other of our intelligences. The memory is sometimes so retentive, so serviceable, so obedient – at others, so bewildered and so weak – and at others again, so tyrannical, so beyond controul! [*sic*] – We are to be sure a miracle in every way – but our powers of recollecting and forgetting, do seem peculiarly past finding out.[36]

Writers have long tried to capture this peculiar power. Yet where the order and linearity of sequential memory characterises the nineteenth-century novel, as the twentieth-century dawns temporal order disintegrates. For Marcel Proust, whose twelve-volume *Recherche du Temps Perdu* is a long drawn-out struggle to recall and thus to transcend a painful past, the trigger which evokes the entire history is the taste of a madeleine cake. For contemporary writers the issue is even more problematic. In the Canadian writer Margaret Atwood's novel *Cat's Eye*, a complex, painful and fragmentarily recalled childhood is only grasped entirely when, almost at the end of the book, the heroine rediscovers and holds in her hand an emblem of her childhood, a mysteriously coloured cat's eye marble. Janet Frame's autobiographical novels are described by a reviewer as a 'meditation on the deceptive layers of memory: "where in my earlier years time had been horizontal, progressive, day after day, year after year, with memories being a true personal history;" as time progresses order and linearity disintegrate.[37]

For the converse, the excess of memory, we can turn to one of the more extraordinary short stories of that haunting Argentinian word-magician Jorge Luis Borges, a tale of a young man, Funes, who, in a manner not dissimilar to Simonides, seemed to be able to remember everything:

We in a glance perceive three wine glasses on the table; Funes saw all the shoots, clusters and grapes on the vine. He remembered the shapes of the clouds in the south at dawn on the 30th of April of 1882, and he could compare them in his recollection with the marbled grain in the design of a leather-bound book which he had seen only

once, and with the lines in the spray which an oar raised
in the Rio Negro on the eve of the battle of the Quebracho
. . . These recollections were not simple; each visual image
was linked to muscular sensations, thermal sensations, etc.
He could reconstruct all his dreams, all his fancies. Two
or three times he had reconstructed an entire day. He told
me: *I have more memories in myself alone than all men have had
since the world was a world*. And again: *my dreams are like
your vigils . . . my memory sir, is like a garbage disposal . . .*

A circumference on a blackboard, a rectangular
triangle, a rhomb, are forms we can fully intuit; the same
held true with [Funes] for the tempestuous mane of
stallion, a herd of cattle in a pass, the ever-changing flame
or the innumerable ash, the many faces of a dead man
during the course of a protracted wake. He could perceive
I do not know how many stars in the sky . . . [38]

It is no accident that, in the story, Funes dies young – of an over-
dose of memory, so to speak.

The problem of the freezing and fixation of artificial memory is even
greater when we move from the individual to the collective in
memory. Is it possible to create a space in which we can both assimilate into our own experience the meaning of the ever-screaming,
ever-napalm-burnt child on our television screens without simply
freeze-framing it, fixing it for ever and thus losing the dynamic of
real, biological memory? Such a fixation of images produces a peculiar and novel form of artificial memory, of the sort of which Plato
was particularly distrustful. What were once individual experiences,
to be made and remade in our imagination and memory, like my
own recollections of a birthday party, or the recovery of a long-lost
Mr Goss from some recess of the brain, are now public. They are
memories which we all share as part of collective experience, even
generations not yet born at the time of Vietnam or the Nazi death
camps. This makes them peculiarly powerful as a way of providing
social coherence. They have become part of our shared history. But

equally, we can no longer make and remake them in our own minds, assimilate them fully into our lived experience and consciousness, because they are for ever fixed by the video. Further, the power of the camera and the film-maker allows history – that is, collective memory – to be remade at the behest of a Big Brother with a memory hole, a revisionist holocaust historian, or a Minister of Education with a belief in the importance of not forgetting 1066.

The new technologies offer unrivalled prospects for, on the one hand artificial memory, and on the other, the production of completely fabricated memories of the Woody Allen type, or even a sort of social amnesia, the public erasure in which the Stalinist airbrushes remove Trotsky from the photographs of the makers of the Bolshevik revolution, or post-communism in the former Soviet Empire once again rewrites its past, turning those once considered heroes into demons and vice-versa. And small wonder that social movements find a continued need to rescue their own memories – and that they are as continually traduced – as Margaret Atwood found out when her novel *The Handmaid's Tale*,[39] an attempt to create a feminist history of the immediate future through the memories of a youngish woman in a masculine, fundamentalist-Christian domi-nated world, was turned into a big-budget movie which sold itself by eliminating memory and flaunting just that sexual availability Atwood's novel had criticised.

The collective in memory transcends the methods and models of the neurobiologist just as much as it does the metaphors of the computer, and will inevitably elude my attempts to grasp it here. I can go no further yet, but must return to the relatively safer shores of individual brain memory, and await the final chapter of the book to attempt to approach the collective as well as the individual. My conclusion from this chapter must echo the words of Fanny Price, only rejecting her view that the mechanisms of memory are 'peculiarly past finding out', for it is here that I wish to stake the claim for neuroscience. The matter cannot be left to the novelists any more than to the modellers, though the former certainly make a better fist of it, and are easier to read. Our task is to show that, just as there is more to memory than a wax tablet, a willing office boy or an ensemble of artificial neurons, there is also more than just the taste of a madeleine or the feel of a cat's eye. As the psychologist Dalbir Bindra put it:

> Psychologists have for too long tried to escape the reality of the brain in favour of physical, chemical, literary, linguistic, mathematical and computer metaphors. Now they must face the brain.[40]

Chapter 5

Holes in the head,
holes in the mind

Memory as image-making

SOME 2,000 YEARS AFTER CICERO WROTE *DE ORATORE*, AND NOT LONG
after Jorge Borges, in Argentina, was writing *Funes the Memorious*, the
Russian neuropsychologist Alexander Luria encountered a patient
with a peculiar problem, an apparent inability to forget. Luria
followed the case of the man he called simply S – Shereskevskii –
for some thirty years, from the 1920s to the 1950s, before describing
it in his book *The Mind of a Mnemonist*. During one particular session
in 1934, Luria gave Shereskevskii an elaborate nonsensical mathematical formula to recall:

$$N \cdot \sqrt{d^2 \cdot x \frac{85}{vx}} \cdot \sqrt[3]{\frac{276^2 \cdot 86x}{n^2 v \cdot \pi 264}} \, n^2 b = sv \frac{1624}{32^2} \cdot r^2 s$$

After seven minutes' study, he was able to reproduce it without
error – and fifteen years later, in 1949, with no advance warning,
Luria asked him to recall it. Shereskevkii did so perfectly, and here,
in his own words, is how:

Neiman (N) came out and jabbed at the ground with his cane (.). He looked up and saw a tall tree which resembled the square root sign ($\sqrt{}$), and thought to himself: 'No wonder the tree has withered and begun to expose its roots. After all, it was here when I built these two houses (d^2).' Once again he poked with his cane (.). Then he said: 'The houses are old, I'll have to get rid of them (X); the sale will bring in far more money.' He had originally invested 85,000 in them (85). Then I see the roof of the house detached (—), while down below on the street I see a man playing the Termenvox (vx). He's standing near a mailbox, and on the corner there's a large stone (.), which has been put there to keep carts from crashing up against the houses. Here, then, is the square, over there the large tree ($\sqrt{}$) with 3 jackdaws on it (^{3}). I simply put the figure 276 here, and a square box containing cigarettes in the 'square' (^{2}). The number 86 is written on the box. (This number was also written on the other side of the box, but since I couldn't see it from where I stood I omitted it when I recalled the formula.) As for the x, this is a stranger in a black mantle. He is walking towards a fence beyond which is a women's gymnasium. He wants to find some way of getting over the fence (—); he has a rendezvous with one of the women students (n), an elegant young thing who's wearing a grey dress. He's talking as he tries to kick down the boards in the fence with one foot, while the other (^{2}) – oh but the girl he runs into turns out to be a different one. She's ugly – phooey (v) . . . at this point I'm carried back to Rezhitsa, to my classroom with the big blackboard . . . I see a cord swinging back and forth there and I put a stop to that (.). On the board I see the figure $\pi 264$, and I write after it $n^2 b$.

Here I'm back in school. My wife has given me a ruler (=). I myself, Solomon Veniaminovich (sv), am sitting there in the class. I see that a friend of mine has written

down the figure $1624/32^2$. I'm trying to see what else he
has written, but behind me are two students, girls (r^2)
who are also copying and making a noise so that he
won't notice them. 'Sh,' I say. 'Quiet!' (s).[1]

As Luria describes it, when he gave Shereskevskii an exercise of
this sort, his behaviour was always the same:

He closed his eyes, raised his finger, slowly wagged it
around and said: 'Wait . . . when you were dressed in a
grey suit . . . I was sitting opposite you in a chair . . .
that's it!' and then and there quite rapidly he reproduced
without hesitation the information which had been given
to him many years before.

How did he do it? Luria got the impression that it was as if Shereskevskii
were reading through material, almost as if he had an open book in
front of him. Does this seem reminiscent of the prescriptions for memorising offered in the ancient arts of memory? Surely Shereskevskii had
not read them. Is there an echo of Funes in Luria's patient? Did Borges'
art imitate nature? Certainly, Shereskevskii's fate was not much more
cheerful than that of Funes. His marvellous gift held him in thrall. It
was hard for him to make normal human relationships because he
found it difficult to merge his recollections of people; a face full frontal
and a face in profile became for him two separate memories. It was
just as hard when he tried to hold down an ordinary job, as his memories of each new event kept intruding on the work in hand. There is
true irony in the way Shereskevskii ended his days, as a sort of music-
hall entertainer, a professional memory man.

Shereskevskii is an exceptional case, but he is not unique. In
1932, a broadcasting company in the United States hired a 'calculating genius', the Pole Salo Finkelstein, to tally the returns for the
presidential election of the year, because he was claimed to be faster
than any then extant calculating machine. Finkelstein was subsequently the subject of a study by the psychologists W. A. Bousfield
and H. Barry, who described his technique. When he was calculating,

the numbers appeared to him as if they were written in his own handwriting in chalk on a clean blackboard. He could then move the numbers around, add, subtract and manipulate them, with the results of the manipulation themselves appearing on the board.

People like Shereskevskii and Finkelstein have what is commonly, if inaccurately, described as a photographic memory. The description is not really accurate, in the sense that the images that are held are creatively manipulable by the memoriser, not simply fixed items to be referred to, as is clear from the sorts of mistakes that people remembering in this way tend to make. However, the technical name for this kind of memory – eidetic, from the Greek for 'image' – simply scientises the same concept.

Scientific knowledge about the world comes from two types of study: the search for underlying regularities in seemingly dissimilar phenomena; and the analysis of the causes of variation – small differences in seemingly similar phenomena. Eidetic memory is interesting for both reasons. First, because it is so different from the ways in which most adults seem to remember things. By its very difference it opens questions about what is for most of us normal memory that we would otherwise not think of asking. The rarity of eidetic memory, coupled with the fact that to possess such a capacity seems not to make for much success in life, suggests that it may not be so beneficial a gift. To be able to synthesise and generalise from past events, to abstract from them, indeed, to forget them, may thus be as essential for survival and effective action in the world as is the capacity to remember them in the first case. If the ancient art of memory were intended to enable the rest of us to learn how to behave in a way that Shereskevskii, Finkelstein – and Funes – could not help behaving, then its success depends on the fact that *except* when we are employing the technique we do *not* remember in this way.

Yet, although eidetic memory is rare in adults, it seems to be much more frequent in young children. Think back to your own early memories, and it is probable that you will recollect them as a series of snapshots, fixed or frozen in time. That is certainly the way those images of my own childhood, which I described at the start of chapter 2, or the account Ingmar Bergman gives of his earliest memory, appear. Nor are such accounts exceptional. Once upon a time there was much interest in eidetic imagery and memory – more than 200 studies on it had been published before 1935, although subsequently it has very much dropped out of the mainstream of psychological research. The early research suggested that whereas eidetic memory is relatively rare after puberty, around a half of the elementary school children who were studied appeared to possess it. In the 1960s and 1970s Ralph Haber, Jan Fentress and their colleagues, following up these studies found a somewhat smaller proportion of United States elementary school children with eidetic memory, but noted that the capacity was widely distributed amongst young children of both sexes, independent of ethnic origin, class or school performance.

In a typical study, Haber would show children a coloured picture of Alice and the Cheshire cat from an illustrated *Alice in Wonderland*. In the drawing, the cat sat on a tree, striped tail curled behind it. Children having been briefly shown the picture could later answer questions in detail about it – for instance when asked how many stripes were visible on the cat's tail, they would behave as if they were counting them off from some sort of mental image.[2] Similarly, children shown a picture with writing on it in a foreign language could subsequently spell out the words as if reading them from an open book.

Many, if not all young children apparently do normally see and remember eidetically, but this capacity is lost to most as they grow up. What is in young children an apparently general capacity has become a remarkable rarity in adults. This change in the quality of memory perhaps also helps to account for the very different ways we remember our childhood experience and our adulthood. This is not simply a matter of time passing. A thirty-year-old man does not remember his ten-year-old self in the same way as a fifty-year-old remembers his thirty-year-old self although the time lapse is the same in each case. For that matter a ten-year-old girl does not remember herself at nine in the same way as a fifty-year-old woman remembers herself a year previously. Memory itself is a developmental process and between nine and ten the quality of memory has changed, whereas between forty-nine and fifty it is seemingly stable. Some time before puberty there is for most of us a transition in how we perceive and remember the world, a transition which means that our adult memories are strangely disarticulated from our childhood ones.

Somewhat to my surprise, the psychologists who have studied eidetic phenomena do not seem to have remarked on the significance of this dramatic change in what would seem to be a fundamental human activity. The way is open therefore for me to speculate on its significance. One of the important differences between human and computer memories that I discussed in the last chapter is the fact that, for humans, to memorise something is an active process. Consciously or unconsciously we select salient information that we need to commit to memory from the blooming, buzzing confusion of the environment around us. To help in this selection, we possess quite elaborate blocking or filtering devices to prevent new information from cluttering up our memories.

For instance, there is a mechanism called *perceptual filtering* which ensures that, of all the information arriving at one's eyes or ears at any given time, only a small proportion is actually registered and even briefly remembered. As I sit and type these words, I do not see the odd debris on my desk, the view from the window, or even my own hands as they strike the keyboard. Yet all these items are within my field of vision; I am just not concentrating on them, and it would have been better to write that I do not perceive than that I do not see the debris. An even better example is of course the well-known cocktail party phenomenon, by which in a room full of babble one can concentrate – more or less! – on the voice of the person talking to one, yet can switch almost completely at will to listen in on other conversations going on around. But it is not as if we have closed our ears to the other conversations – they can easily intrude unasked, as for example if one hears one's own name mentioned elsewhere in the room. So the incoming information, in sound or vision, is entering our brains, and is there being filtered for relevance by processes of which we are largely consciously unaware, but which classify it according to its importance to us by criteria which are clearly rather effective. One of the problems for Shereskevskii and the others may have been that for them this filtering mechanism was only partially functioning.

But one person's relevant information is another's irrelevance. The criteria by which we filter inputs are themselves learned during our own development. Humans, as a species, are characterised by tremendous flexibility; we can survive in many different types of environment, and our survival is made possible by a lengthy period of childhood, during which we have to learn the appropriate survival skills. At birth, we may guess, all types of input may seem to be of about equal relevance; within only the widest possible classification rules everything must be registered and ordered so as to enable each individual to build up his or her own criteria of significance. At this time eidetic memory, which doesn't prejudge the importance of inputs, is vital, because it gives the greatest possible range over which inputs can be analysed. But as we grow up we learn to select from key features of our environment.

Think of the difference betweeen an urban and a rural childhood. The urban child learns street wisdom, to classify vehicles and estimate their roadspeeds, to sum up neighbours and strangers for potential threats or promises. The rural child can tell a bull from a cow and knows that not all trees are identical brown trunks and branches surmounted by green leaves. Even the seasons affect the rural child more profoundly than the urban. These classifications based on experience permeate our adult life, how we perceive and what we remember, in ways that we all recognise, though – except for the novelists and film-makers amongst us – we may find it hard to articulate them.

During the major proportion of human evolution – indeed until the last few generations – it was a good bet that the environment in which one grew up would be virtually identical to that in which one spent one's adult life. Hence the eidetic memory of childhood, enabling rules of perception to be developed, could smoothly transpose at the approach of puberty into the more linear forms of adult memory, while incorporating, for each individual, a uniquely tailored set of such rules which would order their later experience. Even today, at times of rapid environmental change within each person's life experience, such a transition, which must reflect some fundamental biological mechanism, is of survival value. Meanwhile, each of us is left with the fragmentary eidetic images of our own childhood.

The time-course of memory

The transition from childhood to adult memory is dramatic, from imaged and timeless to linear and time-bound. In most adults memories seem to be formed in orderly sequence and undergo a series of transformations from the time when they are first acquired to their later, more permanent form. Only a few individuals seem to retain in adulthood the eidetic memory of their childhood, a sort of arrested development, like the tadpole that won't metamorphose into a frog. We marvel at their talents, perhaps recalling our own childhood eidetic capacities – but usually fail to see at what cost such talents are bought. *Idiots savants*, miracle calculators or professional stage performers impress us all – but no one would wish to suffer the fate of Shereskevskii or Funes. Better to submit to the linearities of time that the transition to maturity brings.

This time dependence can be demonstrated by a version of the number-recall example described in the last chapter. Show a person a string of, say, seven digits, for thirty seconds, and ask them to repeat the numbers back a few minutes later. Most people can manage such a task without too much difficulty; clearly they have remembered the numbers. If we are asked the numbers again in an hour or so, most likely we will no longer be able to recall them – they seem to have been forgotten. But if for instance one were told that it is important to remember the numbers – perhaps the seven-figure string is a telephone number – then there is a strong chance that we will be able to recall them not merely after an hour or so, but days later.

This sort of observation can be made more precise, in the type of experiment favoured in the days before psychologists had to look over their shoulder at biologists. The great pioneer was Hermann Ebbinghaus, whose book *Über das Gedächtnis* (*On Memory*), published in 1885, heralded the arrival of a new breed of psychologists, those who rejected the previously favoured methods of speculation and introspection in an endeavour to study mental processes with the same type of quantitative precision that had long characterised physics and was beginning to be reflected in biological sciences such as physiology.

Ebbinghaus was concerned in this, his most enduring study, done whilst he was relatively young and before he retreated into a world of provincial teaching, first to classify memory, as between voluntary and involuntary remembering; second, to explore individual differences in the content and quality of memory; third, to try to identify similarities amongst individuals in the forms of memory – that is, to ask whether there were general laws of memory formation. To do this, he explicitly rejected the metaphorical approach to memory which dominated chapter 4 of this book:

> It is because of the indefinite and little specialised character of our knowledge that the theories concerning the processes of memory, reproduction and association have been up to the present time of so little value for a proper comprehension of these processes. For example, to express our ideas concerning their physical basis we use different metaphors – stored up ideas, engraved images, beaten paths. There is only one thing certain about these figures of speech and that is that they are not suitable.[3]

To explore these general laws, he invented the simple technique which in various forms has been a staple psychologist's tool ever since – that of the nonsense syllable, a series of three letter sets each composed of a vowel between two consonants, as for instance: HUZ; LAQ; DOK; VER; JIX. Using himself as subject Ebbinghaus then explored the conditions required to remember such lists; numbers of readings, spacing and so forth, until he could make two errorless readings of the entire list. Once the list was learned, he could then test how successful he was at recalling it at various subsequent times from minutes to days. To quantify this process of recall, all that he had to do was to note how many readings of the list were necessary, at any given time after it had been learned, to once again be able to repeat it without error.

A number of general rules could be derived from such observations. For instance, in any such list of, say, a dozen nonsense syllables, some are easier to remember than others – in particular, those at the beginning and at the end of the list. These are the so-called primacy and recency effects. They may seem obvious when described so simply, but what Ebbinghaus did was to demonstrate clearly that in this case at least common sense could be proved right – the first step in any science. In addition, he showed that if a list is once learned, it becomes easier to relearn subsequently. A comparison of the number of trials required to learn it the second time with those required first time round provides a calculation which has become known in the psychology literature as *savings* – the measure of memory.

The use of the savings score enables one to specify more precisely the loss and stabilisation of memory with time. Thus in one set of experiments, involving a series of eight different 13-syllable lists, Ebbinghaus found that the savings when he tested himself 20 minutes after training were 58 per cent, at one hour 44 per cent, at 24 hours 34 per cent and by 31 days 21 per cent. Thus most of the memory loss occurred within the first minutes after training; once the memory had survived that hurdle it seemed much more stable.

This and other studies led to the view that there are a series of processes involved in memorising an item; once it has successfully traversed the perceptual filter and immediate memory it is located in some type of short-term 'store'. (Note the quasi-computer terminology of the concepts of short- and long-term memory 'stores' – although the origin of the term, perhaps in harvesting, long pre-dates that technology too. I shall continue to use the term in this chapter as a convenient shorthand, though I would not like it to be misunderstood; please treat the word and its associated metaphorical baggage with some caution!) Much of the material in the short-term store becomes lost – and presumably functionally so, as it is really not biologically adaptive to remember everything for ever that we need to recall for only a few minutes. But what does manage to survive the short-term filter gets located in some much more permanent store, where it seems to last indefinitely. The attrition of savings in the Ebbinghaus experiment in the days after learning could be the result of forgetting – or it could merely be that he had been doing other things, perhaps learning similar lists, in the interval, and there was some sort of interference going on. As I said earlier, once something has been permanently stored like this it is very difficult to distinguish between such possibilities.

William James, another of the pioneers of modern psychology, saw short- and long-term memory rather differently from Ebbinghaus. For him, short-term was *primary* memory, that which is currently being attended to; 'it comes to us as belonging to the rearward portion of the present space of time and not to the genuine past'. This primary memory James contrasted with *secondary* memory,

> The knowledge of a former state of mind after it has already dropped from consciousness; or rather, it is the knowledge of an event, or fact, of which we have not been thinking, with the additional consciousness that we have thought or experienced it before.[4]

In the decades following Ebbinghaus, psychologists refined and extended his tests. Endless variations explored the effects of interference on memory, by presenting similar or conflicting material either before the learning task (*proactive*) or between learning and recall (*retroactive*). There were comparisons of learning by hearing and learning by seeing, and, in a famous series of experiments by Frederick Bartlett in Cambridge in the 1930s, on strategies in learning complex material. It became clear that practice in memorising tasks led to improvement, often because the learner found ways of grouping the material together into sections (chunks) that could be more easily remembered – like my 48-digit task in the previous chapter. In the 1950s the United States psychologist G. A. Miller noted with some amusement that, just as seven seemed the maximum number of digits in a span that could be recalled readily, so also could seven chunks; he called his now classic paper 'The magic number seven plus or minus two'[5] and used this observation to derive an information-processing model of memory (an approach whose limitations I have already discussed).

Thus over the seventy years between Ebbinghaus and Miller psychologists painstakingly catalogued the capacity, scope and limitations of human memory as studied in the controlled context of the laboratory. But short of ways of actually looking inside the brain during learning and memory formation, there are strict limits to what could be learned about the brain mechanisms of memory by such experiments and observations in normal adults. It became time to turn again from trying to identify general rules of memory based on similarities to exploring the consequences and meanings of differences between individuals. In particular, could disorders of memory – memory loss, or amnesia – help in the understanding of its normal mechanisms?

Diseases of memory

Just as there is a transition in memory between children and adults, so at the other end of the life cycle, memory capacity seems to change again, and people often suffer losses or lapses of memory. It is a common enough experience amongst elderly people that although they can remember well episodes from their own past and childhood, they often seem not to be able to remember what they had for breakfast that morning. Such lapses are often ascribed to the inevitable processes of ageing or to a failing brain – perhaps, in today's computer analogy, the brain memory store is full and cannot add more input. I am not wholly convinced that this is necessarily the case. What one had for breakfast that morning is often not a very interesting item of information; if one has had many breakfasts in one's life, and now lives in a rather restricted and uninteresting environment where one breakfast is pretty much like another, why bother to remember it at all? Far more interesting, in taking stock of one's life, to struggle to put one's earlier and perhaps richer experiences into some sort of perspective. On the other hand, other commonly quoted complaints, such as forgetting where everyday objects such as keys have been left, or failing to remember a person's name, are harder to dismiss in quite so cavalier a manner.

The modern study of diseases of memory begins in 1881, at almost the same time as Ebbinghaus was publishing his experiments, with a book by the French clinician Theodule Ribot entitled *Les Maladies de la memoire*. (The philosopher Ian Hacking dates the 'scientisation of the soul' from this key moment.)[6] 'If, following Ribot, the study of memory failure is removed from the anecdotal world of our own lived experience into the ostensibly more objective environment of the laboratory or the consulting room, it becomes possible to be a little more precise about such matters. There are conditions in which short-term memory seems unimpaired but items cannot be recalled after half an hour or so – that is, they cannot be transferred to long-term memory. In some pathological conditions the loss of memory can be even sharper, so that even a short digit span of three or four numbers cannot be repeated back after more than a few minutes. To evaluate patients who may appear to suffer from lapses of memory, clinicians try to probe the time-course of what may be remembered and what is lost. For instance, by asking patients first to repeat back immediately a short sequence of digits, then to remember a list of items presented a few minutes previously, and finally to describe some past experience, such as what they did last summer, it is possible to identify different classes of deficits in memory mechanisms, often associated with ageing – and in particular with specific diseases of ageing. The first question relies on immediate recognition and memory for the answer, the second requires that events a few minutes old can be recalled, and the third that long past memories can be retrieved. One might expect that the older the memory, the harder it would be for a person with a memory deficit to locate it, but the reverse turns out to be the case; old memories are preserved while recent ones are more easily lost.

The best-known diseases of this type are Alzheimer's disease and Korsakoff's syndrome, both named for the individuals who first described them (Ribot, despite his pioneering role, appears not to have achieved that height of eponymous glory for a clinical scientist, having a specific condition named for him). Clinical research on these conditions is of course primarily directed at understanding their causes, biochemical mechanisms and possible treatments, but there is always a lurking subsidiary hope that an understanding of the memory deficits suffered by Alzheimer's and Korsakoff's patients might cast light on the mechanisms of memory.

Alzheimer's is one of the most debilitating and dreaded of common diseases of ageing – estimates are that it will affect something over a million people in the United Kingdom by 2010, a tragedy both for them and for their family and carers. Until a few years ago it used to be described more generally simply as senile dementia, now sometimes as 'senile dementia of the Alzheimer type', although people at the height of their powers can also be struck down with it. People suffering from Alzheimer's disease experience a frightening loss of the sense of their own identity and access to their store of personal memories – those memories which for all of us form such a central buttress to that very personal identity. The causes of the disease are still unknown – indeed there may be many possible precipitating factors. A tiny percentage – less than 5 per cent – is familial, that is, has a clearcut genetic origin. The presence of some other gene variants are known to be risk factors, as are environmental insults such as concussion. Whatever the causes, the brains of Alzheimer's sufferers shrink and the neurons themselves change their appearance; their internal structures become disorganised, forming patterns which, because of their appearance under the microscope, are called tangles and plaques. I will have a good deal more to say about the biochemistry in chapter 12.

The cause of Korsakoff's syndrome is better understood; it is usually a consequence of deficiency of the vitamin thiamine, resulting from chronic alcoholism (though it can be produced in other ways, for example from viral encephalitis or some sorts of brain tumour), and in the syndrome the sufferer's brain shrinks to only a fraction of the size of a normal person's. As with Alzheimer's, people with the Korsakoff condition suffer characteristic memory losses, especially for recent events. They cannot carry out either verbal or non-verbal memory tasks, find it difficult to remember simple everyday facts and cannot plan quite straightforward tasks, but they have much less difficulty in remembering events from the more distant past. Thus Korsakoff's patients have their greatest problem with recent memory.

But despite the relatively unambiguous nature of the diseases, and the memory deficits associated with them, so general and devastating are the consequences of the Alzheimer's and Korsakoff's conditions that it is difficult to derive any specific understanding of the memory processes from them; and, such diseases apart, there is no real way to be sure what is meant by saying that there is a loss of memory with ageing. Because during ageing there is some loss of neurons from the brain, which also occurs much more dramatically in the disease conditions, it is tempting to assume that the 'normal' forgetfulness of old age is but the disease writ small. This has led to some clinicians speculating that there may be a much more widespread disease of middle age, so-called 'age-associated memory impairment'. A number of pharmaceutical companies have begun to invest heavily in research on this supposed condition and the identification of possible drugs to treat it, and indeed, in the United States, the FDA has somewhat reluctantly come to the conclusion that the 'disease' really exists. In Britain, psychologists and clinicians so far remain more sceptical. But the evidence, either for the existence of a specific disease, or for the popular belief that there is a massive loss of brain cells during ageing, is not good. What is reasonably certain is that mental processes change with ageing, so that a certain amount of youthful speed of processing and reflexes are lost, but instead most of us develop better strategies to cope with and manipulate information. In a culture less obsessed than our own with speed for speed's sake, we might be prepared to regard this change with age as something rather positive – what other societies have called wisdom!

A similar re-evaluation of the biological meaning of the loss of neurons from the brain, if it occurs at all, is also necessary, for the significance of any such loss is not well understood. Early brain development in the foetus and newborn is itself associated first with a massive proliferation of cells, and then by a steady drop in number, but the space once occupied by the lost cells is taken up by an increase in the branching and synaptic connections made by those that remain. Is this a 'good thing' or a 'bad thing' so far as brain function is concerned? We don't know; it seems to be part of a normal developmental sequence, but one should not be seduced into a simple assumption that more means better, and that cell loss with ageing would therefore automatically be deleterious, if it were shown definitely to occur to any significant extent. Indeed, there is now good evidence for something hitherto regarded as impossible – the formation of new neurons in the brain even in adults, derived from a small population of stem cells – that is cells which still retain their embryonic capacity to grow, divide and differentiate.

Nor at the moment are there any effective biologically based treatments for either the diseases or the 'normal' loss of memory with ageing, although, as will become clear as this book proceeds, the prospect is not entirely hopeless. Remembering involves in the first place experiencing and learning something, and subsequently recalling it. Is the life of an elderly or ill person so impoverished that it ceases to be of interest to learn or remember? Do the filtering processes simply block out the trivial day-to-day events so that they are no longer introduced into memory? Are items placed into memory but then cannot be found again, like the memory of a person's name which we 'know' but cannot 'recall' even though it be on 'the tip of our tongue'? Or are they genuinely forgotten – lost irretrievably from memory? The truth is that in most cases it is very hard to distinguish between these possibilities, despite the best endeavours of many experimental psychologists over the years.

Losing short-term memory

The idea, derived from the early psychological studies, that there are immediate and short-term memory processes which last for minutes to hours, followed by more permanent memory, soon received confirmation from observations of the effects of injury. For instance, a blow to the head commonly results in unconsciousness and concussion; on recovering consciousness it is usually the case that a person cannot remember the events immediately leading up to the accident. The coshed movie detective or cowboy hero who staggers to his feet, rubs his head, looks around and says 'Where am I' is not so far off the real thing. Similar consequences – called *retrograde amnesia* – follow anaesthetic coma or the disruption of the electrical activity of the brain involved in electroconvulsive shock therapy (ECT). The 'time out' of memory that such insults produce tends to be about thirty minutes, which is in accord with the loss of savings in the Ebbinghaus experiment.

But as with most of the statements we can make about memory, it is as well not to be too precise about this one. Thus during recovery from concussion people do not regain all their facilities, or indeed all their memories, at once. Rather they seem to pass through a series of stages in which increasingly complex functions are recovered, from simple reflexes through disconnected to purposive movements and speech. At this stage, there are strange gaps in memory and recognition of people and things, which can appear as losses of memory for things that occurred even years before. This memory gap steadily narrows as the patient recovers, as described in a case discussed by the London-based neuropsychologist Ritchie Russell concerning a 22-year-old man who, in the summer of 1933, was thrown from his motorbike, bruising the left frontal region of his brain, though without fracturing his skull:

A week after the accident he was able to converse sensibly, and the nursing staff considered that he had fully recovered consciousness. When questioned, however, he said that the date was February 1922, and that he was a schoolboy. He had no recollection of five years spent in Australia, and two years in this country working on a golf course. Two weeks after the injury he remembered the five years spent in Australia, and remembered returning to this country; the past two years were, however, a complete blank as far as his memory was concerned. Three weeks after the injury he returned to the village where he had been working for two years. Everything looked strange, and he had no recollection of ever having been there before. He lost his way on more than one occasion. Still feeling a stranger to the district he returned to work; he was able to do his work

satisfactorily, but had difficulty in remembering what he
had actually done during the day. About ten weeks after
the accident the events of the past two years were
gradually recollected and finally he was able to remember
everything up to within a few minutes of the accident.[7]

This type of observation suggests that the transition from short-
to long-term memory is not an entirely ordered clock-like process.
There are of course always problems about inferring the nature of
normal processes from the study of the effects of injury, as in the
example offered by Russell. However, a less dramatic example may
indicate the same flexibility about the duration of memory. My laboratory is about four miles from the railway station, and most evenings
during the week I drive from the lab, park my car at the station and
go home. The next morning I collect my vehicle again and drive to
work. The station car park has space for several hundred cars, spread
over a wide area, and where I park my car at night can vary markedly.
Next morning I can with a slight effort normally remember exactly
where I have parked and walk without error to the car. This isn't a
permanent memory, as I cannot under these circumstances remember where I had parked the car two days previously. But nor is it
merely a question of simple longish-lasting short-term memory, for
in the evening when I return to the station in the car I cannot
remember where I parked it that morning, whereas if I leave the car
at the station for a few days while I go off to a meeting, I *can* remember
where I parked it on my return just as clearly as if it had only been
the night before.

Recognition versus recall

Clearly this flexible capacity to remember or not depending on
circumstances makes good adaptive sense – it is the sort of property
we might wish that a memory mechanism could possess – but it
makes clear that, although it seems possible to distinguish conceptually between short- and long-term memory, there is nothing simple

and mechanical about the transition between them (which is why some psychologists now question the legitimacy of making this distinction at all – a matter to which I will come back). But let me return to a different distinction between forms of memory which may cast some light upon how I can find my way around a car park so seemingly unerringly.

As a youngster I was very fond of what we called Kim's Game (after the hero in a Kipling story), though others may know it differently. In this game, someone prepared a tray on which there were a number of small familiar objects – generally about twenty – pencil, toy motor car, egg cup and so on. The tray was covered with a cloth until the start of the game; it was uncovered for a fixed period, generally a minute, then covered again. The task was then to write down a list of as many of the objects on the tray as one could remember. I used to pride myself on my good memory – I could generally manage to get seventeen or eighteen out of twenty of the objects right. In fact, this is a fair average for most people. But the number one can remember does not increase much above this level, even if the number of objects on the tray is increased. Like our memory for digit strings, there seems to be a natural limit to what can be recalled. Memory, it would seem, is readily saturable.

In 1973 the Canadian psychologist Lionel Standing carried out a different version of the same game. Groups of volunteers were shown a series of slides of either pictures or words, in succession, each for some five seconds, at three-minute intervals. Two days later, the subjects were examined on their capacity to remember the slides, by being shown a further series of slides; this time with a double projector so that two pictures were visible side by side; in each case one had been drawn from the previous set and one was novel. All they had to do was to decide which of the two pictures, the left- or right-hand one, looked more familiar to them. Standing's question was: how many photographs could his subjects correctly identify? This seems an easier task than the recall demanded in Kim's Game, because all that is necessary is to *recognise* a photograph, and one has a 50 per cent chance of being right by simple random choice anyhow – so one might expect some improvement on the twenty or so objects which is the limit to recall. But how big an improvement? To Standing's amazement, the difference was dramatic. He went on increasing the numbers of photographs used in the trials up to an astonishing ten thousand, and even then the error rate was very low, and seemingly did not increase at all with the number of items to be remembered. Standing concluded that, for all practical purposes, 'there is no upper bound to memory capacity'; the recognition memory of the subjects seemed unsaturable.[8]

The implications of this extraordinary result are considerable. By contrast with the evidence that there are limits to what is transferred from short- to long-term memory, it would seem that some accessible trace of each of these photographs must have been left, enough to enable a new photograph to be compared with that trace and classified as familiar or unfamiliar. On this basis it could be argued that nothing is forgotten, provided we know how to ask if it is remembered. On the other hand recognition is easier than recall in that in recognising we are presented with a rather limited choice between possibilities; in the case of Standing's photographs, they either had or had not been seen before. The experience is familiar to us from somewhat more complex tasks. Asked to describe the face of a person we know slightly, most of us have some difficulty. (Actually, it is hard enough for most of us to describe the face of someone we know well – though we can instantly recognise that person if he or she comes into the room. Perhaps painters have a special skill – or training – that others of us lack.) But if we are instead asked to build a likeness from a police photofit collection, the task becomes easier; we can create a reasonable convincing likeness from the photofit components. The arrival of the person, or the assembly of the component features, like Standing's photographs, restricts the field over which we have to search.

This makes my capacity to remember where my car is located less puzzling; my choice is not unrestricted, but is limited by the dimensions of the car park and constantly clued by a series of specific contingencies (was it sunny or raining last night; was I going home late or early . . .). Locating my car is in some ways more like a recognition than a recall memory task.

Forms of memory

It is time to take stock. The richly textured world of human memory, out of which I can evoke snapshots of my wartime childhood, the taste of a meal eaten a month ago, images of my son's face, the location of my car in the station car park, the text of a research paper I read yesterday and a telephone number given to me a few moments ago . . . in the century and a quarter since Ebbinghaus and Ribot, all of these have been brought into the psychology laboratory in an attempt to classify their phenomenology, deduce regularities and understand their mechanism by analogy with the already classical methods of physics and physiology. Following Ebbinghaus and some of the deductions from diseases of memory, it seemed as if one could align memory along a temporal dimension, in which, in the periods of seconds to minutes to hours following some new experience, processes of perceptual filtering introduced items into a labile, transient short-term memory and from there into a permanent long-term memory. Short-term memory could perhaps itself be broken down further into immediate and recent memory, but the classification remained within a single dimension, that of time.

But something seems to have gone wrong. Whenever we try to fit the full range of the experience of memory within this single dimension, it escapes us. Not merely does the temporal ordering of forgetting not fit simply into the short-term/long-term frame, but there are other contrasts too. Already this chapter has referred to differences between eidetic and linear memory, between recognition and recall memory. And the widespread feeling that the short-term/long-term distinction imposed as apparent simplicity on a more complex reality has led many psychologists to reject them entirely, replacing them with the less time-bound concepts of 'working' and 'reference' memory, more analogous to James's concept of primary and secondary memory. Of course, in this they are, as James himself was, reverting to the older philosophical tradition, running back through St Augustine to Aristotle, that distinguished between events distanced from the present moment only by a few seconds or minutes, and those positive acts of memory by which we recreate in our minds scenes from the distant past. One of the obstacles in the way of collaboration between psychologists and neurobiologists in this field is that for those of us (neurobiologists) working with animal memory, the short-term/long-term distinction makes experimental sense, whereas for psychologists – especially those studying human memory – working versus reference seems more useful.[9]

The time dimension cannot be ignored, but it is clearly impossible to fit the whole range of forms of memory within it. It is necessary to introduce other dimensions, and create a taxonomy of forms of memory. This is a murky endeavour, for nothing in science raises so much controversy as attempts to classify and order the universe of observables. From the days of Linnaeus, who did this for biological species in the eighteenth century, onwards, taxonomists have been a quarrelsome lot. It is only necessary to think of the bitter arguments amongst palaeontologists about the classification of fossil remains – and at least they have had material objects to argue about. This disputatiousness is partly because the universe is a continuum, and our endeavours to identify discontinuities owe as much to our own human ingenuity and determination as they do to the material reality of what is being classified. Depending on criteria adopted and definitions used, classifications – or taxonomies, to use the biologist's term – can fit as comfortably as a track suit or be as oppressive as a straitjacket.

Nonetheless, the task must be attempted. But what sorts of classifications should we seek, and what sort of evidence might be considered helpful? For instance, we could distinguish between childhood (eidetic) and adult (linear) memory; verbal and visual memories; recent and long-past memories; recognition and recall memories. All these are classifications that this chapter has already discussed. We could look at the 'natural' dissociations of memory in patients like those suffering from Korsakoff's or Alzheimer's disease, and ask which types of memory are lost, and which are spared? Such is the nature of things that it may not be that each approach will give the same answers.

One of the first types of distinction to be extracted was that between memories for *doing* and *naming*. Think of learning about a bicycle. You can learn how to ride it, and you can learn what it is called – bicycle. The two types of learning involve very different processes and are affected in different ways by time and by memory deficit diseases. It is a common enough experience that, once having learned, with however much difficulty, to ride a bicycle, one never subsequently forgets. After many years' lapse of time, one can climb into the saddle, and after an initial wobble, set off. That sense of balance which took so many trials to gain as a child is now almost entirely saved. The memory is there in one's brain and body. I experienced this at first hand a while back when, on holiday in Sweden, I mounted an elderly, gearless – and handbrakeless – machine for the first time in some thirty-five years and took off on a fifteen-mile ride. Within minutes I felt as confident as I had done as an adolescent all those long years before – providing I was on the flat. But never having in the past learned to brake by pedalling backwards as opposed to squeezing a handbrake, slowing down on slopes was an alarming new skill I had to learn rather slowly!

Yet it is possible entirely to forget, as a result of brain damage, that a two-wheeled object which one can sit on, pedal and move about with is actually called *bicycle*. Remembering *how* and knowing *that* seem to be different types of process, and patients suffering from amnesia are generally well able to learn new skills – *how* knowledge – from riding a bike to doing a jigsaw, although they have difficulty in remembering the actual experience of learning.

Although the distinction between these types of memory has long been recognised, it was given prominence and more formal taxonomy in the artificial intelligence community in the 1970s, when 'how' memory was formally named *procedural, skill* or *habit* memory. By contrast, 'that' memory is called *declarative*. The terms were taken over into psychology by the San Diego neuropsychologist Larry Squire*

---

*It will be apparent to those who are familiar with neuropsychology that my treatment of the taxonomy of memory owes much to Larry Squire's poineering contribution to the field; he is also the best-placed of all the neuropsychologists I know to appreciate why, in the previous chapter, I offered to replace the Turing Test by a poker test for computers, which is why I find it especially odd that he should still apparently favour some type of information-processing model for memory.

in the 1980s.[10] Procedural memory thus seems very different in kind from declarative memory. Indeed, the distinctions I drew earlier between short- and long-term memory seem not to apply to procedural memory. Both the way we learn *how*, and the way we subsequently remember *what* we have learned, are different in kind from the ways in which we remember *that*.

But declarative memory is itself not a unitary phenomenon and can be further subdivided into *episodic* and *semantic* memory, terms introduced by the Canadian psychologist Endel Tulving in the 1970s. By episodic, Tulving meant the memory of events in one's own life history; by semantic, he meant knowledge that is independent of that history. My knowing that there was a world war in 1939–45 is in this sense semantic memory; remembering my experience of wartime bombing is episodic.[11]

These multiple taxonomic categories derive initially from psychologists studying what I can best describe as the *phenomenology* of memory, by which I mean the attempt to describe and classify the main features of memory as a phenomenon without raising the question of how these features may be explained in terms of underlying processes. For neuropsychologists or neurobiologists, however, such phenomenology is merely the starting point of our enquiries; it tells us what we are required to account for. This search resolves into two types of question. First, do different types of memory processes involve different brain systems; that is, are memory systems in the brain spatially segregated, or is virtually the entire brain involved in all aspects of memory? And second, if the formation of memory requires changes in the structure, the biochemistry or the physiology of brain cells, do these changes themselves differ depending on the types of memory being stored?

The second of these questions is a present-day preoccupation, very much my own, and hence one to which a good portion of the later chapters of this book is devoted; the first question, however, has much deeper historical and philosophical roots. It is in essence an aspect of a long-running dispute in the brain sciences concerning whether specific mind functions can be localised to particular regions of the brain, a dispute which dates back at least as far as the end of the eighteenth and beginning of the nine-teenth century.[12] This was the period when extravagant claims were made that not merely could everything from mathematical ability to the love of children be localised to different brain regions but that these capacities were reflected in, and could be measured by, studying the very shape of the head. Phrenology, in the hands of its originators, F. J. Gall and J. C. Spurzheim, brought the cause of localisation popular acclaim and ultimate scientific discredit. It was not until the work of Paul Broca in France, David Ferrier in England and other later nineteenth-century pioneers that it became clear that at least some aspects of brain function could be localised, in the sense that damage to them results in more or less specific functional deficits, from motor paralysis to loss of speech. But such studies raised then and still raise today serious conceptual issues: is it legitimate to draw conclusions about the workings of the healthy brain from study of its failures as a result of damage or disease? Many dispute it. Regarding the brain and its functioning more holistically, as the workings of an integrated system, they are unconvinced that meaningful information about normality could be derived from pathology.

Inferring function from dysfunction

It should be apparent that much of the knowledge of both the taxonomy and the biology of human memory is based on the subject of memory dysfunctions. Such studies begin of course with a therapeutic imperative; there are a large number of conditions, ranging from diseases like Alzheimer's to brain damage as a result of injury, stroke, tumour or epilepsy, in which memory impairments occur. Understanding the nature of these impairments might suggest, if not curative treatments – because for most none is known – then at least potential approaches to rehabilitation. But, as I have already emphasised, the memory deficits in disease condi-tions like Alzheimer's and Korsakoff's, where there is a general deterioration of brain and mental processes, are themselves rather general and non-specific. However, there are many types of damage to particular brain regions in which much more specific memory deficits occur. The existence of such deficits associated with brain damage has thus come to be seen as opening a special type of window on the brain. What is more, technological advance has made this type of study still more interesting. In the past, it was possible to study the memory performance of a person known to be brain-damaged, but it was impossible to correlate any deficit in performance with the degree of damage, because this could not be assessed in the living person; looking inside the brain had to wait until post-mortem. Today the existence of scanning techniques such as fMRI (functional magnetic resonance imaging) enables many aspects of a person's brain state – notably the detection of areas of cellular damage – to be examined during life, bringing the goal of an integrated neuropsychology a great deal closer. I will return to what can be learned from these techniques at the end of this chapter.

Before looking in more detail at what such studies have revealed, it is important to spell out clearly the nature of the interpretative problem which affects them all. By studying the performance of the damaged brain, one endeavours to draw conclusions about the functioning of the intact brain. The obvious assumption is that if one identifies a particular brain structure, and notes a particular memory deficit in the person with that damaged brain, then the damaged region is the one which, in the normal person, is responsible for carrying out the missing or deficient function. There are immediate flaws in this logic, however, well expressed by the psychologist Richard Gregory in a famous analogy a few years back.* If I remove a transistor from a radio and the result is that the only sound I can then get out of the radio is a howl, I am not entitled to conclude that the function of the transistor in the intact radio is a howl

---

*Richard Gregory's enthusiasm for puns and paradoxes knows few bounds, as anyone who has read one of his books or seen his TV performances will know. He is the only man I have ever met who can sit up in bed at seven in the morning and make three conceptual puns, still in his pyjamas and *before his first cup of coffee*!

suppressor. When one studies the radio in the absence of the transistor, one is doing just that – studying the system minus a component, not the missing component itself.

What is true for radios is true in spades for brains, because a damaged radio stays damaged – it does not try to repair itself. Yet this is exactly what the brain – and the person who owns that brain – does endeavour to do. Unlike radios, brains are plastic and highly redundant systems. The plasticity means that, although individual brain neurons destroyed as a result of a stroke or brain lesion cannot regenerate, at least in adults, the cells around the damaged area do grow and put out more processes, so that there is some compensatory remoulding of the brain. And equally, it means that the effects of a lesion at one site may be modulated through consequences at another, quite distant one. The redundancy means that if one brain region is damaged, there may be others which can take over at least a part of its function. Further, the fact that the brain is embedded in a person means that the person, seeking to achieve a particular goal – for instance, to remember some item – if blocked in one mechanism will try to develop a new strategy to help in the recall. (Think of the many ways we have of trying to remember a name that escapes us – going through the alphabet, or conjuring up the face, or trying to associate the name with the experience of last meeting the person – all varieties of the memorising devices that were discussed in the last chapter.)

These properties are intrinsic to brains and the humans who possess them; they are part of what makes for life's rich complexity, and should therefore be a matter of rejoicing; that incidentally they put limits on how far function can be inferred from dysfunction is just one of life's – and science's – little problems. Without ever losing sight of these problems, let me turn to what the study of such amnesic patients has revealed.

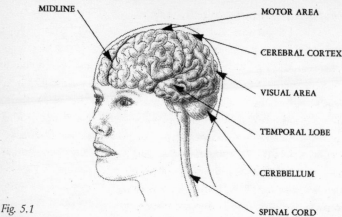

Fig. 5.1
The human brain
Looked at from the outside, so to speak, the brain is dominated by the two symmetrical massive cerebral hemispheres, whose surface is covered by the convoluted cerebral
cortex, packed with neurons ('grey matter'). Different regions of the cortex are responsible for different brain functions, from receipt of sensory information (vision, olfaction,
touch) to speed and motor output. The region which is of primary concern here is the
temporal lobe, which is concerned with aspects of memory – it was the temporal lobe
which was stimulated in Penfield's experiments (pp. 148–9). Behind the cerebral hemispheres lies the cerebellum, concerned, amongst other things, with co-ordination of fine
motor movements. Running away from the brain, through the spine, is the spinal cord,
which both receives and sends nerve fibres out to the body's periphery.

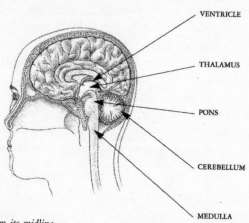

Fig. 5.2
The brain from its midline
This cutaway section enables some of the interior structures lying below the cerebral
cortex to be seen.

Holes in the brain, holes in the mind

Perhaps the most studied single amnesic patient in the history of neuropsychology is a Canadian known to the scientific world simply by his initials, H. M.; several distinguished neuropsychologists have made their scientific careers substantially on the strength of published studies of H. M.'s memory problems. It is worth noting that H. M. is not the victim of stroke or accidental brain damage; his condition is the result of deliberate surgical intervention by those who subsequently began the study of his memory problems. He was an epileptic, and in 1953, when he was twenty-seven, a brain operation was performed with the intention – fairly successful, it is said – of alleviating the effects of this epilepsy. The operation involved the removal of significant areas of the brain, including the anterior two-thirds of his hippocampus and amygdala and part of his temporal lobe – regions whose role in memory were unknown to the surgeons at the time the operation was performed.[13]

The crucial brain structures are shown in figures 5.1–5.3. As these regions of the brain, and especially the hippocampus, a neatly curved structure deep in the centre of the brain, densely packed with nerve cells, will crop up repeatedly in what follows, it is worth noting their location. I am not going to act as a tour guide through

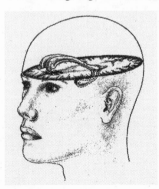

Fig. 5.3
The hippocampus
Sectioning the brain horizontally like this reveals the hippocampus. In this drawing the three-dimensional form of the hippocampus and related structures have been recreated to give a clearer impression of their location, shape and size.

the complexities of the structure of the human brain here, as it is not essential for my purposes, and still less am I going to try to justify the bizarre and fanciful dog-latin names that neuroanatomists have given its several regions. With one exception, which no one who is introduced to the hippocampus for the first time can ever avoid: the word is a bastardised form of Graeco-Latin meaning sea-horse, and the name derives from the imagined resemblance of this brain region to the curved tail of that pretty little creature.

The consequences of the removal of these regions of the brain were catastrophic for the patient. Although his performance on intelligence tests that do not rely substantially on memory remained good and he retained his memory for events long prior to the oper-ation – although with some partial loss for the periods immediately before it – he could not commit anything new to long-term memory. His immediate memory for present events, in the sense defined above, seemed to be unaffected. Fourteen years after the operation, according to Brenda Milner, the psychologist who has given him the closest study over the years:

> He still fails to recognize people who are close neigh-bours or family friends but who got to know him only after the operation . . . Although he gives his date of birth unhesitatingly and accurately, he always underes-timates his own age and can only make wild guesses as to the [present] date . . . On another occasion he remarked 'Every day is alone by itself, whatever enjoyment I've had, and whatever sorrow I've had.' Our own impression is that many events fade for him long before the day is over. He often volunteers stereotyped descriptions of his own state, by saying that it is 'like waking from a dream'. His experience seems to be that of a person who is just becoming aware of his surroundings without fully comprehending the situation . . . H. M. was given protected employment . . . participating in rather monot-onous work . . . A typical task is the mounting of cigarette-lighters on cardboard frames for display. It is

characteristic that he cannot give us any description of his place of work, the nature of his job, or the route along which he is driven each day . . .[14]

Following H. M., several other patients have been subject to similar research. One, N. A., is an ex-airforce technician whose brain injury was caused in an accident in 1959, when he was twenty-one; a miniature fencing foil was driven through his nostril and into the left side of his brain, damaging in particular the thalamic area, a region left intact in H. M. Like H. M., he can remember events well up to the time of his accident, but only spottily for the subsequent thirty years. Larry Squire describes him as living at home with his mother, cheerful and friendly but finding it hard to retain 'the events of each passing hour and day'. He wears his hair in a style appropriate to 1959 and 'recently mentioned the name of Betty Grable, as if she were a contemporary film star'. He 'appears entirely normal on first meeting' but forgets names, meals and that he has new clothes. His life is organised as far as possible to minimise the consequences of his memory loss; as his mother describes it, 'everything surrounding him is a memory thing . . . You've got to have a memory to remember.'[15]

When patients like H. M., N. A. and others are tested for memory in the psychology laboratory rather than on their day-to-day experience, they show a cluster of similar defects. First, it is always declarative rather than procedural memory which suffers. The brain damage does not result in the loss of learned motor skills. Invariably, it seems, well-established, long-term memory is spared, as is immediate memory. It is thus recent declarative memory which seems the problem, as if there is some block in the transmission belt from immediate to long-term storage. Quantifying this loss in more modern and sophisticated versions of the Ebbinghaus type of test shows both general similarities between the patients and also characteristic differences. H. M. is much worse than N. A. at remembering pictures, for example, while they are about equally impaired on remembering words.

One other striking common feature about such observations

relates to the disparity between the sparing of procedural and the loss of declarative memory. Thus a fascinating experiment with H. M. involved training him over a number of days on a task that involved a number of skill and memory elements, the so-called Tower of Hanoi. This is a game in which the player is presented with a board on which there are three spikes, on each of which are stacked a number of rings of different diameters. The object of the game is to shift the rings in the smallest possible number of moves from spike to spike so as to produce a final outcome in which the rings are all stacked on top of one another in decreasing order of size. The constraint is that no ring may ever be placed on top of a ring smaller than itself. Although each time H. M. was introduced to the task he denied ever having seen it before, over the series of trials his performance steadily improved; procedural memory continued to testify to the truth of what declarative memory denied.

At first sight it seems extraordinary that a person's memories could be so decoupled that he can seemingly learn a skill and improve his performance without ever being aware that this is in fact what is happening, that he is rehearsing an experience that he has had before as recently as the previous day; yet this is a common feature of the type of memory loss that goes with such brain damage, provided that the patient's memory can be interrogated with enough ingenuity by the researcher. The inescapable conclusion is that procedural and declarative memory are not merely localised, but localised to different regions of the brain, so that the one, declarative, can be lost, whilst the other, procedural, is spared. (But a word of caution here – to speak of regions implies perhaps a particular circumscribed anatomical area. As will become apparent in later chapters, it is essential to understand the brain in terms not of regions but systems – ensembles of cells perhaps located in different regions of the brain but connected through synapses and also and perhaps more importantly through 'time-locking': phase linked synchronous activity.) As it seems extremely difficult to lose procedural memory and relatively easy to lose declarative, it may also be that the actual biochemical, physiological or anatomical nature of the store is also very different in the two cases. But this is something to come back to at a later stage.

Attempts to explain such similarities and differences in memory capacity focus on the overlapping but different extent of the brain damage in the different cases; thus H. M.'s damage is equally to both halves of the brain, whereas N. A.'s lesion is to the left hemisphere – the side associated more with verbal than pictorial capacity. Both H. M. and N. A. – and the other patients who have been studied – show damage to the set of brain pathways including the hippocampus and thalamus, known collectively as the limbic system. But the problem of course is that each person's lesion is clinically and experimentally unique; such brain damage is an accident of nature, not a controlled experiment, and it is too rare an event to permit the sort of generalisations that can be made about Alzheimer's or Korsakoff's patients. As neuropsychologists have striven to provide some coherent explanation of these effects in terms of brain structures and pathways, there have been attempts in the last few years to mimic the specific types of brain damage in humans by experimental studies with monkeys – experiments which have yielded valuable information about mechanisms and brain structures but have run into profound ethical conflicts about the legitimacy of deliberately damaging monkey brains in order to learn more about human brain processes.

Windows on the brain

If one is to advance in the understanding of the brain mechanisms of human memory beyond the detailed study of such accidents of nature and medical mistakes as brain-damaged and diseased patients provide, what is really wanted, of course, is a way of looking inside a person's head whilst they are learning or remembering. Such an idea would until recently have seemed a sort of science fiction pipe-dream or idle philosopher's speculation (they used to talk about imaginary machines called 'cerebroscopes'). Within the last decades, however, what would once have seemed inconceivable is almost reality. New methods of brain imaging have been developed, either non-invasive or nearly non-invasive, which begin to offer just that prospect.

The earliest of such studies was, however, far from non-invasive.

I have already referred to the fact that H. M.'s hippocampus was destroyed surgically in the course of an operation to relieve the effects of epilepsy. This has been a fairly standard approach to treating some forms of the condition (focal epilepsy) for many years now. Focal epilepsy starts with a wave of electrical activity in a relatively small group of neurons, from whence it spreads across much of the brain. The neurosurgical approach is to locate the offending cells and isolate them or remove connections between them and other brain regions, so as to try to prevent the electrical spread. It was a technique associated especially with the Montreal neurosurgeon Wilder Penfield, and in its early period at least, in the 1950s, required that the site of the epileptic focus be identified by exposing the surface of the brain and then probing it with electrodes through which current could be passed so as to stimulate electrically the cells with which they came into contact. This sounds a more formidable undertaking than it is in actuality. The brain has no pain receptors, and thus once the initial incisions of scalp and skull have been made, it is pain-free; the operation could be performed under local anaesthetic, and patients were able to report their experiences as a result of the stimulation to the surgeons whilst it was in progress.

In the course of the treatment of more than 1,100 patients with this technique, mainly during the 1950s, Penfield and his colleagues explored much of the surface of the cerebral cortex with their electrodes. Stimulating some regions resulted in sensory experiences, others gave motor responses, and Penfield was able to show that the entire body surface is 'mapped' onto the cortex in an orderly fashion; such maps are today an obligatory illustration in any book about the brain. However, it is not these maps which are my main concern here, but the experiences that patients reported when Penfield stimulated particular regions of the right or left temporal lobe of the brain (see figure 5.1). These included auditory, visual and combined auditory-visual effects, such as present voices, music, people and scenes, as well as thoughts, memories of past experiences and visual flashbacks. Not all the patients reported such experiences, and not all stimulations of the regions produced them, but what immediately captured the researchers' imagination were the claims that otherwise hidden memories were being evoked.

The apparent memories had a hallucinatory or dream-like quality. They often began with indistinct or partial images, but as the stimulations proceeded they became clearer, until entire episodes began to be replayed, almost as if the still photographic pictures of eidetic memory were being run through as movies. Sometimes patients described them as 'dreams' of 'people's voices talking' or 'a lot of people . . . in the living room. I think one of them is my mother.' Sometimes they were very general: 'a dog is chasing a cat' and sometimes much more specific: 'I hear people laughing . . . two cousins, Bessie and Ann Wheliaw'; 'my mother was telling my aunt over the telephone to come up and visit us tonight.'[16]

Although these observations were immediately regarded as demonstrating both that memories were localised in the temporal lobe and that they could be evoked by stimulation of the appropriate region, there are real problems of interpretation. Firstly, the fact that stimulation of a particular region invokes a particular memory does not mean that the memory is 'stored' in that region. It might for instance be the case that the electrical stimulation of one site excited its cells, causing them to communicate in turn with other regions which were the 'real' site of memory storage. Secondly, and more important, there is the problem of deciding whether what is being elicited by such stimulation is a 'real' memory for some event which has actually occurred, or, like a dream or hallucination, some type of confabulation. The very nature of the records means that one can never be sure about this; the Penfield studies remain fascinating, challenging, but ultimately uninterpretable.

It is the dramatic advance in recording and imaging technology of the last two decades that have really begun to offer the possibility of a 'cerebroscope' that might move research forward from these rather messy beginnings.

The brain as a physiological system is in a constant state of cellular and chemical flux. Nerve cells in activity are extraordinarily hungry for glucose and oxygen, and the brain's rich blood supply is called upon moment by moment to channel them both to where they are most required. Nerve cell activity is electrical, and biolog-ically generated current flows through the brain in patterns as simultaneously regular and varied as the waves of the sea. And, finally, it is in the physical nature of electricity that when current flows, a magnetic field is created at right angles to the direction of the current. All of these properties offer prospects for brain imaging. Each imaging system requires that an array of detectors is mounted around the head – the smaller the detectors, the more can be included in the array. Each detector records some signal from within the brain and the nearly but not quite simultaneous arrival of signals at the different detectors makes it possible to focus on their sites of origin. Computerised systems then generate maps of 'slices' through the brain at defined coordinates and the black-and-white images are false colour-coded by computer to help visual resolu-tion.

The several methods differ in the nature of the signal, the degree of resolution in space, and the time taken to obtain a signal. One of the earliest to be developed was PET (positron emission tomography – 'tomography' simply describes the computerised signal retrieval and 'slicing' technique). This depends on injecting very short-lived and presumably non-hazardous radioisotopes into the bloodstream and then detecting, by means of their emission of positrons, where they are in the brain at times up to half an hour or so after injec-tion. It is somewhat like the radioactive isotope labelling methods I use in my own lab and described in chapter 2. Depending on the isotope it is possible to map the distribution of blood, of oxygen or of glucose in the brain. If the person being studied is then given a mental task to perform, such as doing simple mathematics or remem-bering a piece of poetry, regions of the brain which are more active in the task than in a quiet condition 'light up'. PET pictures are quite slow to obtain, but they can resolve volumes of brain down to about half a cubic centimetre, which is better than many other of the methods (though this volume still includes many tens of millions of cells). However, they are of course not quite non-invasive; they do require an isotope injection.

By contrast magnetic resonance imaging (MRI) is relatively

non-invasive and detects signals based on the atomic properties of
individual molecules in the brain based on their magnetic properties; the person is placed into a tunnel-like apparatus which generates a strong magnetic flux through the head, and using tomography
to locate the molecules that respond magnetically to the force field.
It depends on passing a strong magnetic field through the brain. It
can only measure molecules present in great abundance – like water.
Originally developed for clinical use – for instance to detect the
regions of brain damage, a variant called f (for functional)MRI even
permits action pictures when for instance there is greater utilisation of oxygen under one condition than another. MRI and PET
imaging, however, are now in fairly routine use in hospitals and
research facilities. Clinical use of the scanner enables the detection
of sites of brain damage and also compensation, when a brain-
damaged person learns again a skill which has been lost and different

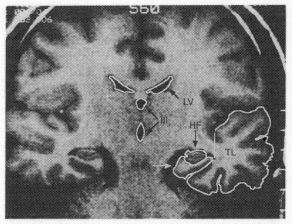

Fig. 5.4
Nuclear magnetic image of the brain
*The image is of a coronal section through the brain of a normal young adult, made by
the non-invasive MRI technique. Black regions in the centre marked III and LV are ventricles. HF – hippocampus; PHG – parahippocampal gyrus; TL – temporal lobe. In amnesic
patients such as H.M. the hippocampus is damaged; in Korsakoff's patients this region
of the brain may be very shrunken and the ventricles enlarged. In some other forms of
amnesia the temporal lobe is damaged or shrunken. (Courtesy Larry Squire; source
L. R. Squire, D. G. Amarel and G. A. Press, J. Neurosci 10, 1990, 3106–17.)*

brain regions take over a task once associated with the damaged area.

One of the most interesting sets of memory experiments using fMRI has been done by Eleanor Maguire and her colleagues at Queen Square in London, asking London taxi-drivers to visualise driving a complicated London journey, and noting how their hippocampus 'lit up' during the process. She went on, also using MRI, to measure the size of taxi-drivers' hippocampi by comparison with controls and found that the posterior regions were slightly enlarged![17] (When I mentioned this to a cabby once he was sceptical – he said his father had been a cabby before him, and he himself had been driving for twenty years, so if it were true his brain would have grown so big it would have exploded!)

In our lab we have been using a different method, called magne-toencephalography (MEG). This measures the tiny magnetic fields around the brain, no larger than a billionth of the earth's field, using devices called SQUIDs (superconducting quantum interference devices). This method doesn't have such good spatial resolution as does for instance MRI, but it can measure in real time (milliseconds) and is completely non-interventive. All it requires is for subjects to sit in a magnetically shielded room with something like a hairdryer on their head. Inside the helmet is a thermos of liquid helium, and a set of up to 344 tiny SQUID detectors to pick up the magnetic fluxes. There aren't many of these machines around yet, and we have been working with the biggest of them, in Helsinki, developed by the Finnish physicist Riita Hari. With my Open University physicist colleague Steve Swithenby and his team, we devised a unique memory experiment.

One of the problems about memory research is of course that everyone's memory is individual, so it is hard to think of a way of testing memory that is reproducible from individual to individual but doesn't involve the artificiality of Ebbinghaus-type lists of words. I wondered whether there was any memory experience which most adults might share, and I hit on the idea of supermarket shopping, which most of us do, whether we enjoy it or not. So we took our volunteers – myself included – on a 'virtual shopping expedition' through a well-known supermarket. In the experiment, each person, sitting in the MEG machine, watched a video tour of the supermarket, stopping at various aisles, at which they were required to choose products from sets of three (for instance, wines, tinned foods, or whatever), based on their preference and past experience. The results were quite fascinating. It takes people on average about two and a half seconds to make their choice (by pressing a key). But during that period the brain is very busy. The first detectable event is about 80 milliseconds after the choice is presented, when the visual cortex becomes active. After about 250 milliseconds, the left inferotemporal cortex lights up – a region known to be associated with semantic memory. A little later, especially if it is hard to make a choice, the 'speech regions' – Broca's area – is active, as we silently verbalise the choices in front of us. And finally at about 850 milliseconds, and *only* if we have a strong preference for one or other of the choice items (Coke versus Pepsi for instance, or for that matter Pepsi versus Coke), then very unexpectedly a region in the rear right quarter of the brain – the parietal cortex – becomes active. It is known, from studies of people who have damage to that region, that it is involved in decision-making especially if it involves some emotional as well as cognitive component. (I was hoping to put in some pictures at this point to show the different regions of the brain lighting up, but sadly they don't work in black and white, so I'll just have to leave them to your imagination.[18])

When we published these results, in 2002, it was hard to resist the press's enthusiasm for announcing that we had discovered the shopping centre in the brain! But of course, nothing is quite so simple. Admittedly, the pictures – and those obtained by fMRI -are dramatic, beautiful and offer astonishing windows into the brain, inconceivable only a few years ago. It is hard not to feel when one first sees them that such techniques will surely answer all the questions we might have about brain function and its relation to mental processes (animal-rights activists are especially quick to jump to this conclusion). But it isn't quite so simple. Showing that a brain region is active when a person is learning or remembering is not the same as showing that the memory 'resides' in that part of the brain. The 'store' might be somewhere quite

different, somewhere that doesn't need a great flurry of glucose utilisation to activate it; we might be looking at the peripherals to the
engine rather than the engine itself. Nor can studies of increased
glucose or oxygen use, or even magnetic fluxes, by themselves tell one
about the detailed molecular mechanisms of what is going on. And,
even more importantly, at the less mechanistic level, they cannot tell
us about meaning, about the translation rules between mind and brain.
They tell us, at best, where we should look in the brain to find the
Rosetta stone; they cannot decode the hieroglyphs themselves, nor read
their message. For this a different approach is needed.

Chapter 6

Animals also remember

Human memory and animal memory

WHEN I SAY I REMEMBER MY MOTHER'S VOICE, OR MY FOURTH BIRTHDAY party, or how to ride a bike, everyone knows what I am talking about. But what does it mean to say that a monkey, a cat, a chick or a sea slug remembers? Can I transfer a word which to most of us describes a uniquely human – even a uniquely personal – activity to a description of something going on in non-human animals, to whose minds I have no direct access and with whom communication is strictly limited? Obviously, if I can devote much of an entire chapter to discussing computer memories, there is nothing inherently illegitimate about speaking of animals as having memories. Perhaps the question therefore needs to be split in two and rephrased: first, is it possible to define and demonstrate activities in animals about which it is reasonable to use the term memory; and second, can the study of such memory-like activity in non-human animals say anything about the mechanism of human memory – or is the human–animal gulf too great to bridge?

Anecdotally, of course, no one who lives or works with animals has any doubt that many species behave in ways which make it seem as if they have good memory. Dogs recognise their owners and distinguish them from strangers; cats, having once after much effort discovered how to open a door, will do so again with ease a few days later; performing animals in circuses run through complex routines of tricks taught them by their trainers. The key phrase, however, is 'seem as if'. Because we cannot interrogate non-humans directly about their experiences, we have to observe them and draw conclusions about the behaviour that we observe. If a hungry rat, once having been placed at the start of a maze and having eventually found its way to the goal at which there is a 'reward' of food, makes fewer wrong turns when placed in the maze again, we are entitled to define this as showing memory for the maze. If, however, the next time the rat is placed in the maze it makes many more wrong turns, we are not entitled to say that it remembers the maze but is not interested in running it correctly, or that it is bored, or inquisitive about the false routes. Because we cannot observe such internal processes, we can only observe the behaviour, record the errors and conclude that the rat does not remember, or has forgotten, the task. Animals have to show us their memories by directly behaving, by acting or responding in some way that we define as appropriate to their environment.

This means that the study of animal memory, once it moves beyond the act of simple observation, inevitably becomes experimental. To interrogate an animal about its memory we have to place it in a situation in which it is required to reveal it – and this means encouraging it to perform some task. Such encouragement can be positive, in which performing the task achieves something desired, like food or drink, or it can be negative, in which performing the task is a way of avoiding something unpleasant or painful. These are the methods of experimental psychology which are inescapable if we wish to study behaviour, and yet which give rise to concern for the welfare of the animals being studied. In describing in this and subsequent chapters some of these methods and their findings I am conscious of this dilemma, but in making a judgement on such experiments and their findings, I want to re-emphasise the position on 'animal rights' that I spelled out in chapter 2.

Learning, remembering and experience

We can observe and measure learning in animals, by setting them some task and watching their performance on that task improve as they attempt it repeatedly; and we can observe recall, or remembering, by setting them the same task again some hours, days or weeks later and comparing their performance now with that in their previous attempts. If they perform no better than on the first occasion they experienced the task, then we assume that they cannot be remembering it; if, however, their performance is nearly as good as it was at the end of the first set of trials, then we can say that they are indeed recalling their past experience in performing the task again. In this way we have defined two active processes: learning and remembering (or recalling). What goes on inside the animals between these two processes of learning and remembering is what we call memory. It may seem a bit late in this book to put the matter of the definition of learning formally; up till now it has been almost implicit. Even memory has not been rigorously defined, and I have contented myself with the odd quotation from James and others. But precisely because of the limits to how we can interrogate non-human animals, it is necessary to be a bit more precise at this stage.

Learning, in this operationally defined form, is a response by an animal to a novel situation such that, when confronted subsequently with a comparable situation, the animal's behaviour is reliably modified in such a way as to make its response more appropriate (i.e. adaptive). Note that it is important to a definition of this essentially operational sort, that the behavioural modification (a) occurs as a result of experience whether deliberate or accidental, (b) is reliable – that is, will occur repeatedly in a single animal or in a similar manner in a group of animals – and (c) is adaptive. If we do not introduce the criterion of adaptiveness, then scar tissue could be described as a form of memory! On the other hand there may be some types of experience in which it is not easy for an animal to find an appropriate adaptive strategy because the situation is too complex, or in which the strategy that it adopts in fact turns out to be mistaken, and yet we would not wish to say that learning had not occurred – there will be examples of this later in this chapter. Terms such as experience and adaptiveness thus each carry a depth and richness of meaning – even of ambiguities – which cannot be ignored.

Recall (remembering) is the expression of the modified behavioural response at some time subsequent to the initial learning. Memory is thus not something we can observe or measure directly; it is something we infer must exist as an intervening process to connect learning to recall. If an animal's behaviour is changed as a result of the learning, and this change is expressed in recall, then it is necessary to assume that something has changed in the biology of the animal to produce the changed behaviour. That is, there must be some type of record inside the organism, by which the information acquired during the learning is stored in such a form that it can be made available to modify subsequent behaviour. This hypothetical record inside the organism (the Rosetta stone, to revert to my earlier analogy) has been given many names over past decades. Ivan Pavlov, one of the earliest of physiologists to research on learning, in St Petersburg/Leningrad in the first decades of the last century, referred to a conditioned reflex. More recently, for reasons the last two chapters have made clear, memory researchers have tended to speak of a memory store or memory trace; or an engram (a term introduced by another of the pioneers of research on the biology of memory, the zoologist J. Z. Young, in the 1950s); the most fashionable terms in the literature today describe animals as forming representations or even cognitive maps. In general this terminological progression has reflected current metaphors and theoretical models for the nature of the memory storage process, from the very rigid 'hard-wired' (to adopt an equally fashionable computer metaphor) Pavlovian concept through to the very fluid ('soft-wired') ideas current today.

I will come back to the significance of these alternative concepts later. For the moment all that matters is that memory is to be measured in terms of changed adaptive behaviour as a result of experience. This is a very broad and essentially biological definition – probably not one that would be recognised at first sight by psychologists, though they might be persuaded, somewhat reluctantly, to accept it; for the moment, however, I want to remain with that seemingly nebulous concept of experience. For whilst we might have difficulties defining precisely what is meant by learning, it is obvious that day-to-day life, for anything from amoebae to rose bushes and humans, involves experience, and that one definition of life itself must involve the capacity to adapt to experience by changing behaviour.

Plasticity and specificity

Experience is a term in the behavioural lexicon. Its translation in the language of biology is plasticity. To function effectively – that is to respond appropriately to its environment – all living organisms must show two contradictory properties. They must retain stability – specificity – during development and into adult life, resisting the pressures of the endless buffeting of environmental contingency, both day-to-day and over a lifetime. And they must show plasticity – that is, the ability to adapt and modify this specificity in the face of repeated experience. Whereas once biologists used to speak of organisms as the product of the interplay of nature and nurture, or in modern language, genes and environment, today this dichotomy is recognised as simplistic, for it is an individual's genes, expressed during development, which provide the basis for both specificity and plasticity. If we didn't have the genes which are instrumental in producing a brain which could learn by experience, we wouldn't survive. But equally, if we didn't have the genes which ensure that our brains become wired up correctly, so to say, during development, we also would fail to survive. To unravel the dialectic between specificity and plasticity and to understand its mechanisms form some of the major tasks of modern biology. The changes that occur in the brain as a result of experience are a form of plasticity, and memory is one major aspect of that experience. So it turns out that to understand the mechanisms of memory, of plasticity, it is also necessary to understand the mechanisms of specificity. If the brain were not, most of the time, invariant, unmoved by experience, we would be unable to survive, as indeed the fates of Funes and Shereskevskii attest. Ordered and restrained variance can only make sense against a largely invariant background.

To achieve stability in a fluctuating world requires stable systems to recognise and respond to that world – systems in which the brain is a key player. Specificity is the property that confers on brains the ability to make sense of their environment, to recognise invariance and to respond in an orderly way to regularities in the external world. Without such specificity we would live in a constant state of blooming, buzzing confusion. For instance, a major part of the information that the brain receives about the external world comes from vision, and therefore through the eyes. The position of our head and eyes with respect to the outside world is constantly varying, yet we see the world stably and can interpret our position within it in three dimensions. To achieve this interpretation requires a remarkable degree of order and constancy in the way in which light, falling on the retina, is converted into signals which can be passed to and processed within the brain, an order which must be retained throughout a lifetime in which cells die, molecules are replaced, and multitudinous experiences are remembered.

How can this be achieved? The human eye, whose retina contains more than 110 million light sensitive cells, is connected to the brain by way of the optic nerve tract, a pathway along which run a million individual nerves. The fact that there are many more retinal cells than optic nerves to which they connect means that each nerve integrates information from many individual cells. The optic nerves terminate in a region deep in the brain called the lateral geniculate; there they make synapses with a set of many million neurons from which run a set of further connections to a region of the cerebral cortex, the visual cortex (this set of pathways, or wiring diagram, is more-or-less identical in all mammals). Thus visual signals processed and converted into nerve impulses by the retinal cells are first compressed in the optic nerves and then expanded again in the lateral geniculate. If the brain is to interpret images arriving at the retina of the eye these pathways, with their compressions and expansions, have to be organised in an orderly manner – and indeed it can be shown that there is a precise topographic mapping of the retina onto the neurons of the lateral geniculate and a further mapping of these cells onto those of the visual cortex. That is, there is a type of map of the retina in the lateral geniculate, and a further map, albeit transformed at least as much as Mercator's projection transforms the globe of the world into a two-dimensional plan on a classroom wall, in the visual cortex. At each level of mapping, from the retina to the geniculate to the cortex, analysis of the information occurs so that it is classified in terms of signals for edges, angles, movement, light of different wavelengths (colour), etc. In the cortex further mapping and classification by increasingly complex criteria occurs to generate the pattern of neural activity which we define as vision and perception.

Actually this is a great oversimplification. Just as there may be many types of geographical map (atlases contain geological maps, climatological ones, linguistic ones, political ones, etc. each of which can be overlaid on the others but occupies a separate atlas page), so the visual cortex contains many maps, analysing visual input in terms of shape, colour, movement, etc. Such cortical maps, which are composed of three-dimensional chunks of cortex each containing many tens of millions of neurons, are sometimes referred to as *modules*, a term that can be somewhat confusing as it is used in a variety of ways. One of the peculiar consequences of this modular organisation is that brain damage can remove one module without necessarily affecting others. This means that is is possible for a person with such damage to perceive a world without colour, or even more bizarrely, a world in which motion can be perceived without the objects in motion being 'visible'. Such cases have been described by the neurophysiologist Semir Zeki.[1] The existence of such modules also raises the question of how what we normally actually perceive is a single coherent image. How are the activities of the modules integrated? This, which has been called the 'binding problem', is one of the great current neuroscientific mysteries, to which I will return much later.

The difficulties of achieving an orderly mapping, classification, analysis and finally synthesis are increased by the fact that in mammals such as humans both eye and brain change rapidly and disproportionately in size and cell number from birth through the first years of life. As a consequence, the mapping cannot be done once and for all, with individual cells assigned functions at birth and connections that will last them a lifetime. To retain the mapping relationship the connections involved in the pathways from eye to brain have continually to be broken and remade. Yet the infant human can see from birth, rapidly develops stereoscopic vision and the capacity to focus, and soon learns to recognise objects even when they are seen at fluctuating angles and distances and in different forms of illumination. The 'wiring' that connects eye to brain must therefore be remarkably stable, connecting and reconnecting according to some internal developmental programme by rules relatively little affected by environmental contingency.

How is such stability achieved despite the tremendous changes that occur in size, cell number and experience during development? There is a lot of evidence that during early phases of brain development there is a huge overproduction, a veritable efflorescence, of synaptic connections which are later steadily pruned back in number; redundant or irrelevant synapses are discarded and the remainder in some way stabilised. This pruning and stabilisation process seems to be an important way in which experience modifies brain structure during development, and just how this synaptic stability – specificity – is achieved has become a major research theme for developmental neurobiology. Indeed, so intertwined is the dialectic of specificity and plasticity that it may eventually turn out only to be possible to understand each in the context of the other.[2]

For environments are only partially regular and invariant. They also change in both the short and long term. Organisms, and therefore their brains, need to be able to adapt to such changes. Hence, plasticity, which implies this capacity to adapt. In fact the term is used rather loosely in the neurobiological literature. If during its development a young animal is deprived of key hormones – thyroid hormone, or oestrogen or testosterone – the nervous connections in parts of its brain grow characteristically askew. If it is malnourished, or otherwise environmentally restricted during development it may end up with a reduced brain size. Both of these are regarded as examples of plasticity – but the term is also often used to refer to the capacity of the adult nervous system to recover from damage, either by the growth of new connections or by an undamaged part of the brain taking over the functions of the damaged region.

Enrichment and impoverishment

There is clearly some danger that the term plasticity will outgrow its usefulness, implying a similarity or even identity of mechanism between phenomena as varied as recovery from injury, the effects of malnutrition and hormonal imbalance, and even memory, where perhaps no such identity really exists. Nonetheless, if in behavioural terms memory is a special case of experience, it is at least worth considering the possibility that the brain mechanisms of memory may be special cases of neural plasticity. This idea lay behind the experimental programme, begun in the late 1950s by the psychologist Mark Rosenzweig, biochemist Ed Bennett and anatomist Marian Diamond in Berkeley, California to study the effects of different rearing environments on the brain structures of young rats. In the basic experimental design (which was in the years that followed sometimes modifed to include other types of control groups), some of the rats were kept in isolation, in individual cages with plenty of food but no contact with their fellows, and with sensory stimulation kept to a minimum – constant low-level illumination, continuous mild 'white noise' and so forth. The second group was reared communally, in large cages filled with toys and, in the early experiments, regular periods of training on a variety of learning tasks. The first group was referred to as 'impoverished' and the second as 'enriched' – although such terms are surely only relative. By comparison with the way real rats live their lives outside the laboratory walls, it might have been better to describe the conditions as 'relatively impoverished' and 'very impoverished'.

The Berkeley group was able to show that following several weeks of rearing the rats in these different conditions, characteristic differences could also be found in their brains. Typically the animals of the 'enriched' group had a thicker brain cortex; there were more synapses in some parts of the brain and increased quantities of several enzymes, especially those concerned with the synthesis and breakdown of the chemical transmitters that carry signals across the synapses between one nerve cell and the next. These experimental findings may not today seem very surprising, but they generated considerable interest at the time they were published, partly because they represented a then novel collaborative effort between a noted psychologist and two neurobiologists. The impact of the Berkeley results was the more striking perhaps because the dominant view of the brain, at any rate at that time and within the United States tradition, was that specificity far outweighed plasticity and that brain structures were very hard to modify by environmental contingency and experience. Over the subsequent two decades the group went on to refine its measurements somewhat, to show that relatively short periods of enrichment even in adult animals could produce similar changes, and to measure more precisely some of the anatomical changes that occurred. This was clear evidence that experience could modify brain structure and biochemistry.[3]

Nonetheless, the problem with interpreting the results is that the types of experience the rats were subject to vary in so many ways, some very obvious, some much more subtle and hard to pin down at either the behavioural or the cellular level. Were the observed differences due to the fact that the animals in the 'enriched group' had more social interactions, perhaps affecting their hormonal state? Or perhaps those in the 'impoverished group' were exceptionally stressed by their isolation? Did the 'enriched' animals really learn more than their 'impoverished' counterparts and if so, how could one tell? If the questions one wants to answer are about the causes of the observed changes, it is hard to design experiments, based simply on the antithesis between such 'enrichment' and 'impoverishment', to discover just which aspects of these complex environmental situations are crucial for the observed brain differences to occur. For all the efforts of the Berkeley workers to find ways round this problem, it continues to represent, I believe, a clear limitation to their approach – insofar as we are interested in more specific questions of memory and not just the range of possible plasticity. It is necessary to find a more controlled and limited way of modifying experience.

Light and dark

One more specific approach is to try to limit an animal's sensory input during development. Perhaps the easiest such input to control is that of vision, for instance by comparing the effects of rearing animals in light or darkness, or subjecting them to different types of visual stimulation. My own first, crabwise approach to research on memory, described in chapter 3, was based on this type of thinking, as I studied what went on in the visual cortex of a rat during its first experience of light. A similar rationale has guided much such research from the 1960s to the 1980s. The visual system is an attractive model because it is relatively easy to manipulate – animals can be reared in the dark or in environments with restricted illumination much more readily than they can be raised in a sound- or smell-proof environment. However, some researchers have tried to limit or modify other senses, such as that of smell, by altering the 'odour environment' of animals during their development – less easy experimentally, but perhaps more biologically relevant to animals like rats or mice, which depend much more on their sense of smell than on vision.

It is physiologists rather than biochemists who have made the greatest use of such experimental designs. The animal of choice has been the kitten, as cats are highly visual creatures which are born with a still very underdeveloped brain and visual system. The first three weeks or so of the kitten's life, before and during the time when its eyes first open, are critical periods for the proper development of the visual pathways. In a series of experiments made in the 1960s, Richard Held and his colleagues showed that if kittens are kept in complete darkness over this period their visual system does not develop properly; although not blind they cannot coordinate their movements by eye, avoid obstacles while walking or discriminate depth. Complete darkness is not obligatory – rearing a kitten fitted with a large paper collar so that it could not see its own body had similar, though less marked, consequences.[4]

These gross behavioural effects can be shown to have their correlates in the brain. Colin Blakemore, at Cambridge (and later at Oxford) was the first to show that if kittens were brought up in an environment in which the only visual input was a pattern of black and white vertical stripes, then the neurons of the kitten's visual cortex became particularly responsive to vertical patterns but would scarcely respond at all to horizontal ones.*

Experience thus modified the 'wiring' of the kitten's visual system during development in such a way as to expand the proportion of cells responding to types of stimulus which were met frequently (vertical stripes) and decrease the proportion of cells responding to infrequent stimuli (horizontal stripes). More recent experiments have tended to make use of the fact that most neurons in the visual cortex receive, and can respond to, input from both right and left eyes. If, however, during critical periods of brain development the input is limited to one eye – for instance by rearing a young animal with an eye-patch – then the visual cortex neurons reorganise so as to become responsive almost exclusively to input from the open eye. Such methods have been used to show that the visual system is capable of a considerable measure of what is formally referred to as activity-dependent self-organisation. What this means is that the selection of just which synapses become stabilised and which are pruned back during the development of the visual system is partly determined by experience. And what is true of the visual system turns out to be the case for other senses as well. (Among the best studied are the nerve cells in the cortex which receive inputs from the sensitive whiskers on the muzzles of many mammals, notably rats and mice.) Brain cells that can

---

* Colin Blakemore has received much obloquy as well as threats to himself and his family from animal-rights activists for such experiments, yet they have proved important in developing procedures for correcting human developmental visual problems. And he himself has very courageously spent a great deal of time and energy discussing with those animal-rights activists prepared to take part in reasoned discourse the issues surrounding the ethics of animal experimentation – a discussion that has helped move both sides forward.

change their properties in this adaptive way in response to environmental contingencies are clearly showing something rather close to memory as I defined it earlier.[5]

Conditioning and associationism

Nonetheless, the approach to the phenomena of memory by way of neural plasticity and development is a biologist's rather than a psychologist's route. For experimental psychology, the first steps to the experimental study of animal learning and memory are generally attributed to Pavlov (although he himself drew on an extensive set of earlier traditions in neurophysiology and psychology). He was not a psychologist by training, and came late to work on learning, following the experiments on digestion for which he was awarded a Nobel prize. As is well known, it was laboratory practice during these experiments, which were made on dogs, for a bell to be rung as a signal that the animals were to be fed. Pavlov observed that the animals, which would of course salivate when the food was put in front of them, came instead to salivate when the bell was sounded. They had learned to associate the sound of the bell with the imminent arrival of the food, and as a result their behaviour was subsequently changed. Pavlov's physiology was determinedly mechanistic, and he interpreted these observations very straightforwardly.

Sit someone crosslegged and tap his or her kneejoint sharply with a hammer; the result is a quick jerk of the leg, an act seemingly without conscious intervention or control. Nineteenth-century neurophysiologists, studying the determinist arc that connects such a sensory input, via the neurons of the brain or spinal cord, to a motor output, described the phenomenon as a reflex. Reflex theory to account for the motor activity of the nervous system was strongly built into neurophysiology by the time Pavlov came to make his experiments. He thus interpreted the dog's response, of salivating to the taste or smell of food, as just such a fixed reflex of the nervous system, built into its wiring. Because it occurred 'naturally' without the intervention of the experimenter, Pavlov called it an unconditioned reflex.

As a result of the dog's repeatedly experiencing the conjunction of food with the sound of the bell (the conditioning stimulus) a new reflex or nervous pathway was established, connecting the registration of the sound of the bell directly with the salivation response – this was the conditioned reflex.

The essence of Pavlovian conditioning is thus the pairing of a normally evoked response with a previously unconnected stimulus. For the next thirty years Pavlov, and following him generations of his pupils, continued to explore the nature of this pairing.* They showed, for example, that the timing of the association was vital – if the bell was rung after the food was brought, or too long beforehand, the dogs could never learn the association. On the other hand, it was possible to build up chains of conditioning – for instance if a dog had learned the pairing 'bell–food', and then a light were to be switched on just before the bell, the animals would eventually come to salivate to the light alone. Such studies led Pavlov to a narrowly mechanical general theory of learning based on sequences – 'chains' – of conditioned reflexes. In an effort to explain more complex human cognitive functions ('higher nervous activity') he developed as an extension of his reflexology the idea that, as well as the simple conditioned reflexes available to other animals, humans also possessed a higher order 'second signalling system'.

Pavlov's experimental approach and theoretical formulations were received with enthusiasm by the Bolshevik government. There was an almost boundless enthusiasm for science among the Communist philosophers and political leadership of the 1920s.[6] Pavlov's materialist approach to psychology was attractive in that it offered to abolish any possible ghost in the brain-machine and, as a Russian Nobel prize winner, the international acclaim for his work was seen as

---

* During a visit in the 1990s to Leningrad, then just getting used to being renamed St Petersburg, I visited the institute that bears Pavlov's name, and saw his laboratory, preserved as a museum. To my amazement I found that adjacent to it there was an identical laboratory in which minute variations of Pavlov's experiments were still being conducted, the only difference being that the introduction of the food and the quantity of salivation were now being controlled and recorded by computer. Alas, this was not intended as an animated demonstration of a classic experiment; rather it was an example of mummy-worship masquerading as science.

bringing prestige to the fledgling Soviet Union. Lenin himself authorised the building of a major new institute so that Pavlov's work could continue even under the direst conditions of civil war and economic crisis of the early years of the regime, though, almost until his death in 1936, he declined to join the Party or even to support publicly its social or political goals.

Despite Pavlov's own formation as a physiologist and his committedly materialist – even narrowly reductionist – approach to behaviour, he remained oddly indifferent to the study of any actual neurophysiological processes that might be associated with the formation of reflexes; he was content to study the phenomenology of conditioning and to draw abstract flow diagrams of brain events that might underlie it. For several decades during the Stalinist period and its aftermath, research on brain and behaviour in the Soviet Union became shoehorned into Pavlovian orthodoxy, despite the presence of new generations of researchers who, while prepared to give Pavlov credit for his undoubted achievements, sought to break loose theoretically.*

From the 1920s on, the study of Pavlovian conditioning also became the stock-in-trade of Western psychology. Here, however, it was soon complemented by an equally arid orthodoxy deriving from a United States school of psychologists calling themselves behaviourists. The behaviourist manifesto was launched by John B. Watson in 1913. Before that time, much Western psychology was strongly non-experimental in its approach, and psychologists concerned themselves, rather in the tradition of Descartes, with acquiring knowledge about the mind by thinking about their own

---

*Vygotsky, Luria – whose book *The Mind of a Mnemonist* I have already discussed – Anokhin, Rubinstein and Beritashvili are amongst the most significant of these figures. Vygotsky died young, and his promising, more socially oriented approach to psychology was lost in the dark Stalinist years of the later 1930s, to be resurrected only in the 1960s. The others, though each experiencing long periods of personal hazard and difficulty, survived long enough to found schools and institutes of their own which even today continue to dominate neuropsychological thinking in the former Soviet Union and its successor republics.[6] Despite my scepticism about aspects of the Pavlovian tradition, major advances in analysis and experiment on neuropsychological processes were made in the Soviet Union over these decades, albeit often ignored in the West, for reasons discussed more fully in chapter 12.

(the method known as introspection). In reaction, the behaviourists were determined to banish ideas of mind and the techniques of introspection from psychology. They therefore restricted themselves firmly to the observables of behaviour; to make assumptions about unobservable internal mental or brain processes was, for the behaviourists, unscientific, merely an act of faith.

Watson and his protégé B. F. Skinner argued that the child, or animal, was at birth virtually a *tabula rasa*, physiologically competent, but behaviourally a blank slate on which experience would cut the grooves that would determine all subsequent patterns of function. All behaviour was seen as a sequence of stimuli and responses. Organisms learned by experience which stimuli were followed by pleasant (rewarding, or positively reinforcing) responses and which were followed by unpleasant (punishing, painful or negatively reinforcing) ones, and subsequently responded accordingly. The formal terminology of negative and positive reinforcement was part of behaviourism's deliberate attempt to create itself as an abstract science devoid of any language that might resonate with normal day-to-day experience. Equally, just as experimental subjects no longer thought, nor did they even behave or act; instead they emitted pieces of behaviour or operants. The model science to which these psychologists aspired was clearly physics rather than biology, a graphic example of the powerful attraction this rather untypical science holds for misguidedly envious non-physicists.

Watson and Skinner's theories were parodied by Aldous Huxley in his dystopic novel *Brave New World*, in which *in vitro* bred and genetically engineered babies are firmly moulded to their future place in life. In Huxley's dystopia infants already genetically destined to become low-grade workers are electrically shocked as they crawl towards books or flowers. As a result they grow up with a profound distaste for either reading or natural beauty.

Watson soon left academic research for a presumably more profitable career in advertising, and it was left to Skinner, at Harvard, to carry the torch of behaviourism, which he continued to do throughout his long career as experimental psychologist, educational adviser, philosopher and novelist, until his death in 1990. Skinner's approach offered a theory and soon a technology that seemed to go beyond Pavlov's. Whereas in Pavlovian – or as it became known, classical – conditioning the experimental animals were the passive recipients of the unconditioned and conditioning stimuli which they had to learn to pair, the Skinnerian approach was to put the animals into a situation in which they had to do something positive, to act on their environment – to emit an operant. In the standard form of learning studied by the behaviourists, an animal, say a hungry rat, is placed in a box equipped with a pedal. When the animal stands on the pedal, initially by chance, it receives a food pellet, and it soon learns to press the pedal deliberately for food. Beyond this point the animal can be 'shaped' – to use yet again the Skinnerian terminology with all its ideological undertones – to press only one from a choice of pedals or levers, or to press it a predetermined number of times, or to press it only when a light appears. This is known, by analogy with, and in distinction from, Pavlovian or classical conditioning, as instrumental, or operant conditioning. A great range of permutations is obviously possible, and just as the animal may learn to perform certain acts to obtain a food reward, it may also learn to perform others or similar acts to avoid punishment – say to escape a mild electric shock (negative as opposed to positive reinforcement).

Rather as classical conditioning became the Soviet orthodoxy, so generations of psychology students in the United States and in Britain learned their psychology by way of watching rats in Skinner boxes. Skinnerian acolytes spent thousands of person-years of research minutely manipulating the variables in such boxes; comparing the capacity of rats required to press the lever not once but several times or repeatedly for a given number of seconds to receive food, or measuring how long it takes the animal to cease pressing the lever if food fails to appear (the response is extinguished). Obsessed with formalising these effects, of comparing different 'schedules of reinforcement', the behaviourist journals became full of trivial phenomenology, mere catalogues of seemingly objective facts about the behaviour of rats in boxes as tidily classified as the stamps in a fanatic's collection, but as indifferent to either the biology or to the broader behavioural range of the animals they were ostensibly studying as were the Pavlovians. Skinner's rats were empty black boxes, being stimulated and emitting responses like the most rigidly wired computer chips. Like Pavlov, Skinner was a materialist, anxious to eliminate mind from his psychological equations, and his materialism, also like that of Pavlov, is mechanical and reductionist. However, while Pavlov was at least content to leave the implications of his theories for human education and society to his acolytes, Skinner knew no such inhibitions and, seemingly without even noticing the relevance of Huxley's parody, repeatedly argued the case for a psychologically engineered society based on his theories of behaviour and its formation.[7]*

Even more than Pavlov, what Skinner was doing was emptying organisms of their biology, turning them into little more than artefacts, robots to be manipulated – trained – by the psychologist. True, Skinner found when shaping his animals that they sometimes behaved inappropriately. Occasionally they would form associations their trainer had not intended – for instance approaching a lever from a particular direction or performing a particular gesture before pressing it – a behaviour Skinner regarded as 'superstitious', the result of a one-off adventitious association from which the animal generalised mistakenly. And it was hard to deny that animals could learn about their environment by exploring it without a specific association of a particular behaviour with a particular reward – what might be called incidental learning – as a result of merely attending to what was going on around them.

---

* Skinner went so far as to claim that children acquire language because their elders reward them when they use words correctly and punish them when they make errors – a theory which was effectively demolished in a famous debate with Skinner by the linguist Noam Chomsky. Chomsky's view is that language and indeed grammar are programmed into the brain's 'deep structure' – that is, they are an aspect of what I would describe as the brain's specificity. Just as, during development, the eye becomes wired to the brain to provide ordered vision, so, for Chomsky, the speech centres of the brain become wired up so as to produce ordered grammar. A child is thus developmentally programmed to learn to speak grammatically, although the language in which that grammar is expressed is determined by experience – the linguistic environment in which the child grows.[8] However, Chomsky's innatist views on how language is acquired have been taken to biologically determinist extremes beyond his intentions by evolutionary psychologists such as Steven Pinker[9] but also heavily criticised by more sophisticated evolutionary biologists and psychologists.[10]

But it became almost a source of embarrassment to behaviourism to be confronted with the facts of animal life (one psychologist was to write a paper entitled, only half ironically, 'The misbehavior of organisms'). For instance, rats readily learn to run mazes and distinguish odours. Pigeons, on the other hand, are not very good at learning odours and a pigeon-maze, while not impossible, is hard to imagine. Pigeons are, however, very good at learning to distinguish colours and patterns, and they can easily be trained to peck at a red light for food reward rather than a green one, say. And although they cannot run a maze, many species of bird – including homing pigeons – have an almost incredible capacity to form internal maps of their environment, to navigate over thousands of miles or to remember sites at which they have nested or stored food. Such biological constraints on learning are clearly related to the evolutionary history of rats and pigeons outside the artificial conditions of the laboratory. Rats are nocturnal creatures, live in tunnels and work by sound, smell and touch much more than by vision. Birds explore their world visually and by pecking, not by sniffing, and can therefore make some types of association much more easily than others.

I spent some time teaching myself this rather obvious fact when, having found how easy it was to train a chick to avoid a bitter bead once it had pecked it, I tried to pair the bead-pecking with a different form of discomfort. I arranged the experiment so that every time the chick pecked a dry, tasteless bead, it felt a mild electric shock to its feet. However many times they experienced the pairing of bead-peck with foot-shock though, the chicks would not avoid the bead – if anything they pecked at it more vigorously and somewhat aggressively. In nature they are scarcely likely to have to learn the peck–pain-in-the-foot relationship, whereas the peck–nasty-taste relationship must happen quite frequently.

Hebb synapses

Nonetheless, the main thrust of the Pavlovian/Skinnerian theory, that learning occurs by association, that is, by the regular conjunction of two phenomena closely associated in time, be they bell and food, or lever-pressing and food, became the mainstay of learning psychology, and remained dominant in the Anglo-American, Soviet and Eastern European traditions for several decades. (The domination was much less marked elsewhere in Western Europe, where alternative schools flourished.) What is more, associationism, initially a psychological theory, became a neurobiological one as well in the hands of the Montreal psychologist Donald Hebb, whose book, *The Organisation of Behaviour*, published in 1949, explicitly offered a cellular, neural version of associationism.

Hebb's argument was as follows. Consider the classic Pavlovian reflex, of the dog learning to salivate to the sound of a bell. Suppose that there is in the brain a simple pathway by which the sight or smell of the food activates receptors in the eye and nose; these pass messages up the optic (and olfactory) nerve and eventually, via several synaptic connections, arrive at particular neurons in the cortex. The output of these neurons in turn eventually runs to the salivary gland, signalling for the salivation to start. This is basically a reflex arc, like the knee-jerk, although involving rather more complex circuitry. But, Hebb supposed, the inputs from the visual system will not be the only ones on these particular neurons in the brain; there are also others, originating say from the auditory system, which make weak or even non-functional synapses with them. An input from the auditory system alone would not normally be sufficient to cause the cortical neuron to respond by firing. Now consider what happens when the bell rings just before the food is presented. Auditory and visual signals now arrive at the cortical neurons more or less simultaneously. The visual input is already sufficient to cause the cortical neuron to fire, and the core of Hebb's idea is that the frequent pairing of the two inputs results in the neuron becoming sufficiently activated as to strengthen the auditory synapse. Subsequently, the auditory signal alone is enough to cause the brain neuron to fire and the gland to salivate. Synapses could be strengthened like this by biochemically modifying their structure, or by changing their electrical properties, or by simple growth. (It must of course always be understood that although one conventionally speaks of 'the' neuron, 'the' synapse, or whatever, in practice any such process is likely to involve many hundreds of neurons and perhaps, granted that each neuron may have anything up to a hundred thousand synapses on it, many hundreds of thousands of synapses. Learning even a simple task is a process likely to involve whole populations of cells.)

In Hebb's own words (and indeed his own italics):

> Let us assume then that the persistence or repetition of a reverberatory activity (or 'trace') tends to induce lasting cellular changes that add to its stability. The assumption can be precisely stated as follows: *When an axon of cell A is near enough to excite a cell B and repeatedly or persistently takes part in firing it, some growth process or metabolic change takes place in one or both cells such that A's efficiency, as one of the cells firing B, is increased.*
>
> The most obvious and I believe much the most probable suggestion concerning the way in which one cell could become more capable of firing another is that synaptic knobs develop and increase the area of contact between the afferent axon and efferent [cell body]. There is certainly no direct evidence that this is so . . . There are several considerations, however, that make the growth of synaptic knobs a plausible perception.[11]

The whole process can also be shown diagrammatically (see figure 6.1).

Hebb's idea wasn't new (no idea in science is ever completely new, I suppose). It can find precursors in the writings of the great Spanish neuroanatomist Ramon y Cajal at the beginning of the twentieth century, and assiduous graduate students wanting really to assign priority refer to an article by an Italian, Tanzi, in 1893 – though how many have actually read the article they cite is a different matter. Tanzi's formulation, composed at a time when it wasn't even fully recognised that neurons were individual cells or that synapses existed in the brain, is extraordinarily prescient:

> Perhaps every representation immediately determines

a functional hypertrophy of the protoplasmic processes and axons concerned; molecular vibrations become more intense and diffuse themselves, momentarily altering the form of the dendrites; collaterals originate and become permanent . . . Each fresh occurrence of the same conscious phenomenon in the form of recollection, fantastic representation, or identical repetition, strengthens the mnemonic capacity, because it determines the formation of new impressions that are substituted for or added to those that were there . . . Where this takes place the neurons momentarily associated in the action ultimately become united in a stable functional solidarity . . . by a progressive process of functional hypertrophy, which leads to a more or less permanent increase of the neurodendrons that connect the nervous elements . . .[12]

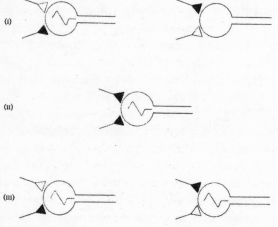

Fig. 6.1
Hebb synapses
According to Hebb, memories could be formed by modification of synapses. In this diagram, two incoming neurons synapse onto a third. At first (i), when the lower synapse is active (shown as dark), the third neuron fires, but the upper synapse is too weak to cause it to respond. However (ii) if both synapses are active simultaneously, the third neuron fires, and sets in train biochemical processes which strengthen the previously weak upper synapse. As a result (iii) the upper synapse becomes capable in its own right of causing the third neuron to respond.

Of course, if one is looking for historical precedents one can go back even further. The founding father of Russian neurophysiology, Sechenov, whose work was so important for Pavlov and his successors, argued something similar thirty years before Tanzi (and one could of course go back to Descartes and his fine hairs, or even to St Augustine and Aristotle). Speculations however plausible are fine, but the problem remains to test them. By contrast Hebb's formulation came at a time when it was beginning to be possible actually to test some of these theories experimentally, and 'Hebb synapses' and 'Hebbian rules of association' have become, as I discussed in chapter 4, the raw material for modellers and theorists of memory ever since his book appeared. Most of us working experimentally on memory think that something like what Hebb suggested must occur – the problem is to prove it and to show just how, in biochemical, physiological or structural terms, the change occurs.

The nice thing about the Hebbian model of course was that it seemed then, and still indeed does seem to many, to offer a direct 'translation' of the behavioural phenomenon of association – classical conditioning – into a neural mechanism. It can accommodate species differences in the possibility of learning, because a basic postulate is that at least the potential synaptic connection must exist (in the Pavlovian example above, that between neurons which receive auditory inputs and the neurons responsible for triggering the salivation response) for such synaptic strengthening to occur. This is an important constraint on learning any particular association. Presumably in my chicks there would not be such a ready set of synaptic connections beween the nerve cells responding to foot shock and those responsible for preventing pecking.

Bringing back the organism

Nonetheless, despite its popularity with psychological theorists, modellers and neurobiological experimenters, association cannot be the only way in which memory occurs, as many critics of Skinnerian and Pavlovian learning theories pointed out from the 1920s on. For instance, on Skinner's theory, rats ought to learn to run a maze correctly by learning each correct turn (first left, second right and so on) individually and sequentially as a chain of stimuli and responses. But it was quite straightforward to show, by rearranging the maze or altering the cues within it, that the animals are not so inefficient; instead they seem after a few trials to be able to form some sort of a global image of the maze, a map if you like, in their brains, so that wherever they are placed in it they can deduce where the goal may be and adopt the most efficient route towards it without being excessively confused by the rearrangement of the maze. Animals use strategies when they learn; they can create concepts. To understand such mechanisms it is not adequate to reduce them to linear sequences of stimulus–response, positive and negative reinforcement.

In the 1920s and 1930s this less impoverished view of how animals behave formed the basis of the gestalt school of psychology which flourished particularly in Germany and Austria. If Skinner and Pavlov were reductionists, the gestaltists were holistic. Their protoype learning experiment was provided by Wolfgang Köhler's chimpanzees, who, when placed in a cage with bananas visible but out of reach, and a couple of sticks each in itself too short to reach them, eventually hit on the idea of joining the sticks together so as to capture the food. Köhler and the gestaltists regarded such problem-solving as evidence of creative, conceptual thinking, not the linking together of interminable chains of stimuli and responses.[13] This alternative tradition, together with the ideas of genetic epistemology developed over the same period by the Swiss psychologist Jean Piaget, became a serious competitor to associationism, especially in Western Europe. The competition was strengthened by the emphasis within European and British research on the study of the behaviour of animals in more natural environments than the artificial constraints of the psychologist's cages and mazes – the science that became known as ethology, and which began to rediscover the richness of animals' lives, which Skinnerian or Pavlovian reductionism had endeavoured to eliminate from their experiments and cancel from their equations.[14]

It was this contrast which led to the comment – I believe due to

Bertrand Russell – that the difference between German and American animals was that in America rats rush about energetically learning by trial and error, while in Germany they sit and think before acting. But even less than the other schools of pre-biological psychology did the gestaltists have much interest in the physiological processes going on inside the brains of their thinking apes.

This is why I would argue that for neurobiologists concerned with learning and memory, the legacy of this period of experimental psychology, always excluding Hebb, is not its theoretical constructs, its painstakingly accumulated phenomenology, the minutiae of schedules of reinforcement or of conditioning chains. If Alexandrian fires were to consume all the thousands of metres of library space devoted to the archive of Behaviourist and Pavlovian journals from the 1920s to the 1960s, I doubt that much of more than historical interest would be lost. For all their concern with theory, the residue of Pavlov and Skinner is to be found in techniques; some experimental models of learning using classical conditioning, and the Skinner boxes which were nearly ubiquitous in experimental psychology laboratories until the 1980s but now at last have fallen out of use in favour of training tasks involving more clearly cognitive or map-making types of learning, of which more later.

If asked to single out one particular turning point in the history of American psychology's approach to learning I would unhesitatingly point to experiments made by John Garcia in the late 1950s and early 1960s. Garcia was interested in bait-shyness, a phenomenon known to anyone who has ever tried to get rid of plagues of mice by poisoning or trapping. If an animal finds poisoned food, eats it and becomes sick for a few hours before recovering, it will subsequently take avoiding action. If the food is of a familiar kind but located in an unfamiliar place, it will later avoid eating food in that place. If the food is in a familiar place, but is itself unfamiliar in some way, in taste or colour, the animal will afterwards avoid eating food of that taste or colour. Garcia was able to mimic this bait-shyness experimentally if instead of using poisoned food, he allowed the animals to eat normal food, either located at an unfamiliar site or coloured in an unusual way, and subsequently made them sick by injecting a small dose of lithium chloride into their gut. The result of doing this is to produce the equivalent sensation to a mild case of food-poisoning, which lasts for up to three hours. When they had recovered from their sickness, Garcia's animals avoided food at the novel site or of the unfamiliar colour which they had clearly learned to associate with the subsequent sensation of sickness. (This phenomenon occurs in humans too; patients undergoing chemotherapy in hospital often form aversions to the hospital food.)*[15]

This behaviour is so obvious to anyone who has studied or worked with real animals – or humans – that it might seem surprising that it should so disconcert associationist psychologists. The problem was simple though. The painstakingly worked-out rules of both classical and operant conditioning insisted that learning an association depends on a close contiguity in time of action and effect, conditioning and unconditioned stimulus. The conditioning stimulus has to precede the unconditioned by only a quarter of a second; the operant behaviour has to precede the reinforcement by a similar short time, or the association cannot be made. Yet in conditioned taste-aversion, as the Garcia effect became known, eating the food can precede the feeling of sickness by some hours. The associationists, as might be expected, struggled hard to save their theory. Perhaps what was being paired was some residual sensation of the taste of the food in the stomach at the time when the feeling of sickness was experienced? Yet the fact that place aversions could also be produced, to eating even familiar food in an unfamiliar site, made this argument hard to sustain.

A few years ago I blundered inadvertently into this debate by developing a version of the Garcia test for my chicks in the course of some other experiments with which I was concerned. I offered them a green bead, dipped in water, and, as I had expected, they pecked it eagerly. Half an hour later I injected them with lithium

---

*Actually Garcia's original research was sponsored by the United States Navy, interested in the effects of ionising radiation such as that produced by nuclear explosions, which are rather analogous to those of chemotherapy or radiotherapy, on eating.

chloride, they got sick, recovered, and three hours later I offered them the green bead again. Not surprisingly, they avoided it. No problem for the associationist theory here; the chicks were learning that 'green water' tasted bad. But then I modified the experimental design and instead of a wet bead, offered them a dry green bead. They pecked and got sick as before, and later avoided a similar green bead, although they would go on pecking at a red or chrome one. Clearly in this case it could not be any residual taste which was being 'paired' with the sickness. The only possible explanation could be that the chicks, when they pecked a novel bead for the first time, formed some sort of a 'representation' of the green bead, a model, in their brains, which they could preserve for at least half an hour without it being paired with anything obviously either aversive or pleasant, but simply neutral. Later, they could form an (inappropriate, of course) association between this pre-existing representation of the bead in their brain and the sensation of sickness, and hence avoid the green bead when offered it again.[16]

Although I hadn't intended to get involved in a debate about associationism, once I began to think about the implications of this experiment I realised that, straightforward enough though the effect may be, it clearly does not conform to any simple associationist theory derived by an extension of Pavlovian or Skinnerian conditioning theory, whose essence is the immediate linking in time of stimulus and response. True, associationists could (and some did, when I discussed the experiment with them) argue that what is being paired in remembering the bead-peck is not the association of peck and taste but peck and green colour, or maybe bead-shape and green colour, or maybe . . . Clearly one could invent many such possible pairings, and there is no real way of disproving them. But all of them miss the point of associationism as a theory. If Pavlov or Skinner are to be believed, what should become paired are not simply two properties of the same object (green and round, say), but some property of an object with a particular reinforcement, either a pleasant taste or an unpleasant experience, or an action such as pecking with its immediate consequences such as distaste. Simply to say that an animal is pairing the neutral act of pecking at something which is both green and round, unaccompanied by any such rewarding or aversive experience is to say nothing other than that the animal, in remembering the bead, can recall various aspects of it, including colour and shape. Indeed, in chapter 11, I shall go on to show how our own experiments can prove that this is the case – with implications far different from those the associationists wish to draw from them.

There are three lessons from this: first, simple experiments can often have further-reaching implications than one thinks when one starts to make them; second, theories, as many philosophers of science have pointed out, die hard; and third, animals, even day-old chicks, learn in many ways, and our theories will need to be broad enough to accommodate them. Associationism is not wrong, it is just not the whole story.

But rather than pursue further the twists and turns of psychological theory-making about memory, I want to return to the biological issues involved. I have argued that memory is a particular, rather specialised form of adaptation to experience, itself a general property of living organisms. To explore the mechanisms of human memory, I have suggested turning to the behaviour that can be defined as learning and remembering in non-human animals such as cats, dogs, rats, pigeons or chicks. All these are vertebrates with large and complex brains. Does memory depend on having such big brains, or can animals with much simpler nervous systems also learn and remember? Can some animals learn better than others, and if so might there turn out to be some specialised structure or particular form of biochemistry or physiology within their brains compared with those that learn less well? Answering such questions may cast light on the mechanism of memory by helping define those types of brain properties that are necessary for it. To approach such questions, it is time to consider the evolution of memory, and the forms that it may take in organisms with smaller, less complex nervous systems – and even with no nervous systems at all. This will be the task of the next chapter.

Chapter 7

The evolution of memory

IT IS A COMMON ENOUGH LINGUISTIC TRICK – AND ONE OF WHICH neurobiologists are themselves often guilty – to speak as if there were some sort of evolutionary scale or ladder of complexity, along which all the living forms found on earth today can be arrayed to form a series of 'more evolved' and 'less evolved' organisms. An even more extreme form of this way of thinking is to assume that there has been some type of directive force in evolution such that humans represent the 'pinnacle' of progress. Such ideas, which derive from views about the place of humanity in nature that long pre-date Darwin and the birth of modern biology, very much misunderstand evolutionary theory. A short evolutionary sermon is therefore necessary before the argument can run on.

All of today's living organisms derive from primitive forms of life which first appeared on earth not long after the birth of the planet itself perhaps some four billion years ago. Evolution, which literally means unrolling, or development, is, for biologists, the process by which there has occurred a steady change in the form of organisms across generations. By form here is meant anything from their biochemistry and internal structure to their behaviour. These changes in form thus occur at a multitude of levels, from the molecular to that of an entire population, and are conserved by genetic transmission.* When the change of form becomes too great for the old and the new form to interbreed successfully, one is entitled to say that a new species has been generated.

The most generally accepted mechanism of evolutionary change is the modern version of Darwinian natural selection, based on the simple propositions that (a) like begets like, though with minor, essentially chance variations; (b) all organisms are capable of producing more offspring than actually can survive to maturity and reproduce in their turn; (c) those offspring that do survive to reproduce must in some way be variants that are better adapted to their environment than those that fail; and (d) those favoured variants are likely to reproduce the favourable variation in their own offspring. Thus better-adapted forms will tend to replace less well-adapted forms.

Organisms can only adapt to their presently existing environments, they cannot predict future ones. The fact that future generations may find themselves living in a warmer planet cannot be built into today's selection processes. Thus evolutionary change in a population of organisms can only track and respond to the changes in the environment, continually being wise after the event. Evolution has no foresight. Yet environments are never static; they are always changing. Physical, non-biological forces modify climates, raise mountains and sink seas. And living organisms themselves constantly change their environment, altering its chemical composition by eating, breathing, excreting; altering its geography by building and destroying (this, relatively obvious, concept has been raised to almost metaphysical status by James Lovelock and his devotees as the grandiosely named Gaia hypothesis). Above all, living organisms are part of one

---

*I have given a very general definition of evolution. My description of form includes both what geneticists call an organism's phenotype – that is, everything about it except its genes, and its genotype, or the full complement of genes that it possesses. Population geneticists tend to work with a much more rigorous and reductionist definition, a gene's eye view of evolution, in which the essential phenomenon is a change in gene frequency – the rate at which a particular gene occurs among a population of organisms of the same species – within such a population. But I don't wish to get embroiled in this debate here, which many will recognise as part of an ongoing dispute that I and many others have with such exponents of sociobiology as Richard Dawkins and E. O. Wilson.[1]

another's environment, as sexual partner, parent, offspring, prey or predator. And finally, organisms are not passive recipients of their environments; they can – animals to a greater extent than plants – choose their environment by moving from a less attractive to a more attractive location. Thus as in development, the interplay between organism and environment during evolution is not a one-way process, by which environments constantly set challenges to organisms which they either pass (reproduce successfully) or fail (die out). Organism and environment are dialectically interlocked.

While Darwinian processes are likely to be only one of several mechanisms responsible for evolutionary change in form (there is much debate, which need not concern us here, about the relative contribution of these other mechanisms), the point for the present is that all forms of life on earth today are clearly the results of comparable evolutionary pressures over the whole of geological time. It is thus wrong to assume that any one form can be 'more evolved' or 'better adapted' than any other. To be here today by definition means to have survived, whether as dung beetle, mushroom or human. There are many ways of surviving to produce offspring, and there is little consensus between organisms about what counts as evolutionary 'success' in this sense. Only a very biased, brain's-eye-view of the living world would assume that the bigger the brain the more successful the species. In sheer bulk of biomass, organisms without brains or even without central nervous systems far outnumber those possessing these desirable features. Even amongst those organisms with well-coordinated nervous systems, a good case can be made for the dominance of beetles over mammals.*

Nor are evolutionary processes progressive or purposive; the primitive single-celled creatures which were our early forebears were not possessed of a burning imperative, or driven by a mystic higher force, to evolve into sentient humans. True, there are sharp constraints on what can evolve and what cannot, given by the physical

*I can't refrain here from quoting the response of J. B. S. Haldane, the Marxist physiologist and geneticist who made a major contribution to the new synthetic theory of evolution in the 1920s and 1930s, when asked what attributes he might imagine a god to have. 'He must be inordinately fond of beetles', Haldane replied.

properties of the planet we inhabit and the carbon chemistry on which our molecules are based. But within these constraints, chance events, the contingencies of particular organisms present at particular times in particular environments, provided the push towards the living forms we are today; there was no master plan. Studying the fossil record enables us to show what must have happened, but, as Stephen J. Gould has put it most cogently,[2] were we to be able to wind the clock of evolution backwards towards the origin of life on earth, and then run it forward once more, quite different outcomes would most likely have occurred; not only would there probably have been no humans, but perhaps not even any brains. The study of evolution is the study of history, not of the working out of some mathematical programme of progress towards complexity or a mystical urge to evolutionary perfection.*

In that historical sequence, some organisms, such as humans, have appeared only very recently, and their past ancestry seems to have been one of rapid and quite dramatic change. Other forms of life on earth today seem much more closely to resemble what one knows, based on the fossil record, to have been their evolutionary ancestors of many million years past. It is as if having once found an effective way of surviving and reproducing, and a relatively stable environment in which to go about their business, there was little pressure to change their way of life. By studying their biology and behaviour, therefore, it becomes possible to draw reasonably probable

---

*A final footnote on evolutionary matters. Some may recognise several sentences in this last paragraph as a coded reference to a friendly debate which has been going on for some years now with my one-time Open University colleagues Brian Goodwin and Mae-Wan Ho, who, along with Gerry Webster and Peter Saunders have been strong advocates of just such a rational as opposed to historical approach to the understanding of biological form.[3] I share the view of most biologists that nothing in biology makes sense except in the context of history, although agreeing that nothing in biology is possible except within the framework of structural limits imposed by physical principles. Thus I occupy a position somewhat midway between Gould and Goodwin et al. For the former, chance is all, random mutations, or externally unpredictable accidents like mass extinctions due to changing environments determine future history, only understandable with hindsight ('for the want of a nail . . . the war is lost'). For the latter, history is trivial – Webster calls it 'mere antiquarianism' – because if one understood enough about the physical and chemical forces that shaped form one would see that the number of possible types of organism is strictly limited, and humans are maybe predictable after all.

conclusions about how their – and thus our own – ancestors lived.

If one goes far enough back in evolutionary time, all present-day living organisms share such a common ancestor – or group of ancestors. Along the evolutionary path that led to humans there were, three million or more years ago, creatures which were also the ancestors of today's apes; earlier, those who were the ancestors of all modern mammals; earlier still, of all today's vertebrates; yet earlier, of all multicellular animals; and so on. Rather than drawing some evolutionary ladder or tree, the best representation is a sort of multi-twigged bush. Because the only records of the ancestral forms are their fossils, without any soft tissues like brains and nerves, the brain structures and behaviour of such ancestors can only be inferred by studying their present-day descendants. From this comparative psychological approach one can begin, however, to draw some conclusions about the evolution of nervous systems, of brains, behaviour and eventually of memory.

The origins of behaviour*

Although all organisms show adaptive behaviour, for it is a necessary condition of existence, many highly successful life forms have managed very well without the capacities to learn and remember – even without brains. Think of a sunflower turning its flower head towards a source of light – and therefore of energy. The sunflower does not learn by experience to turn its head more effectively as it matures, or not to turn at all if it is repeatedly electrically shocked every time it does so. The flower-head turning behaviour, adaptive or not, is a 'given' property of the organism, fixed within its genetic and developmental programme. Such a given behaviour (innate is the conventional term, but I prefer to avoid it if possible, for it carries along with it a load of redundant ideological baggage) ensures that appropriate responses are made to particular stimuli without the need

---

*I discuss these questions at much greater length in my forthcoming book on the future of the brain (Jonathan Cape, 2005).

for trial and error learning, but at the expense of limits to both the range and the flexibility of the response. Changes in behavioural responses occur not in an individual but over many generations as a consequence of evolution. By contrast learned behaviour requires experience and practice. The advantage is that responses are highly flexible and can be modified in response to environmental changes and perceived outcomes within the lifetime of the organism. However, it is crucial to recognise that such learned responses are not genetically transmitted to offspring. What is transmitted is the plasticity and capacity to learn which are themselves 'given' by the genetic and developmental programmes of the organism. This is why it is important to reject tired old debates about nature versus nurture. To understand brain and behaviour means rejecting that dichotomy and instead trying to interpret the intertwined dialectic of specificity and plasticity.

A plant's mode of existence is shaped by the fact that it can survive, biochemically, by making use of the sun's energy to trap carbon dioxide and convert it to the foodstuffs such as sugar that it requires. All it needs to do therefore is to stay with its leaves out in one place, photosynthesise, and avoid being eaten until it has managed to reproduce. (The price it pays is the need to develop quite elaborate and inherently uncertain reproductive methods, making use of other organisms or natural forces such as wind to spread pollen and seeds.) Animals cannot make a living so simply; not being able to photosynthesise, they have to hunt for prepackaged food, either directly by eating plants or indirectly by eating the animals that have eaten the plants. Thus they need to be continually on the move, to develop specialised ways of finding food sources, avoiding danger and locating mates. For non-photosynthesisers, the premium is therefore on an increasing range and flexibility of adaptive behaviour, on developing sense organs that can detect food or predators at a distance and motor skills to move towards the desired – or away from the desiring.

Single cells

Even the simplest, single-celled organisms can show examples of this type of behaviour. As long ago as the 1880s, W. Pfeiffer, in Germany, dipped a thin capillary tube containing glucose into a drop of liquid containing bacteria, and observed that the bacteria tended to collect at the mouth of the capillary. The bacteria were behaving as if they 'knew' the glucose was there and were responding adaptively. The explanation is relatively straightforward; the glucose slowly diffuses out of the capillary into the surrounding liquid, creating a gradient of sugar concentration. The bacteria have, embedded in their cell membranes, receptor molecules – that is, proteins whose molecular architecture enables them to recognise the glucose molecules – and tiny whip-like projections (flagella) which, by beating in unison, can row the bacteria up the glucose gradient. Mutants which lack the glucose receptors fail to show such behaviour, even though they will continue to make use of the glucose as an energy source if placed directly in contact with it.

More complex single-celled animals, such as paramoecium, have similar projections, called cilia, connected at their bases by a system of fine threads, which enable the beat of the cilia to be coordinated. Except when it is actually feeding (it eats bacteria), the paramoecium is in constant motion, frequently bumping into obstacles in its path, and, when it does so, reversing by switching the beat of the cilia on one side. It thus steers itself like a toy motor car, peregrinating around obstacles and into food-rich areas. As well as physical obstacles it will move away from regions of heat or cold or irritating substances such as acids. These goal-seeking and avoiding reactions represent, within an organism no bigger than 0.2 mm in length and with no sort of nervous system, the beginnings of a pretty sophisticated range of active behaviours.

Multicellularity

Single-celled organisms have to be self-sufficient, packing all their behavioural repertoire into a tiny compass. But at least communication between one part of the cell and the other is not a major problem. With the evolution of multicellular organisms, both the range of possible behaviours and the organisational problems the organism has to solve increase. There is a need for some form of rapid internal signalling system so that the activities of cells in different parts of the same creature can be coordinated. It seems likely that in very primitive multicellular forms the main mode of communication was chemical – a substance released by one cell, say signalling for the cell to contract, could fairly quickly diffuse to other cells, ensuring that they too contracted. Such chemical signals would have been the forerunner of modern-day hormones. One of the most intriguing of evolutionary clues is the close chemical similarity between many hormones and the substances that function within the nervous system as neurotransmitters, suggesting that perhaps the second group, the neurotransmitters, may have developed evolutionarily from the first.

Multicellularity makes possible specialisation. Different functions and properties can be distributed between different cells; some for instance become contractile, whilst others specialise in synthesising and secreting the chemical signals. Yet others, on the surface of the organism, concentrate receptors responsive to particular chemicals, like the bacterial glucose receptors, or even to light.

Signalling by diffusion can work well for small organisms, but it is limited; diffusion of chemical signals over long distances is slow and inefficient as it cannot be specifically targeted to ensure that the signal arrives only at the cell it was intended for. If, on the other hand, the signalling cells grew into such a shape that they could come into direct contact with their target organs the chemical signal could be discharged directly at site across the 'synaptic' gap between the cells. This would ensure directionality, but it would still leave the problem of getting the message from one end of the signalling cell to the other over what would now be a relatively long distance.

Here, the electrical properties of cells must have become important. It is a universal property of living cells that their external cell membranes are electrically charged. This charge, called the membrane potential, comes about because cells contain large numbers of dissolved salts in their internal fluid (the cytoplasm),

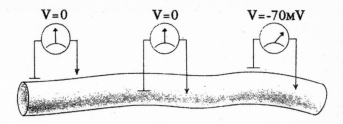

Fig. 7.1
Resting membrane potential
If two recording electrodes are placed on the surface of a nerve fibre (or indeed any other cell), no voltage difference is registered between them. The same is true if the electrodes are both placed inside the nerve. But if one is placed outside and the other inside, a voltage of some 70 millivolts negative to the inside is measured. This is the resting membrane potential and results from the biochemical and physical properties of the membrane, which ensure an unequal distribution of sodium, potassium and other ions across it.

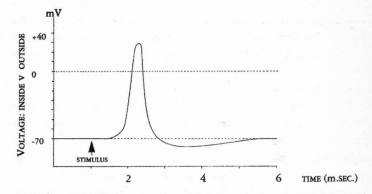

Fig. 7.2
Action potential
It is the action potential which is the unique property of excitable cells such as neurons. If a nerve axon is stimulated by an electrical, mechanical or chemical signal, the properties of the membrane at the site of stimulation are briefly changed; sodium ions enter the axon from the outside medium and the voltage rapidly swings from 70 millivolts negative to some 40 millivolts positive, before the membrane properties change again, sodium entry ceases and ions are exported until the original potential difference is restored. The graph shows the brief pulse of the action potential, which travels down the axon like a wave, normally commencing at the site where the axon leaves the nerve cell body and terminating at the synapse, where it triggers the release of neurotransmitters which diffuse across to the post-synaptic cell, carrying the depolarising signal to it.

including sodium, potassium, calcium, chloride and others. These salts, in solution, form electrically charged ions (for instance sodium chloride, NaCl, forms positively charged sodium, Na+, and negatively charged chloride, Cl-). However, the inside of the cell differs from the outside in that there is a high internal concentration of potassium and a low concentration of sodium. Inside the cell as well are proteins, whose constituent amino acids are also electrically charged. The ionic composition of the inside of the cell is thus different from that outside, and as a result the inside of the cell is some 70 millivolts negative with respect to the outside.

Nerve cells, neurons, resemble all other types of cell in possessing such a resting membrane potential. Where they differ is in the unique properties of their cell membranes, for the nerve cell membrane is excitable – which means that in response to a signal such as a small local fluctuation in ion concentration across the membrane, it can rapidly become permeable to the ions outside it. Sodium ions enter and the membrane becomes depolarised, turning from 70 millivolts negative to as much as 40 millivolts positive. This change results in a wave of electrical activity passing down the nerve cell membrane – a wave called an action potential that in a few milliseconds passes from the cell body along the axon to the synapse. The action potential serves as a signal in its turn for the synapse to release neurotransmitters which trigger the response in adjacent neurons. The evolution of cells with action potentials and chemical signalling processes at their termination points may have provided the basis for the evolution of a modern-day nervous system.

One example of an organism possessed of such a rather basic nervous system is the tiny, pond-living hydra, which sits at the bottom of ponds and streams attached to rocks or water plants and waving its tentacles above its mouth (figure 7.3). Like a sea-anemone it will close down to a blob if touched, and feeds on small organisms which brush past its tentacles, by shooting out special poisonous threads which paralyse its victim and can then be collected by the tentacles and drawn into the hydra's mouth. This complex behaviour requires mechanisms to register the presence of prey or danger and to decide on and make the appropriate response, of attacking or contracting into a blob – sensory cells, secretory cells, muscle cells and above all a network of electrically connected cells running right across its surface which can coordinate the hydra's responses.

The individual cells of this network are not quite like neurons in more complex animals, though, for the network lacks specificity or directionality. Stimulate the hydra at any point on its body and a wave of excitation runs from that point in all directions so as eventually to involve every part of the nerve net. The hydra's nervous system is like a telephone exchange whereby whatever number you dial you are eventually connected to all other subscribers. By contrast the key feature of a fully developed nervous system is its specificity, the precise set of connections by which a signal beginning at a particular sensory cell runs in a defined route, ending in some effector cell, a private line, essentially insulated from the multitude of other neurons within the system.

Private lines and nervous systems

Private lines of this sort, and therefore real nervous systems, appear in rather more complex organisms than hydra, the planaria, or

Fig. 7.3
Hydra
Note the diffuse network of nerves running through the body.

flatworms. Put a piece of raw meat into a stream and within a few hours it will be covered with small, flat, black worms feeding on it. These are planaria. Unlike hydra, they have clearly defined head and tail ends, and a much more elaborate behavioural repertoire. Most importantly they have a specific set of connections between their neurons, so that if a cut is made in a set of nerves leading to a muscle, that specific muscle becomes paralysed. What is more, whereas in hydra the proto-nerve cells are more or less evenly distributed throughout the animal's body, in planaria the distribution of cells is not symmetrical. Neurons are concentrated in clusters with short interconnecting axons and dendrites between the cells of the group and defined nerve tracts leading in and out. Within each cluster there are some cells which receive inputs from the incoming nerves, some from which the outgoing nerves run, and others (interneurons) which connect inputs and outputs. These clusters of neurons (called ganglia) thus contain all the essential elements of a complete nervous system (figure 7.4).

Planaria, because they have a head and a tail end, also have a clear sense of direction, of going forward and going backward, which is not apparent in hydra. It is obviously advantageous for an animal to receive more detailed information about where it is going to than about where it has come from, and it is therefore not surprising that

Fig. 7.4
Planarian
This organism has a ladder-like nervous system with a clearly defined head end.

as well as the mouth at the front end of the planaria there is a concentration of sense organs, such as light-sensitive eyepits, and to process the information arriving from these sense organs there is a group of ganglia concentrated in the head – forming at last the forerunners of real brains. This much more elaborate nervous system is associated with a much greater range of what we can recognisably call adaptive behaviour. Planaria tend to shun the light and move towards the dark, they are sensitive to touch and keep the undersides of their body in contact with solid objects, move towards food, and tend to move upstream in water currents.

Habituation and sensitisation

They also show another form of behaviour. Touch a planarian firmly with a glass rod and it will curl itself into a ball – its normal response to a threatening situation. After a few minutes it will slowly and cautiously uncurl itself again. Touch it again, and it will curl up again, and then stretch out once more. But repeat the procedure often enough and the response will diminish and eventually the animal will no longer curl itself into a ball when touched – as if it had become accustomed to the stimulus and no longer regarded it as dangerous. This phenomenon, in which an animal responds to a repeated stimulus by eventually disregarding it, is familiar to everyone. As one dresses in the morning one is very conscious of the feel of the clothes on one's skin, but after a fairly short time one ceases to be aware of the sensation. This is called habituation; the nervous system has become familiarised with the particular set of stimuli which are the pressure of the fabric of the clothes on touch receptors in the skin, or of the glass rod on the planarian's body; the stimuli can thus be disregarded; no longer a factor to be taken into consideration in assessing the current state of the environment.

However, if the planarian is left undisturbed for a period, the original response returns. This could be regarded as nothing more than fatigue and recovery from fatigue, but it is not. If during the period of habituation, the touch of the glass rod is coupled with some other stimulus – say a bright light – the original response at once returns in full, so it cannot be merely because of exhaustion that it disappeared in the first place. The recovery is thus itself an active process; the system has become dishabituated. Habituation and dishabituation, which thus fulfil the criteria for the definitions of learning given at the beginning of this chapter, can be regarded as very basic and simple forms of short-term memory, adaptive mechanisms which economise on unnecessary responses and hence help avoid fatigue. In general habituation and dishabituation follow certain clear rules: the more rapid the frequency of stimulation, the stronger the habituation; the stronger the stimulus, the weaker the habituation. Repetition of the same stimulus following a sequence of habituation and disahabituation produces a more rapid habituation on successive occasions.

Planaria also show another form of short-term learning, called sensitisation, which is in some ways the antithesis of habituation. Where habituation is the weakening of a response, sensitisation is the strengthening of a response to a particular weak stimulus if that stimulus is coupled with an unpleasant one – an electric shock, say. Give planaria a mild electric shock and they will respond by curling up; touch them very gently with a rod, or squirt a mild jet of water at them, and they will not show any response at all. However, if the gentle touch or water squirt is always coupled to the electric shock, the planaria will eventually respond to the otherwise unnoticed stimulus itself. Sensitisation is a generalised process, for after such a shock the animal's responses are not specific to the touch or the water squirt; instead reactions to a wide variety of mild stimuli all become exaggerated. Thus sensitisation lacks the specificity which is the hallmark of truly associative learning in which a particular pairing of stimuli is achieved. In practice sensitisation and dishabituation are rather closely related processes.

Habituation and sensitisation – often grouped together as non-associative learning – can be seen as the first steps towards full-fledged memory, and are universal properties of all organisms with nervous systems. Indeed, there is some evidence that they occur in creatures without proper nervous systems at all – even paramoecium. But 'real' memory must last for a longer time than such short-term effects as habituation and sensitisation, and must show greater specificity. The key property should be the ability to form associative – that is, specific – memories. Whether planaria can carry out these more complex forms of learning is much in dispute, as will become apparent in the next chapter. Nonetheless, with the evolution of planaria-like organisms appeared both the rudimentary forms of a nervous system and the basic behavioural building blocks out of which fully developed memory-processes are eventually fashioned. Discovering just how much creatures with nervous systems of this degree of complexity can remember, and whether they can meet the rigorous criteria laid down by association psychologists as to behaviour to be counted as learning, classical or operant conditioning, becomes a matter of the ingenuity of the experimenter in designing appropriate, biologically relevant tasks. The capacity of such nervous systems to demonstrate learning and memory cannot be in doubt.

Bigger brains, bigger memories

The concentration of neurons in ganglia is perhaps the first step towards building a brain, but even the appearence of a large head ganglion does not ensure that brains of the sort that humans and other mammals possess are the only design solution that can result. In the nervous systems of arthropods, for example – including insects and crustacea – the head ganglion – the brain – may be important but only so in the sense of King John among the barons; there are many other ganglia distributed throughout the arthropod body, each with a considerable degree of autonomy. Consider, for instance, the ability of the male praying mantis to continue copulating whilst the female steadily devours him from the head end downwards, or for the head end of a wasp to continue eating even when severed from the abdomen.

This independence means that, even without their head ganglia, insects can show some behaviour which could be called learning. In the 1960s, Gerald Kerkut, in Southampton, described a series of experiments in which he suspended a headless cockroach above a bath containing a salt solution. The normal tendency of such a 'preparation' (biologists' speak for an animal to which they have done something nasty, akin to the use of the term 'sacrifice' that I commented on earlier) is to extend its legs into the liquid bath. Kerkut arranged it such that whenever the leg touched the liquid an electrical circuit was completed and the cockroach was shocked; when it withdrew its leg from the liquid the circuit was broken and the shock ceased. The residual cockroach, even without its head, eventually ceased putting its leg into the liquid – it had 'learned' how to avoid being shocked.[4]

All animals more than a few thousand cells big need to have some sort of scaffolding to keep their bodies in shape. Vertebrates do it by means of a backbone and internal skeleton, arthropods achieve structural rigidity by means of a tough external skeleton or shell. This produces a fundamental design limitation to the size to which an insect or crustacean can grow (for all the ingenuity of fishermen, breeders and genetic engineers, with however enthusiastic support from Livebait, Legal Seafoods or Hammer Films, a lobster the size even of a small dog isn't on; it would implode under its own weight). Such a design also strictly limits the size of ganglia and brains.

Such a limitation has not obviously affected the evolutionary 'success' of this type of body pattern, as witnessed by the vast number of species of insects and crustacea alive and flourishing today – but it certainly affects the extent of their behavioural repertoire. To a certain extent this limit is transcended in social insects – bees or ants for example – which live in a highly organised and cooperative community. Some have indeed argued that such a community can be regarded as a sort of superorganism which adapts to the environment, regulates its numbers, stores information and behaves in a strongly cognitive manner. When bees swarm, for instance, which happens when any given colony gets too big, a new queen is produced and departs the hive with a group of followers. When a suitable site for a new hive is found the bees have to learn its location and get rid of their earlier learned behaviour of flying back to the old hive. They can not merely learn and keep maps in their head of the location of hives and sources of nectar, and recognise colours and patterns, but they can communicate such orientations, distances and directions to their colleagues by means of the famous 'waggle dance'. Even before Karl von Frisch's well-known studies on bee orientation and his interpretation of the waggle dance in the 1910s and 1920s, students of bee behaviour, from fable-writers and apiary-keepers to entomologists, had marvelled at the skills, apparent intelligence and capacity to learn of the bees, which seemed very far from being rigidly programmed robots.

Bees do have, however, a very tightly defined repertoire of skills and learning ability. Thus, although they can learn colours as signals for food and as markers for the entrance to their hive, they have difficulty with other types of colour learning, such as using colour as a cue to finding their way out of a closed space. The extent of these learning skills has been ingeniously studied for individual bees by Randolf Menzel in Berlin over the past three decades, and he has been able to clarify the role of many regions of the bee's head ganglion. It remains extraordinary though that with a brain of only 950,000 neurons – less than a thousandth of those in the human retina – bees can learn colours, textures and smells as well as motor skills, and when set appropriate tasks can show most of the features of conditioning, associative and non-associative learning and relatively long-lasting memory found by mammalian psychologists in organisms with many-fold larger brains. Particular regions of the brain, the so-called mushroom bodies are crucial structures in such insect learning and memory.[5]

But perhaps if they had listened to what folk-legend had maintained for thousands of years, neurobiologists would not have been so surprised at the learning and memory capacity of the bees. It may come as a greater surprise to discover that much humbler, less social insects than bees can also show features of learning and memory. For most of the twentieth century a favoured organism for geneticists to study, because of the ease with which it can be maintained, its rapid breeding cycle and the possibility of studying populations of many thousands, has been the tiny fruit fly (sometimes called vinegar fly), *Drosophila melanogaster*, which gathers like specks of coaldust, seemingly magnetically attracted to over-ripe fruit. Exposing fruit flies to X-rays or certain chemicals produces mutations, many of them lethal, as a result of which the flies cannot survive at all. Some mutations, however, permit the fly to survive, although in an altered and normally less adaptive state. Study of these mutations, for example in the colour of the eyes, the patterning of the wing veins, the number of bristles on the abdomen and many other minor variants, has provided vital clues to the mechanisms of genetic inheritance and the control of development in the fly, and by generalisation, in other life forms too. It wasn't until the late 1960s and early 1970s, however, that one of the most experienced of fly geneticists, Seymour Benzer, and his students, notably Chip Quinn at Yale and Yadin Dudai, now in Jerusalem, began studying Drosophila behaviour in some detail as the opening shot in a programme to try to identify mutants with abnormal behaviour, and in particular, abnormal capacities to show forms of learning and memory.[6]

The reasons for this hunt, and the role that the flies have played in elucidating learning mechanisms will become clearer in the next chapters; for now the point is that even flies, which might traditionally have been regarded as stupid, and with tiny brains of no more than 20,000 neurons, turned out to be able, under the appropriate circumstances, to learn and remember. The way to teach them is to work with the behaviours they show naturally; like moths, they are for instance attracted by light, but find the fruit on which they feed primarily by smell. The first reasonably reliable and convincing learning task for Drosophila involved training them using just this sense of smell. Groups of flies were attracted by light into a test-tube containing one of two differently smelling substances. When they entered the mouth of the tube containing one of the odorants, they were shocked. The flies were then tested by being offered the choice of the two odours, and the proportion avoiding the shock-associated odour was compared with that avoiding the control odour. In experiments of this sort, which provide a sort of population index of learning, about two-thirds of the normal (what geneticists call 'wild-type') flies avoid the shock-associated odour, and only one-third avoid the control odour. From this and other types of experiment, it is beyond dispute that, even by the most rigid of the criteria used by mammalian psychologists, Drosophila show not merely habituation and sensitisation but classical and operant conditioning based on visual, olfactory and even touch cues.

Slugs and other molluscs

In terms of having bigger brains, arthropods are, as we have seen, limited by their external skeletons. A way out of this design limitation is offered by molluscs, whose best-known land-living forms include slugs and snails. In water, shell-less molluscs can grow to large sizes, and include squid and octopus. Like those of arthropods, the molluscan nervous systems are organised in a series of ganglia, mainly distributed around the animal's gut. In the 1930s and 1940s research on one large mollusc, the squid, revealed that it had truly giant nerve axons, which could be dissected out individually and were big enough to insert electrodes into. A favoured saying amongst biologists is that for any biological question God has created an ideal organism in which to study it. The squid axon certainly conformed to that belief, and became the preparation in which details of the ion movements and changes in electrical activity which occur during nerve conduction and the action potential could be explored.

One of the group of researchers who first explored the merits of the squid giant axon was John Zacharay Young, of University College, London. Young's lifetime passion for the large molluscs led him from the squid to the octopus. Octopodes offer no great advantages for the study of nerve transmission, but they do have relatively large brains. Working at the Marine Station in Naples in the 1950s and 1960s, Young began to study octopus behaviour, and especially their capacity to learn and remember. Octopus feed on small crustacea, and Young designed experiments in which the octopus would be shown a large black or white shape – say a cross – at the same time as it was offered a small crab. If the cross was black, the octopus would be shocked as it touched the crab; if white, it would not. The octopus, he discovered, could learn to distinguish such shapes and patterns and avoid those coupled with the unpleasant experience. They could learn by sight, and also through their remarkably sensitive tentacles to distinguish rough from smooth, heavy from light cylindrical shapes.[7] The brain regions responsible for storing the memory could be located to one of the major lobes into which the octopus brain was divided. But there, essentially, the search ended. The octopus brain is a mass of small neurons and surrounding cells, whose connections are not well understood and the mapping of which will require as many lifetimes of research as has gone into mammalian studies in the last century. Beautiful as the creature is, and instructive as the study of its behaviour has been, the octopus did not prove to be God's organism for the elucidation of memory. Despite Young's enthusiasm, in this context his research programme became an experimental dead-end.

It required the next generation of researchers and a simpler mollusc to come closer to finding God's favourite. In the 1940s, Angelique Arvanitaki, and in the 1950s and 1960s Ladislav Tauc, in Paris, began studies of the sea mollusc Aplysia, a slug-like hermaphrodite creature which lives on the sea floor close to the beach and grazes on seaweed. Aplysia can grow up to about 30 centimetres long and weigh a couple of kilos, and of the several species the biggest and experimentally most popular (I don't say the best) comes from California – *Aplysia californica*. Its central nervous system consists of a few ganglia, no more than about 20,000 neurons in all. Four of these ganglia are arranged in a ring round the gut, interconnecting with a big abdominal ganglion by large nerve tracts. The great experimental merit of Aplysia, by contrast with Drosophila, which has as many neurons, or the octopus, which has far more, is that many of the Aplysia neurons are very large – up to a millimetre or so in diameter – and are located in characteristic and recognisable patterns, which are reproducible from animal to animal. This means that the 'same' cell can be studied in preparation after preparation, its connections and the effects of stimulating or excising it followed in detail – something that is quite impossible in any of the organisms yet discussed in this chapter (figure 7.5).

The advantage that this property conferred on Aplysia for those

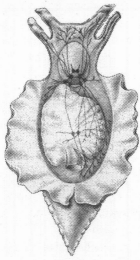

Fig. 7.5
Aplysias
A mollusc with a well-organised nervous system, containing large, identifiable neurons, especially in the abdominal ganglion, shown here in cutaway form.

studying neurons is as great as that of the squid giant axon for the study of the action potential, and was early recognised by Arvanitaki, Tauc and their collaborators. It has, however, been Eric Kandel, initially jointly with Tauc in Paris and later in New York, who has over the past quarter century made the study of Aplysia learning and memory so especially his own, and which contributed to his award of a share in the Nobel prize for physiology and medicine in 2000. Much of this work will be discussed in chapter 9. The question here, however, is one that occupied him and his colleagues for many years of sometimes acrimonious debate with mammalian psychologists: can Aplysia learn?

It was relatively easy to show that the animal could habituate and show sensitisation, and much work was put into the study of a particular set of reflexes, by which, if it is touched, the Aplysia withdraws its breathing organs, the gill and siphon, which normally stick out from its body surface, into its body cavity. Repetitive touching of the body surface results in a decrease in the amplitude and probability of withdrawal of the gill and siphon, a decrease which can persist for weeks. This may be a form of habituation, but it is very long-lasting and thus can reasonably be regarded as at least a form of non-associative learning. It wasn't until the early 1980s, though, that convincing evidence for classical conditioning of the gill and siphon withdrawal reflexes could be obtained. In these experiments the unconditioned stimulus was a strong shock to the tail, which produces a strong gill and siphon withdrawal response, paired with, as conditioning stimulus, a mild tactile stimulus to the siphon, which normally produces only mild withdrawal. After pairing, the conditioning stimulus produces a strong withdrawal as well.[8]*

The vertebrate solution

The design problem of disentangling one's guts from one's brain was solved for vertebrates by the invention of an internal skeleton built around a backbone. The skull cavity could now house an enlarged head ganglion or brain, while the nerves from the brain to the rest of the body run inside the backbone down the spinal cord; those ganglia that remained outside this central nervous system became reduced in significance and autonomy. Nonetheless, despite these radical changes in design, the basic cellular organisation of the nervous system, with its neurons, synapses and ensembles of interconnecting cells is the same for vertebrates as for invertebrates, as is much of their biochemistry. It is a bit like consid-

---

*An alternative molluscan species, Hermissenda, has been championed as a rival to Aplysia by the Woods Hole-based neurophysiologist Dan Alkon. Hermissenda, like Aplysia, has a simple brain with a small number of relatively large and identifiable neurons. It responds to rotation, or shaking, by contracting the muscular foot with which it normally clings to the surfaces on which it crawls, so as to minimise disturbance to its position. By contrast, it responds to a weak source of light by moving towards it and to do so needs to extend its foot. Pairing of light with rotation, Alkon showed, will eventually result in Hermissenda responding to light alone, as it does to rotation, by contraction.[9] Once again, this type of behaviour meets all the criteria the mammalian psychologists lay down to count as associative learning. Over the past decades there has been a degree of rivalry between the two research groups which has spilled over into public contention over the relative merits of the two molluscs, and priority disputes about the research findings they have produced, rivalry which has even formed the basis of a popular book.[10] But of this, more in subsequent chapters.

ering the multiple forms that vehicles based on the internal combustion engine have taken since its invention at the end of the nineteenth century. Weird and wonderfully designed cars, motorbikes, planes, improved engines, and year-by-year variations in efficiency, finish and fashion-features there may have been, but the principle of the piston and cylinder internal combustion engine, oil-based products as fuel and wheels for movement along the ground all remain.

What changes with the appearance of the vertebrates are not its building blocks or basic sources of energy utilisation and transmission, but the organising principles of the nervous system as a whole – a system that now contains the fully-fledged learning and memory capacities that are the properties of all mammals, including, in their greatly expanded forms, primates, amongst them of course humans. Whether despite this revolution in design the cellular mechanisms required for learning and memory in invertebrates are similar to or radically different from those in vertebrates is an issue which is still unresolved in research terms, and one which will occupy some of the discussion of the following chapter. But the task of this chapter, of tracing the evolution of learning and memory-like phenomena in non-human animals, is done.

Chapter 8

Molecules of memory?

Why biochemistry?

WHEN, IN 1929, THE SWISS AMATEUR PHYSIOLOGIST HANS BERGER reported that by taping a set of recording electrodes to the human scalp he could record continuous bursts of electricity pulsing through the brain he was at first not taken seriously. As I've mentioned already in the context of metaphors of memory, the phenomenon of 'animal electricity' and its relation to neural activity had been known for a long time – at least since Galvani's demonstration in Bologna in the 1790s that electrical pulses caused a frog's legs to twitch. In 1875 Richard Caton, the professor of physiology at Liverpool, had shown that electrodes placed on the exposed brain of a rabbit could record electrical pulses, but Berger's records came through the skull and could easily be dismissed as artefacts until his results were systematically vindicated by the Cambridge neurophysiologists Adrian and Matthews in the mid-1930s. The brain's ceaseless electrical activity showed characteristic waveforms that varied with sleep and wakefulness, mental activity or tranquillity.

The electroencephalogram (EEG) seemed for a while to carry the secret of the soul in its multiple waveforms.[1] Could it also hold the key to memory mechanisms? Perhaps when memories were made they were stored in the form of continuous reverberating circuits, endless electrical loops made by opening or closing synaptic connections? Alas, the brief popularity of this idea could not survive the demonstration that long-term memories persisted even if the total electrical activity of the brain was disrupted, by epileptic fits or electroconvulsive shock for instance, or was brought virtually to zero by coma or concussion. Thus, though it might well be that the very short-term phases of memory are dependent on the continued electrical activity of the brain – and there will be more to be said about this in due course – in the longer term any persistent record or trace must demand some more permanent incarnation.

What form, though, might such a trace take, and at what level should it be sought? Hebb's view, as described in Chapter 6, was that memory formation should involve some element of synaptic growth or reconstruction, thus providing a new pattern of connectivity between neurons which can subsequently be preserved, and this indeed remains the consensus, though by no means the only view. But how many synapses and cells might be involved for any single memory? Indeed what constitutes a 'single memory'? Could one write an equation: one association = one synapse? Or are multiple cells and synapses involved? Are such cells and synapses localised to a particular brain region, or are they diffused across many sites within the brain? Are memories multiply represented? Such a debate over localisation reflects the conflicting evidence from human data but now translated into a cellular language. Does the same set of cells embody the memory for all time, or is remembering a more dynamic, less fixed process? Even granted Hebbian principles, all such questions would remain to be addressed, and answering them helps decide at what level of cellular complexity memories are represented within the brain.

Answering them also requires the development of experimental models in which to test different hypotheses, and measuring techniques refined enough to be able to detect any postulated changes with learning. Until the last few years, the thought that one might have microscopic techniques sensitive enough actually to see tiny changes in the structure of neurons and their synapses as a result of learning seemed improbable – to start with one would need to have a very good idea where to look and what to measure in the brain. However, an alternative approach might argue that, if learning does involve making structural changes at synapses, and the synapses are built of proteins and packed with molecules of neurotransmitter, then learning must itself involve the synthesis of new proteins and transmitters. Might it be easier to measure this rather than to look directly for structural changes?

Protein synthesis

Living organisms are much more stable than the molecules of which they are composed. No molecule of our body survives unchanged for more than a few weeks or months; over that period, even in adults, it is synthesised, plays its part in the cellular economy, and is then discarded, broken down and replaced by another more or less identical. The extraordinary feature of this ceaseless flux is that structures of the cells and the body they compose remain constant whilst their components are replaced. Bodies are not even like cars, in which every so often a faulty exhaust, spark plug or body panel is removed and replaced by another, identical one. They are more like brick houses in which a demented builder is steadily, day and night, pulling out individual bricks and slotting others in their place. The overall appearance of the house is unchanged in this process though its components are continually being replaced. Like the bricks in such a house, the protein molecules of the body are replaced ('turned over') so that on average, about half of all the protein molecules are changed every two weeks. Synthesising any new protein molecule from scratch takes a matter of minutes. Having been synthesised it has to be transported to the part of the cell in which it is required; there it will remain for its lifetime of hours, weeks or months until it is due for renewal, when it is pulled out of its place in the cell and broken down by enzymes as quickly as it was previously synthesised, its building blocks (the amino acids) being recycled in the synthesis of other proteins.

Now normally in an adult the rate of synthesis and breakdown of proteins is equal. Every brick inserted into the structure of the house is matched by one removed. But suppose that the builder decides to add a new chimney to the house. To do so, it is necessary briefly to increase the rate of insertion of bricks at a particular point in the house – the roof – without changing the rate of removal; bricks thus accumulate and the chimney can be constructed. Once it is in place, the rate of insertion of bricks can decrease once more to its previous level, in balance with the rate of removal. This brief flurry of brick-building will leave one with a house plus chimney, to be maintained just as before. As with chimneys, so with synapses; if they are constructed – or even reconstructed – during learning, one might expect a brief increase in the rate of synthesis of proteins over the time when an animal was being trained and memory was being formed. And reciprocally, if memory formation requires the synthesis of proteins for the construction of synapses, then if one could stop the proteins from being synthesised around the time of learning then the memory should not be formed; an animal trained on a task and prevented from synthesising proteins should behave as if it has no memory for the task – is amnesic – when it is subsequently asked to perform it.

This was the state of biochemical thinking about the study of memory in the early 1960s. And fortunately, there were simple techniques available both for the measurement of the rate of protein synthesis and for preventing such synthesis. Proteins are built by joining together long chains of individual units – amino acids – which are either themselves made in the body or are present in the diet. Measuring the rate of protein synthesis then becomes a matter of measuring the rate at which amino acids are incorporated into proteins. If one of the dietary amino acids is made radioactive (as in the experiment I described in chapter 2) and fed to or injected into the animal, it is incorporated into the protein just as its fellow, non-radioactive amino acids are, and the protein becomes slightly radioactive in its turn by virtue of containing the radioactive amino acid. The amount of radioactivity in the protein is proportional to its rate of synthesis – and this can be measured both easily and extremely sensitively. There are twenty different naturally occurring amino acids in proteins, and any individual type of protein is a unique sequence of up to several hundred such amino acids. The precise assembly of the amino acids into the appropriate sequence depends on another giant molecule in the cell, ribonucleic acid (RNA), whose synthesis is in turn directly under the control of the cell's genetic material, deoxyribonucleic acid, DNA. (It is in this sense that protein synthesis is often, though somewhat misleadingly, said to be 'directed' by the genes.) Increased protein synthesis thus also may require increased RNA synthesis, which can be measured by exactly analogous procedures using a radioactive precursor to RNA.

As for preventing protein synthesis, it was discovered many years ago that many antibiotics – which of course work by preventing bacteria from growing and multiplying – achieve this goal by preventing bacterial protein or RNA synthesis. Injected into the brain in high enough doses, such antibiotics will also stop most of the brain's RNA or protein synthesis for a period of some hours. More specific methods are now available based on gene technology for either temporarily or permanently preventing the synthesis of particular proteins. Thus two types of experiment are in principle possible. In the first, sometimes called a correlative approach, one can inject a radioactive precursor of protein or RNA synthesis into an animal, train it on a task, and ask if the amount of radioactivity in the protein or RNA has increased by comparison with that in appropriate control, 'non-learning' animals. In the second approach, described as interventive, one injects an inhibitor of RNA or protein synthesis, trains the animal and asks if it can still remember the task.

The early 1960s saw both these types of experiment being done. I have already described the tremendous impression that Hydén's experiments – in which he measured increases in RNA and protein synthesis in tiny cellular regions from the brains of rats trained to balance on wires to reach for food – made on me as a young post-doc. In later experiments he somewhat altered the behavioural design. Noting that rats tend naturally to be either left- or right-pawed in reaching for and picking up their food, he constrained them to reach for their food with the non-preferred paw and reported changes in RNA and protein synthesis in the region and side of the brain responsible for the motor coordination of the 'learning' paw compared with the 'non-learning' side.[2]

Meanwhile perhaps the earliest of the inhibitor experiments was made by Wesley Dingman and Michael Sporn, in Rochester, New York, in 1963.[3] They taught rats to swim a water maze, and injected them with an inhibitor of RNA synthesis. They first showed that the inhibitor had no effect on the rats' ability to swim in general, nor, if the animals had already learned the maze by the time it was injected, did it prevent them swimming it correctly. However, if the inhibitor was injected at such a time that RNA synthesis was inhibited while the animals were being trained on the maze, then they failed to remember it when tested on it later. This experiment was quickly followed up by others using protein synthesis inhibitors, all essentially leading to the same conclusion – that if protein synthesis was prevented during the period over which an animal was trained, or for up to about an hour subsequently, then although the animal could learn the task, when tested on it some time later – say the next day – it behaved as if it were naïve. Long-term memory, it would appear, required protein synthesis, and it seems, despite some earlier doubts, that this is universally true for all species and forms of memory that have so far been studied, in organisms from flies, bees and slugs to primates.[4]

My own immediate reaction when hearing of such results was incredulity. The brain synthesises protein at a rate higher than that of any other tissue in the body. The antibiotics were being injected into the brain at doses sufficient to block all protein synthesis for several hours, and yet it appeared that no other aspect of the animal's behaviour was affected – not its capacity to perform already learned tasks, to see and respond to the world around or to act otherwise 'normally'; the only thing it seemingly couldn't do was to memorise new tasks. Surely not all that protein synthesis going on in the absence of the inhibitor could be about learning and memory; some other fundamental aspects of behaviour must be affected? But no, it appeared not; report after report, in many different learning tasks and in species as diverse as rats and goldfish, came to the same conclusion. To convince myself I eventually performed the ultimate in doubting Thomas exercises, trained some chicks using the inhibitors, and got the same result. It had to be true!

There is a methodological issue worth drawing out here, for this business of replicating – or failing to replicate – someone else's research findings is of course what 'the scientific method' is supposed to be about, at least according to the standard philosophy-of-science textbooks. A researcher reports a particular result, and to verify it other scientists repeat the same experiment in their own labs. If they agree, the result is provisionally true. If they fail to agree, it is necessary to decide where someone has gone wrong, in experimental design or theory-making. This is what is meant by claiming that scientific knowledge is 'public' knowledge – that is, that it is in principle testable and verifiable by anyone/everyone and not merely a matter of private belief.[5]* Even in so-called 'basic' or 'pure' science, in practice, except possibly in some areas of physics, direct attempts to replicate reported experimental findings are very rare. Firstly, there is no prestige to be gained simply from repeating someone else's experiment; you are very unlikely to get a grant to do it, and the main scientific journals are not normally interested in publishing 'replications' of experiments unless they are on a particularly controversial topic. Even failures to replicate are not very interesting to the journals; experiments with negative results therefore rarely get reported.†

What people do tend to do if a result someone else reports interests them is to repeat it with variants – that is, they test it in their own favourite animal or experimental situation. This is of course

---

*In practice, things are a good deal more complicated than this. Replication demands that one has the laboratory facilities to replicate; labs are expensive and access to them privileged.[6] The possibility of questioning such ostensibly 'public knowedge' is distinctly restricted, as many environmentalist action groups have discovered when wanting to challenge expert pronouncements about such-and-such a procedure being safe.

†The furores in the late 1980s and early 1990s over Pons and Fleischman's claims on cold fusion and Benveniste's results apparently supportive of homeopathy are among the few, apart from the 'transfer of learning' experiments I discuss shortly, where direct replications have been attempted – and failed.

what I did – rather than repeat exactly experiments done on mice or rats that other researchers had already reported, I asked instead what would happen if I tried a similar procedure with my chicks. Such a roundabout way of replication means that if one gets similar results in a different species, they are worthy of publication in their own right. Even if one gets different results they are still publishable without necessarily running into head-on confrontation with the earlier claims. Granted the rich diversity of the phenomena of the biological world, a disagreement over results can generally be put down to differences between animals ('species or strain differences' for instance) or to subtle alterations in experimental conditions, and can therefore be fudged or ignored. Thus rather than direct refutation, a controversial or dubious result can remain 'in the literature' publicly unchallenged but generally disregarded. Those in the know – the core group of researchers in any field who spend a lot of time at conferences and seminars chatting with one another about the state of the art – will simply disregard the anomalous result, or they will have gossiped it away in the bar after the meeting. The reports of the amnestic effects of protein synthesis inhibitors were at first ignored, often on the same sorts of *a priori* grounds that had led to my initial scepticism, and only after some struggle accepted by these definers of the field.

Thus it wasn't for many years after the first inhibitor experiments were reported that I actually got around to test their effects myself. At that stage my priorities lay elsewhere and to start playing with inhibitors seemed a diversion – when I have turned to using them, in the late 1980s, as I will describe in chapter 10, it was with rather more specific goals in mind. For by the late 1960s and early 1970s, our own imprinting studies had their own strong momentum. Essentially the experimental design involved exposing day-old chicks to an imprinting stimulus, injecting them with radioactive precursors to RNA or protein, and measuring the amount of radioactivity in protein or RNA extracted from different brain regions. If that description is excessively dry and abstract, let me spell it out in a little more detail.

First, what do I mean by an imprinting stimulus? In nature, so to say, chicks soon learn – within the first three days after hatching at the latest – who their mother is, and follow her thereafter. But their definition of 'mother' begins by being pretty flexible. When they hatch they will try to follow and come close to the first slowly moving object that they see which is the right sort of colour and more-or-less chicken-sized. People studying imprinting had used a stuffed hen, or even a red ball on a rotating arm. Pat Bateson stripped the learning down to its essentials. The chicks are placed in pivoted running wheels, a bit like the treadmills one can buy for pet hamster cages, placed in front of a red or yellow rotating, flashing light. The flash gives the light the appearance of moving and the chicks, in the wheels, try to follow. After an hour or so they are given a brief rest by switching the lights off and then tested by being given a choice between attempting to follow the light on which they had been imprinted (say the red) and another, unfamiliar light, say the yellow. The extent to which they try to run towards the red compared with the yellow is taken as a measure of how strongly they have been imprinted.

At a chosen time during their exposure to the flashing light, the birds would be injected with the radioactive precursor, the training would continue, the chicks would be tested and then killed. Because we didn't know where in the brain any changes might occur, nor very much about chick brain anatomy (nor it should be said did anyone else at the time) we divided the forebrain arbitrarily into two regions which we called simply 'roof' and 'base' – later, as we focused in on where the changes occurred, we were able to subdivide much further and according to more meaningful anatomical criteria. Pat and Gabriel Horn would code the brain samples and send them down to me for analysis. In the very first experiments it became clear that, compared with 'control' birds that stayed in the dark or had been exposed simply to diffuse overhead light there was indeed increased synthesis of RNA in the roof region in the hours after training in birds which had been imprinted on the flashing light. We repeated the experiments with a precursor for protein and found that an increase in protein synthesis occurred also. With two of my first graduate students to work on the chick, I went on to explore some of the more detailed biochemistry of these changes.

However, the main problem that concerned Pat, Gabriel and myself at the time was not so much the details of the biochemistry, but a theoretically more important matter. True, we had shown that protein and RNA synthesis increased when chicks were exposed to the stimulus, but how could we be sure this was due to the fact that they had learned something about it? Maybe it was because they were just more active, running busily in their treadmills, than their hatchmates in the dark or in subdued lighting. Or maybe they were more aroused because of the flashing light, or maybe the light was affecting their visual system in some way. All these could be alternatives to the explanation that the changes in RNA and protein synthesis were due to learning. Designing experiments to check these possibilities isn't easy, and it took us several years, through the early seventies, to try to eliminate one after the other. By 1973 we were finally able to convince ourselves, and, I believe, the rest of the research 'community' interested in the problem, that the biochemical changes were indeed an aspect of learning and not any of what we had come to call the concomitants of learning, such as motor activity or visual experience. In one key experiment, for example, we trained over a hundred birds, measuring their imprinting preference score and motor activity (that is how much they ran in the wheels) as well as RNA synthesis. The amount of RNA synthesised was unrelated to motor activity or any measure of stress that we could identify (for instance how much the birds peeped or twittered), but was strongly correlated with their preference score; that is, the more the chicks had learned about the flashing light and preferred it to any alternative, the more RNA was synthesised in their forebrain roof.[7]

Thus it seemed, from our work and that of many other labs at about the same time, that brain RNA and protein synthesis were necessary for learning and memory formation, and that the field of biochemical memory research had taken a flying leap forward. Sadly it was not to be. What happened was a mixture of biochemical hubris and technical muddled thinking which between them were to confound memory research for more than a decade. As the errors remain instructive, it is worth spending a little time considering what went wrong.

Memory molecules and artefacts

Molecular biology's own favourite philosopher, Gunther Stent, has called the late 1950s and early 1960s the classical period of molecular biology.[8] Since 1953, when Watson and Crick had solved the structure of DNA and recognised that embedded in its famous double helix lay a mechanism both for genetic transmission of information and the directed synthesis of proteins, the detailed mechanisms of protein and nucleic acid synthesis had been unravelled and more and more aspects of its exquisitely precise cellular controls were becoming clear. Nothing seemed impossible; the whole of biology was about to become transparent to this wondrous new science. Where once biologists and biochemists had been concerned with questions about where and how cells got and used their energy, the new molecular biologists had a different language for what was important. Not energy, but, drawing on the then equally new computer sciences, information was what mattered. Controlling and reproducing the cell was, it seemed, all about controlling and reproducing information; and what distinguished the molecules that embodied this new idea, proteins, DNA and RNA, from the much more boring small molecules that until then biochemists had worked with was that these giant molecules seemed to embody information; they were, it appeared, 'informational macromolecules'. As the brain was a machine for processing information, what more logical than to assume that it did so by utilising in some very special way these informational macromolecules?

Furthermore, could not the very reproduction of the species itself be regarded as depending on a form of memory – genetic memory, the apparent capacity of the DNA, transmitted between parent and offspring, to carry the rules for the future accurate development of the new organism? If DNA was the carrier of genetic memory, why could not it – or RNA or protein – also be the carrier of brain memory?

This punning logic became extended by parallel developments in the field of immunology. Antibodies are proteins which, once synthesised by cells of the immune system to counter and inactivate 'foreign' molecules, enable the body to retain the 'memory' for the intruder and hence the capacity rapidly to inactivate it on subsequent invasion. And as immunological memory too depended on proteins, could there not be a grand convergence of mechanisms operative here? Forget structure, the intricately intertwined pattern of 100,000 billion synaptic connections within the brain. Perhaps memories were carried by the very macromolecules themselves? Sure, the DNA was perhaps a little preoccupied with carrying the genetic memory, but could not the memories of a lifetime be readily encoded in the myriad of potential unique protein sequences? Such was (and remains) the power of molecular biological rhetoric that many who should have known better were swept along with it. The sloppiness of such thinking-by-pun affected many of the leading molecular biologists and immunologists of the period (two who were swept up in this early enthusiasm but stayed on to become wiser neurobiological theorists were Gerald Edelman and Francis Crick). The misguided enthusiasm spilt over into the most prestigious of journals. A few examples will give something of the flavour of the times:

> There are known to be three types of biological memory: (a) genetic memory, the discovery and unravelling of which has been due to molecular biology: (b) conventional memory, which is a function of the brain; and (c) immunological memory. In spite of the apparent differences between these types of memory they probably have much in common, and presumably a single mechanism is responsible for the working of all three.[9]

Even sober-minded mathematical modellers fell under the spell, as witness the mathematician J. S. Griffith, who had helped Watson and Crick solve DNA back in the early 1950s, writing jointly with one of the doyens of biochemistry, Henry Mahler, and offering what they called, for reasons I have never quite understood, a 'DNA-ticketing theory of memory'.

> DNA does have an intuitive appeal as the depository

of learned information as well as of genetic information
... we are impressed with the possibility that the peculi-
arity that nerve cells possess of not dividing has been
devised so as to avoid disturbing the learned information which is somehow stored in their DNA.[10]

Enthusiasm for such ideas has even now not completely evapora-
ted; voices advocating them can be found through the 1970s and
1980s and, in only marginally more sophisticated forms, in the
1990s:

Individual molecules are the fundamental decision-
making elements in the brain . . . the function of the
neuron is to allow the elements to communicate with
one another.[11]

In this paper, animal behaviour, in particular learning
and memory, has been reduced to the behaviour of
proteins, whether individual or assembled in super-
structures . . . the interplay of billions of such molecular
events, ensured by appropriate wiring, brings about
complex forms of learning in animals and in man.[12]

Experimental results showing the involvement of RNA and
protein synthesis in memory formation could readily be accommoda-
ted to the new molecular thinking, but what really raised the
temperature of the whole enterprise were reports that began to appear
of bizarre memory experiments involving planaria. The originator of
this research was the maverick James McConnell, at Ann Arbor,
Michigan, who in a series of papers during the 1960s, first in
conventional scientific journals and then in his own publication, the
exotically named *Worm-Runners Digest*, reported experiments in which
flatworms, trained by pairing light with electric shock, were chopped
up and other, 'naïve' (that is, untrained) worms allowed to
cannibalise them. McConnell claimed that the cannibal worms
behaved as if they remembered the conditioned response their food

had learned, whereas worms allowed to cannibalise other, untrained worms showed no such change in behaviour.[13] The experiments hit the scientific and popular headlines over a number of years before falling into disrepute when others found it quite hard even to train flatworms reliably on this pairing, let alone repeat the later steps in the procedure.

But by that time it no longer mattered, for reports of similar effects in mammals had begun to appear. Some of the first of these were made by a pupil of McConnell's, Allan Jacobson, in Los Angeles, who announced in 1965 that if he trained rats to approach the food dispenser of their cage when a light flashed or a clicking sound was made, then killed the animals, extracted the RNA from their brains and injected it into the gut cavity of untrained animals, these then tended to approach the dispenser when the appropriate stimulus – click or light – was given, even though the dispenser was now empty of food and the animals received no reward. Jacobson even managed to 'transfer' the approach behaviour from rats to hamsters in the same way.[14]

Meanwhile, human analogues of such experiments began to appear. Ewen Cameron, a McGill-based psychologist, began to feed large doses of RNA – typically 100 grams of yeast RNA extract, a truly massive amount – to elderly people with memory difficulties. He claimed that this amount did have significant effects on the person's capacity to remember (though presumably it did not cause them to remember their past experiences as yeast cells!).[15] More than likely the explanation for such results lay simply in the fact that the elderly, institutionalised patients who formed Cameron's subjects were sufficiently pleased to be noticed and made a fuss of in experiments of this sort that their memories improved as a consequence.* Or perhaps the subjects of Cameron's study were simply malnourished, as happens in many institutions, and the RNA served as a dietary supplement. Taking RNA as food, as Cameron had given it,

---

* See chapter 5; this type of phenomenon is known as a Hawthorne effect, after the research which first, inadvertently, discovered it whilst trying to assess the effects of different types of work organisation on factory efficiency.

would simply mean that it would be broken down to its precursor molecules in the gut before being absorbed into the bloodstream.*

Doubt was cast on Cameron's results partly by the lack of control data he offered, and, later, after his death, his reputation for scientific integrity was irretrievably damaged by the revelation that much of his experimental work had for a long time been secretly supported by the CIA, including some rather insidious studies of the effects of covertly administered LSD on the behaviour of unsuspecting people.[17]

Claims for the memory-enhancing effects of RNA provoked a fierce controversy in the research literature, with many laboratories attempting to repeat them, and mainly failing to do so. In the same year as Jacobson published his experiments, a 'failure to replicate' report signed by twenty-three authors appeared in the major journal *Science*[18] and the matter might have rested there but for the fact that it was noticed that the method which Jacobson had used to extract RNA from his rat brains also liberated a good deal of protein and other contaminants. Perhaps the active material was not RNA at all? By 1967 memory transfer labs were back in business again, injecting a variety of brain extracts and claiming many and varied results – one group of researchers for instance trained some rats to press levers for food with their right paw and others with the left, and found that one behaviour could be transferred but not the other!

The most systematic claims, however, were made by Georges Ungar, of Baylor University in Houston, Texas. His training protocol exploited the fact that, given a choice between staying in a lighted environment or entering a dark compartment, rodents such as rats or mice will go into the dark. Ungar placed rats in a start box which opened onto a lighted arena with a dark compartment in one corner, and shocked the animals electrically when they attempted to enter

---

*It is interesting that several years later, in the late 1970s and early 1980s, a research group in Magdeburg, in what was then East Germany, led by the pharmacologist Hans-Jürgen Matthies, claimed that large doses of a precursor of RNA, orotic acid, would improve memory in rats trained on a variety of lab tasks.[16] Orotic acid is now included in the lengthy list of 'smart drugs' and 'cognitive enhancers' that are increasingly found on sale in the less scrupulous 'health food' stores, despite the lack of evidence for either their need or their effect.

the dark compartment; this punishment rapidly results in the animals refusing to enter the dark box. Material extracted from their brains was then injected into mice, which were then given a a similar light/dark choice, though without being shocked. Mice which had received material from the brains of the trained rats, Ungar claimed, also refused to enter the dark box, whereas animals which received material from naïve animals showed no such inhibitions.

Ungar and his group went on to endeavour to purify the component in the brain extract which, they claimed, carried the information about 'fear of the dark'. As I have already hinted, there had always been a biochemical puzzle about how proteins or RNA might work in these transfer experiments because all such large molecules are rapidly degraded in the gut and broken down into their component amino or nucleic acids before being taken into the general metabolism of the recipient. And indeed Ungar found that his active component was neither a protein nor a nucleic acid but a peptide – a generic name given to a short sequence of linked amino acids anything up to fifteen or twenty units long and administered by injection, so avoiding the digestive barrier offered by the gut. Ungar's peptide was fifteen amino acids long and he called it 'scoto-phobin' – Greek for 'fear of the dark'. After much furore, his results were published in *Nature*, along with, in a development unheard of at the time, a criticism of his data by one of the paper's referees, the chemist Walter Stewart.[19]*

Stewart's *Nature* critique focused not on an analysis of Ungar's behavioural claims, but on the chemical purity and composition of the presumed scotophobin. However, for me, as for other neuro-biologists, there were always other implausibilities about Ungar's results, even assuming they could be replicated (and many labs remained sceptical). How could minute quantities of an injected peptide be guided to and then enter the appropriate neuron so as to

---

*By the 1980s Stewart had became one of the US Congress and the Editor of *Nature*, John Maddox's, favourite scientific fraud-spotters, from spoonbending and homeopathy to priority disputes.

code for the new memory? Why should the same peptide 'mean' the same highly specific memory/behaviour in different animals/species? And if peptides really did code for memories should there not be many more of them in the brain than one actually finds? If there were such 'memory peptides' and each was present in the brain in the concentration of scotophobin, then to code for the memories of a human lifetime would demand that the brain contained a mass of peptides weighing something of the order of 100 kilograms – or rather more than the weight of an average human.

But perhaps our main concern was even more fundamental – could the behaviour of the injected mice really be said to show learning at all? Let me explain. In Ungar's test situation, the mice are released into a lighted compartment and their behaviour observed. The time taken for them to enter the dark box is noted, and if they have not done so by the end of a fixed time, say one minute, the experiment is terminated. The time mice injected with material from the trained rats take to enter the dark box is compared with that for animals injected with 'naïve' material. The difference between the trained and the naïve rats is that the former have been electrically shocked, and therefore certainly stressed and pained, when attempting to enter the dark box. The common response of a rat or mouse to stress is to freeze and become immobile. So suppose that the result of the stress caused by shocking an animal in this way is the production of some hormone – for instance a peptide – which then produces the freezing behaviour. It would be present in the brains of the trained – and shocked – animals in higher concentration than in the naïves – and when injected into mice would in its turn produce freezing behaviour. This behaviour would be measured in Ungar's test as an increased reluctance, or even a refusal, to enter the dark compartment within the minute of the trial, simply because the recipient mice were relatively inactive. Because of the design of the experiment, this inactivity would appear as the learning of a specific response. However, in reality what would have been 'transferred' would not be specific learning, but a rather general emotional stress reaction – a very different matter.

Not long after this controversy, Ungar himself died. There were some irreverent proposals to perform the ultimate experiment and try injecting material extracted from Ungar's brain into his critics – a human trial that I suspect that Ungar himself might have been rather in favour of! In any event, with his death scotophobin disappeared from the research literature, and so did the memory transfer experiments. (Friedrich, a Hungarian enzymologist whose views on molecular memory are quoted above, was one of those whose association with the memory transfer work lasted the longest.)

I do not want to suggest that all this work fell into such simple methodological traps as that I have described above, though I suspect that much of it did. Many unexplained and ignored results still litter the research journals of the early 1970s, unread and unconsidered because as a research paradigm memory transfer is no longer taken seriously; it had become another victim of scientific fashion, though, unlike McIlwain's slices, this time probably deservedly. And as no one takes the paradigm seriously, no one is troubled to try to find an explanation for the seemingly anomalous data. Perhaps, as most of us would assume, it can all be explained by inadequate statistics, faulty experimental design, overenthusiastic interpretation of ambiguous results, or, as I have argued in the case of Ungar's experiment, misinterpreting the biochemical and pharmacological consequences of stress or other, rather non-specific aspects of behaviour. Perhaps there is something else residually there that doesn't fit our present-day models? None of us at the moment is prepared to spend – probably waste – our time trying to find out.

Non-scientists – and anti-scientists – are often troubled by such apparent refusals by scientists to spend time investigating seemingly paradoxical results which don't fit into current experimental paradigms – corn-circles, ESP, UFOs, aromatherapy or whatever. To such critics it shows just how blinkered orthodox science is, and the rather arrogant refusal of most scientists even to take the phenomena seriously, instead waving them irritatedly aside, merely serves to reinforce their opponents' criticisms. What such critics for their part fail to realise, however, is just how difficult scientific research actually is, how complex the testing of any even seemingly trivial hypothesis or hunch may be, and how many paradoxes and seeming mysteries we confront day-by-day in our research, which to us may seem at least as challenging as, but theoretically more relevant than fretting about probably untestable phenomena like ESP. While poets and magicians are concerned to draw attention to the anomalies that break the regularities of our day-to-day world, the strength of natural science has lain in the meticulous and often boring study of its seemingly tedious regularities. To us they seem at least as intriguing and worthy of study as do the signs and wonders which become the obsessive concern of mystics, religious and many others outside the laboratory.

However, I have here rescued the lost experiments on memory transfer from otherwise merciful oblivion not merely to do historical justice to a now suppressed period of my chosen research field but for three rather more solid reasons. First, the episode shows how easy it is to be led astray by one's own rhetoric. As I pointed out in chapter 4, there is a very important way in which science proceeds by metaphor, and metaphors can illumine – or they can mislead. In this case, the metaphor was given by the use of the words memory and information in three different contexts, heredity, immunology and learning. The power of fashionable slogans – such as informational macromolecules – and the search for sensational results to feed press and paymasters swept caution to the winds. The result was a hunt for the biochemical mechanism of memory at what was a fundamentally mistaken level – that of molecules rather than the brain systems in which molecules are embedded. The choice of the right level at which to study a phenomenon is as important a strategic decision in biology as is the choice of the right organism or the appropriate control experiment, and it affects present-day research just as much as it did thirty years ago.

The second point, however, is not so negative. Ungar's experiments were done before other research – which was to lead in due course to the opening up of a major new branch of neuropharmacology and to the making of some very distinguished scientific reputations – had revealed how important many peptides were in the brain. The best known of these peptides are those sometimes described as the body's natural pain-killers, the morphine-like family of the opioids such as enkephalin and endorphin. Dozens of such brain-acting peptides are now known to exist, including many closely related to hormones which act elsewhere in the body. They function as neurotransmitters and as modulators of neuronal activity (neuromodulators) and are associated not only with pain but with pleasure, stress, arousal, attention and many other such global mental and bodily states. Another group is the so-called growth factors, which help direct the development of connections between neurons, especially during periods of brain plasticity. Our own lab and others have shown that one of these factors, the so-called Brain Derived Neurotrophic Factor, BDNF, is intimately involved in memory formation.[20] Remarkably – or perhaps not so remarkably – Ungar's mythic scotophobin had an amino acid composition rather reminiscent of that of the endorphins and enkephalins. He had, in effect, stumbled without recognising it into a major new area of understanding of the chemistry of brain function. But he died without knowing it, and the true significance of the peptides was discovered by others without reference to his premature and misinterpreted finding.

The third lesson is the ease with which artefacts can enter into experiments intended to study phenomena as complex as learning and memory. Just because non-human animal memories can only be expressed in behavioural terms, the possibility always arises that what we are measuring is an aspect of the behaviour rather than the memory. In learning experiments, animals are stressed or hungry, they receive sensory inputs, they perform motor tasks. If we find changes in protein synthesis, say, in a correlative experiment, how can we be sure that such changed synthesis is not the consequence of these expressed behaviours rather than the learning which we presume accompanies them? If we conduct an interventive experiment, and inject a drug which results in an animal not performing some task on which it has been trained 'correctly' (I won't bother putting that word into quotation marks hence-forward; I have already spelled out that what we read as correct in an animal's behaviour is interpreted by our criteria, not necessarily by its own), how can we be sure that what has been blocked or disturbed is the memory rather than the motor or sensory activity on which its expression depends? A drug may make an animal less hungry, less mobile, less sensitive to the pain of an electric shock, just as much as it may make it forget.

Designing experiments to control for all these possibilities is not at all easy, as much of the debate within the research literature shows. To take but one other example from an experimental approach common in the late 1960s and early 1970s, suppose one trains a mouse by putting it onto a small shelf in the wall of a cage with a grid floor which can be electrified. Every time the animal steps down it receives a mild electric shock (yet another of the less-than pleasant learning tasks which experimental psychology finds it necessary to employ). Within a few trials it learns to stay on the shelf (this is called 'step-down avoidance'). In a parallel cage to this 'learning' mouse is another; however, this second cage contains no shelf. Each time the learning mouse is given an electric shock, so is the parallel 'yoked' mouse. However, without a shelf to escape to the yoked animal cannot learn the task. So one seems to have a perfect control; both mice have received the same number of shocks, but one learns how to avoid the shock, the other cannot, for there is no escape; the number of shocks it receives is entirely governed by the behaviour of its learning partner. Thus any differences between them must be caused, not by the shock, but by the fact that the mouse in the cage with the shelf is learning an avoidance response. Many experiments have used this type of design and have shown differences in rates of protein synthesis between such 'learning' mice and their yoked controls, from which it is concluded that the learning rather than the shock produces the biochemical response. Sometimes a third, 'quiet control' group is included in the study, and often differences in biochemistry are found between it and both the yoked and the learning groups.[21]

But wait. Can one really be sure that the yoked animal is not learning anything? Perhaps it is indeed learning that there is no escape from the shock – and such learning may be very important to its subsequent behaviour.[22] Are the differences between yoked and quiet control the result of learning or of stress – and is the learning group perhaps less stressed than the yoked group? Stress will certainly affect the levels of a number of hormones circulating in the bloodstream and could well alter brain metabolism. Biochemical differences between the groups of animal may thus not have such a simple meaning as learning versus non-learning – even if the biochemical effects can be themselves regarded as unequivocal.

For there are similar sources of artefact and error in the biochemical measures. Many of these depend on rather sophisticated biochemical arguments which do not really concern me here; just two must suffice. I have said (in chapter 2) that one can calculate the rate of protein synthesis by injecting a radioactively labelled amino acid into the bloodstream and measuring the amount of radioactivity to be found in the protein of particular brain regions after a given time. True enough, but to be incorporated into the protein the radioactive amino acid has first to be taken up from the bloodstream into the neurons. Changes in blood flow and other such physiological effects can alter the rate of uptake, and hence produce an apparent change in the rate of protein synthesis. There are similar ambiguities in interpreting the effects of inhibitors of protein synthesis. Because proteins are made of amino acids, if protein synthesis is blocked by giving an inhibitor, amino acids which would otherwise have been converted into proteins tend to accumulate in the cell. Several amino acids, as well as being necessary building blocks for the synthesis of proteins, are also powerful neurotransmitters, and their presence in excess can result in disruption of the electrical activity of the nerve cells. The inhibitors might therefore not be causing amnesia because they prevent protein synthesis but because of their effect on increasing amino acid levels.[23]* In the first

---

*This demonstration that even a simple chemical interference with a complex biochemical process results in multiple effects is a graphic example of the folly of the pharmaceutical industry's way of speaking about the 'side-effects' of drugs; introducing an exogenous chemical like a drug into the body has a multitude of biochemical consequences, some anticipated, others unexpected – but never 'side-effects' – the phrase is a misnomer, concealing the reality that such consequences better seen as unwanted effects or adverse reactions are inevitable, even though they are ones that the researcher or the clinician doesn't want or hadn't thought about. No drug is a 'magic bullet' with only a single target.

flush of enthusiasm for the 'molecules of memory' many experimenters failed to take the precautions necessary to control for such biochemical and behavioural ambiguities, and as a result their research – and with it the entire field – became discredited.

Starting again

As the artefacts and problems of the early research became apparent, and bubble scientific reputations were painfully pricked, the rush of researchers into the field ceased and then reversed. For a few years funding for memory research became hard to come by. Those of us who stayed faithful to the project found ourselves isolated, our findings met by at best polite scepticism. When, in the early 1980s, memory came back into neuroscience fashion again, it did so in new forms.

It was at about that time that I wrote a paper entitled – I hoped provocatively – 'What should a biochemistry of learning and memory be about?'[24] The problems of memory research, it seemed to me, were in part the problems of any new research field. A multitude of labs had started in enthusiastically using many different forms of learning paradigms, often taken over very straightforwardly from the experimental psychologists. But it didn't necessarily follow that the sort of training beloved of the psychologists – say shaping a rat to press a lever for food in a Skinner box – was best suited to the study of the cellular and biochemical processes going on within the organism. It might simply be that the amount of learning involved in such a task was not great enough to generate biochemical changes big enough to be measured. Researchers working on the biochemistry of learning and memory needed to develop new model systems, in which the changes that we sought to identify would be big enough to be measured, yet in which we could be sure that they weren't simply artefacts. More than one researcher abandoned the field with the argument that if the biochemical changes really 'coded for' memory in the brain then they would be too small to be measured, and if they were large enough to be measured then they probably weren't anything to do with memory.

The question of the scale of any possible biochemical changes was (and is even now) a serious one. Psychologists and physiologists have always been resigned to using statistical analyses to extract meaning from and interpret their data. Many biochemically oriented biologists, and especially molecular biologists, tend to feel unhappy about any such need; the phenomena they study have often tended to be all-or-none, or at least so large that differences between experimental treatments or conditions produce apparently unequivocal results. If you need statistics to demonstrate an effect, they argue, it may not be real, and anyway cannot be important. Waving to one side the 15 to 20 per cent changes in protein synthesis rates and enzyme activity I had reported, Francis Crick was explicit on this point at a Royal Society discussion meeting in London in 1977 when I presented the results we had by then obtained on imprinting in the chick and the effects of first exposure to light in the rat.[25] If it is less than a hundred per cent change, ignore it; you are studying the wrong system or have designed the wrong experiment, he insisted. Yet to psychologists or physiologists in the audience, the surprise was that simply training an animal on an imprinting stimulus, or indeed any other form of learning, could produce a change of measurable magnitude at all; they would search our experimental designs for sources of artefact just as rigorously as I myself had done for the 'transfer' experiments.

Quite apart from the scale of any observed effects, and of rather more theoretical importance, was the issue of whether any biochemical change we found was unique to, say, imprinting in the young chick, so that memory for other types of behaviour in older birds or in other species would involve quite different processes. Or were we tapping into some kind of general biochemical mechanism relevant to all types of learning? Granted that psychologists have described a whole taxonomy of memory, procedural and declarative, episodic and semantic, working and reference, should one expect similar underlying biochemical and cellular changes to be involved in each, or would every form of memory have its own special biochemistry? Are there universal cellular memory mechanisms found in all mammals, all vertebrates or even all animals, or are they specific to particular species?

In a sense such questions are about the right level at which to study memory. If the key processes are biochemical, then it might be expected that each memory will have its specific representation in terms of the synthesis of unique proteins or other molecules. But if we reject this view in favour of memory as a property of the brain as a system, rather than of its individual cellular and molecular components, then memory will depend not on distinct biochemistry but on just which cells and synapses are showing the changes, where they are located in the nervous system, and which other cells they make contact with.

Think of the front door to an apartment block, with its array of bells. There are two possible ways in which the bells can be arranged so that someone arriving at the door can signal their arrival to a person in a specific apartment. Either each bell sounds different, and each could be heard in every apartment, or each sounds the same but is wired up to sound only in a single specific apartment when pressed. The first method – each bell unique but each ringing everywhere – means that the 'message' of the bell lies in its specific sound; in the second the message lies not in the bell, but in the way it is wired up. These, in essence are the two alternative ways in which the biochemistry of memory might work. For those believing in molecules of memory, the message is in the bell and its unique properties; for those believing that memory is a system property of the brain, the bell is merely a part – albeit an important one – of the system, and to understand the message one must read the wiring diagram, not listen to the sound of the bell ringing.

If this second approach is right – and despite my own biochemical enthusiasms I believe that it is – then the biochemical events that I study are likely to be very general processes of protein synthesis and membrane modification, sometimes described as 'housekeeping'. (Often the term is used in a rather derogatory way, as 'mere housekeeping' – albeit then mainly by male biochemists who may not recognise that housekeeping is a pretty serious business!) The memory lies within the topography – the wiring diagram – and dynamics of the neuronal system. This means that the cellular mechanisms of, say, remembering a telephone number and of remembering how to drive a car wouldn't differ – it would just be that different cells, connected up in different ways with other parts of the brain, are involved.

At the start of the 1980s, it seemed to me that until we knew more about the answers to such questions then it would be hard to compare and make sense out of the varied and possibly conflicting results that were coming from different labs. How many depended on the minor peculiarities of the learning task or organism involved, and were just another almost random item of knowledge to be added to the burgeoning catalogue of phenomena of memory? How many were 'true' generalisations which went beyond the particular species and task and could begin to provide a real cellular and biochemical 'alphabet' of memory – or is such a search a mere will-o'-the-wisp?

The great triumphs of molecular biology have come about because the research groups centrally involved in the experimental programmes of the 1950s and 1960s had concentrated their efforts on a single simple organism, the common gut bug *Escherischia coli*. Indeed, Francis Crick had gone so far as to suggest, at least half seriously, that all work in molecular biology and biochemistry on anything else should stop until *E. coli* was 'solved' – whatever might be meant by such a solution. Other, less molecular, biologists had protested in outrage that what was true for *E. coli* was not necessarily true for E.lephant, that biology has its diversity as well as its universals and that multicellular organisms with complex brains are not merely aggregates of 10^{15} or so single cells; the properties of systems of cells include relationships between those cells which are not inherent in any single unit. Nonetheless, might neurobiologists in general, and memory researchers in particular, not gain something by concentrating on a limited number of model systems that everyone could agree on?*

What is certain is that learning is a complex business involving many aspects of brain activity, and not reducible to a single linear sequence of events. The stress, arousal, motor activity and so on inevitably associated with learning themselves result in biochemical and physiological changes in the brain at the same time as the animal is learning, and they are all important and interesting to study in their own right. Anything that alters the rewarding or punishing effect of the learning task (if you are less hungry, thirsty or fearful, you are less likely to work hard to learn a task which provides food, drink or avoidance of electric shock as a prize for success) will also affect the study of learning and memory.

Arousal, reward and punishment, as I have already suggested in relation to the interpretation of Ungar's experiments, are associated with changes in the amounts of the opioids and other peptides in the brain and bloodstream, so injecting the peptides, or drugs which interact with them, will alter behaviour, including the expression of memory. Such agents will therefore affect the learning

---

*When along with other molecular biologists at the time, Francis Crick's closest Cambridge collaborator over many decades, Sydney Brenner, began to make tentative steps towards neurobiology, he took the route proposed by Crick in opting to turn virtually the entire work of his laboratory over to the study of the anatomy, development and behaviour of one of the simplest organisms to possess a nervous system, the small nematode worm *Caenorhabditis elegans*. *C. elegans* is only some half a millimetre long, has a nervous system consisting of just 302 neurons, wrapped, like that of other worms, around its gut. Its short breeding cycle means that it can be studied for mutations rather like fruit flies. Its rather limited behavioural repertoire is virtually confined to eating, sex and locomotion by means of convulsive wriggles. By electron microscopic mapping of normal and behaviourally mutant worms, and producing a comprehensive 'wiring diagram' of its nervous system, Brenner hoped to be able to reveal the translation between neural wiring and these simple behaviours *C. elegans* is by now probably the best anatomically understood animal in the history of the world and was one of the first to have its entire genome sequenced. Brenner's vision has paid off in that there is now a large international *C. elegans* research community. Finally, in 2002, Brenner received his much-deserved share of a Nobel prize for the work. The worm's behaviour is intimately studied, many behavioural mutants have been generated, and even its social life explored. But it is still probably fair to say that the enthusiastic reductionist manifesto with which the project began has not so far yielded great neurobiological dividends in the form of universal mechanisms of the sort which *E. coli* provided.[26]

process even though they are not directly part of it, in the same way as the tone or volume controls on a tape recorder affect the recording and playing of the tape even though they are not directly part of the message the tape carries. There are now available drugs which, injected prior to or just after a learning trial, improve retention of the memory (that is, increase the 'savings' as defined in chapter 4) in animals tested hours or days subsequently. Other such substances can diminish retention. This discovery, of proactive or retroactive interference in memory formation, has led to the hunt, strongly backed by a number of pharmaceutical companies, for drugs which might improve human learning or memory, especially in elderly people with conditions like Alzheimer's – the so-called smart drugs – a topic I will later return to in the context of our own research.

By the end of the 1970s, it seemed clear to me that if any cellular or biochemical process were to be regarded as forming part of some type of memorial code, it must show just these features that the drug studies could not readily provide, of being both necessary and sufficient to account for the memory. Whether one could go further and show that any particular process was specific to a particular memory in that it represented it and only it within the brain remained to be seen. Granted that it had turned out to be only too easy to make experiments in which training an animal on some task results in large biochemical and cellular changes in its brain, I felt that it was necessary to establish guidelines to help judge whether any particular change indeed has these characteristics of necessity, sufficiency and specificity. This was a need that Pat Bateson, Gabriel Horn and I had hammered out in many long discussions about our imprinting experiments, and which we had tried to meet in practice in the design of the controls we had used in the early 1970s. But in 1981 I tried to go further and to identify a set of criteria which any proposed biochemical or cellular correspondent of memory formation would have to meet if it was to be regarded as a candidate memory process. Because the discussion of these criteria is associated with a fresh and more promising period in the history of memory research, and also symbolises the point in my own research trajectory at which I switched from working on imprinting to an even simpler form of learning in the young chick, they can appropriately form the starting point for the next chapter.

Chapter 9

God's organisms? Flies, sea slugs and sea-horses

Criteria for correspondence

BY 1980, THE MEMORY RESEARCH COMMUNITY WAS REASONABLY CLEAR about the things it didn't believe in. For instance it was pretty clear that there were no unique memory molecules. It was also cautiously convinced that the most promising way to think about memory was along the lines of Donald Hebb's model, which involves changes in the strength of the connections between nerve cells, perhaps by growing new or enlarged synapses, and so altering the physiological relationships between neurons. Such a Hebb-type modification might indeed be detectable by appropriate neurophysiological or biochemical measures. But no one was very sure that any specific biochemical processes, apart from the rather general one of protein synthesis, had yet been unequivocally linked to memory formation, and the community had become distinctly cautious about evaluating any new claims. But what after all should a biochemistry of memory – or at least a biochemistry of learning and memory formation, as these approaches do not yet address the question of recall, of remembering – be about, and how would one know if one were discovered? That is, what sort of biochemical answer might prove convincing to both biochemists and to psychologists? What criteria should any experiment we made try to meet to satisfy the claim that the process we were studying was a necessary, sufficient and possibly even specific aspect of the memory formation process?

This was the question I tried to answer as the decade of the 1980s began.[1] It was becoming easy enough to find biochemical changes that occurred when an animal learned; the problem was to show that such changes were really part of the memory-making process. And I felt I couldn't design rational experiments without having some type of criteria to judge the results against, otherwise I wouldn't be able to see where I was going. I decided that, for anything I could measure to be considered part of a memory trace or engram, it needed to show the following properties:*

Criterion one – There must be changes in the quantity of the system or substance, or the rate of its production or turnover, in some localised region of the brain during memory formation.

If there are more or modified synapses, the chemicals and structures of which the synapses are composed must show signs of change, which might be measurable either by biochemical methods (for instance, an increase in the amount of synaptic membrane protein) or under a microscope (for instance a change in the dimensions of particular synapses, or an increase in synaptic numbers). But if I were to find such a change taking place while an animal is learning, unless the conditions for that change meet all the subsequent criteria, I would be no further forward than the experiments of the 1960s that I criticised in the last chapter. Some change is certainly *necessary*, but by itself cannot be said to be *sufficient* or *specific*. Nonetheless, this criterion is of course fundamental to any

---

*I have slightly modified the formulation of these criteria and the order in which they are presented from the version in the original paper without in any way affecting their substance. I am not trying to rewrite history. More than two decades on, I feel less sure about this attempt, but as it guided my research strategy for many years, it forms an essential part of my story in the next few chapters.

materialist model of memory. The criterion also makes a claim about *localisation*; the changes cannot be all over the brain but must be concentrated to some specific region or at least subset of cells, even if these are distributed between anatomical regions. This of course harks back to the much older debate about whether memories can be localised – something I'll come back to later; much of the next two chapters will be taken up with the question of the localisation of memory in space and time.

Note also that this criterion doesn't say anything about the direction or size of any change. It is perfectly possible to imagine that memories are coded for negatively, by reducing the level of some substance or process, although in practice nearly all research seems to be devoted to trying to identify increases. How about the size of any change? If something is remembered twice as strongly, or if two items are remembered rather than one, should the change be twice as big? Not necessarily, because in our experiments we measure the strength of a memory in arbitrary terms that we, as experimenters, devise; however, we have no way of telling whether the scales we use are the same as those the animals themselves use in making their memories. For example, I can train a chick to avoid a bitter chrome bead, then a bitter red bead, then a bitter blue bead. But the chick may not remember these as three separate items; instead, and more likely, it will adopt a different, and indeed more rational strategy, and generalise along the lines that 'all objects of a certain size, irrespective of colour, are likely to taste bitter and should be avoided', that is, it would be remembering one item, not three.

Criterion two – The time-course of the change must be compatible with the time-course of memory formation.

Clearly memories are not formed instantaneously, as if by throwing a switch, but are built up over a period of hours after the event to be memorised has occurred; during this build-up the form in which any memory is stored changes. At least for declarative memory (chapter 4) there is a transition over a period of minutes to hours from the initial and labile short-term phase to long-term, stable memory.

One could envisage a number of ways in which this transition might occur. At one extreme, there could be a continuous process in which a sort of chain reaction of biochemical processes in a particular ensemble of cells led inevitably from the early, vulnerable phase through to a final fixed form, like the hardening of glue or the developing of a photograph. On the other, there could be two more-or-less independent processes occurring in parallel, such that there were transient changes in the electrical properties and responsiveness of one set of neurons which could 'code for' the memory for a few minutes before gradually fading away. Meanwhile, if the memory were 'important' enough for lasting representation, there could be a steady development of biochemical processes, such as the reconstruction of synapses, which would permanently

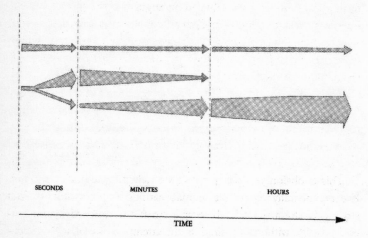

SECONDS   MINUTES   HOURS

TIME

Fig. 9.1
Short- and long-term memory
Two models for the transition between short- and long-term memory. In the upper version the processes are sequential, with a linear transition between the various phases. In the lower version, the processes are independent. An initial cellular activation provides the signal for the start of both a transient short-term memory and the steady build-up of long-term memory. If the sequence leading to long-term memory is disrupted, memory is lost once the short-term phase has declined. Evidence favours the lower version.

represent the memory in another set of neurons, perhaps in another part of the brain. These two types of process, on the one hand serial and on the other parallel processing of memory, are of course extremes, and there are many intermediate possibilities;[2] it becomes quite hard to design experiments which will unequivocally distinguish between them.

However, the experience of human memory points to just this type of separation between short- and long-term forms of memory. Evidence from studies of patients like H. M. suggests that, even though it is not required for the expression of long-term memories, the hippocampus is necessarily involved in the transition between short- and long-term memory. But whether serial or parallel processing turns out to be how the brain works (and there will be more to say about this too in the next chapters), there will be cellular events associated with both the short-term and the long-term phase, and we have to try to distinguish between them experimentally. This of course impacts on conclusions from Criterion 1 – it could be that the changes one detects occur first in one region or set of cells and at a later time in others – that is that the 'trace' is mobile.

Criterion three – Stress, motor activity or other processes which accompany learning must not, in the absence of memory formation, result in the structural or biochemical changes.

This is obvious in theory but immensely hard to test in practice. If experimentally observable animal learning is impossible without stress or motor activity or whatever, is stress or motor activity or whatever possible without learning? Is the attempt to make such a separation meaningful, and above all can one devise the appropriate form of highly reductionist experiment which will enable a distinction to be made between these processes? Are we asking for the experimentally and/or theoretically impossible? Granted that I have spent a considerable portion of my theoretical energies over the years criticising reductionism, should I even be asking this question at all?

Because trying to meet this criterion has occupied so much of my research time in the past two decades it is worth spelling out again that to adopt a reductionist methodology in research strategy – that is, to try to stabilise the world that one is studying by manipulating one variable at a time, holding everything else as constant as possible – is generally the only way to do experiments from which one can draw clear conclusions. Error comes in if one over-interprets the relevance of these conclusions, by forgetting the artificial constraints of the experiment and assuming that in real life, outside the laboratory so to say, such changes involving only a single variable can actually take place, and that it is a simple matter to extrapolate back from the artificiality of laboratory isolation to the blooming buzzing confusion of the real world. It is this procedure, which involves turning a reductionist *methodology* into a reductionist *philosophy* (and which I have written about in more detail in my book *Lifelines*), that is the manoeuvre so popular among molecular biologists and some geneticists, but, fortunately, rather rarer amongst psychologists or neurobiologists. (Indeed, it is psychologists who have been among those who have queried most sharply even the theoretical feasibility of my third criterion.)

Except under artificially reduced circumstances, variables are in continuous interaction, and this interaction is not simply a matter of addition. A commonly held example of such an error comes not from neuroscience but from genetics, where for a long time there was a rather simple-minded assumption that the physiology and behaviour of an organism (its phenotype) could be arbitrarily divided into two components, one given by the genes, the other by the environment. An organism's phenotype was thus believed to be almost entirely accounted for by the sum of these two apparently independent variables. In fact of course, genes and environment interact in highly non-linear ways during development, and attempts to partition the phenotype out into a genetic and an environmental 'component' are doomed to failure. Patrick Bateson's analogy is to consider baking a cake. In cake-making, a variety of components, flour, milk, butter, sugar, spices, eggs, etc., are added, mixed and heated together. Although each component is necessary to the final taste of the cake, to ask how much of the taste is contributed by the flour, how much by the eggs, how much by the time and temperature of baking makes no sense; the mixing and baking have qualitatively transformed the components.

Simple-minded formulae about the additive relationship of genes and environment – although they still appear, often without even a health warning, in standard genetics text-books – bear little relationship to what happens in real life. So too in memory research; if an animal cannot be shown to have learned except by changing its behaviour and this change in behaviour can only be induced by some form of stress or constraint, then the changes in biochemistry that one finds in relation to the learning must include the changes in relation to the stress – including all the types of neuromodulators discussed in the last chapter. And yet, within the artificial world that the laboratory enables one to create, we can and must isolate the variables, and, if we are clever and lucky enough, can discover how to fit them back into some meaningful real-life pattern. We have no choice but to deduce as much as we can about the baking of the cake from studying what happens if we miss out a constituent, alter the temperature or cooking time or whatever . . .

Criterion four – If the cellular or biochemical changes are inhibited during the period over which memory formation should occur, then memory formation should be prevented and the animal be amnesic; and vice-versa.

Clearly this is logically necessary, and in the 'forward' direction is the basis for the interventive strategies making use of protein synthesis inhibitors that I discussed in the last chapter. The problems arise because in practice no inhibitor is a 'magic bullet' with a single target and without so-called 'side-effects', so an experimental finding using an inhibitor is not likely to be unambiguous. There are of course many more specific inhibitors of particular biological pathways that are now in general use. Most potentially powerfully, the new genetic techniques open up an entire new spectrum of possibilities. Thus it is possible, in both fruitflies and mice, to 'engineer' animals in which a particular gene has been deleted either wholly or in part. The protein coded for by that particular gene is thus no longer made, and if it is an enzyme, the metabolic pathway which requires it is also blocked. Such methods were introduced to neurobiology in the 1990s by the Nobelist Susumu Tonagawa, at MIT. Initially they suffered from the deficiency that many such induced mutations were lethal, and others appeared to have no effect, perhaps because of the brain's plasticity that meant that during development the organism finds both biochemical and behavioural strategies to circumvent the lesion. Further ingenious molecular biological tricks have made it possible instead to inactivate the target gene only in specific brain regions and only at particular times, so that the lesion is more specific and reversible. It is also possible to block specific protein synthesis using so-called 'antisense' – nucleic acid strands that bind to and temporarily block the function of the RNA on which the protein is made. I'll describe some experiments along these lines later.

Meanwhile I have also added a behavioural consideration. Suppose one trains an animal on a task that normally results in it learning, but adds some treatment that prevents the learning from occurring, then if the biochemical process under study is really a process which is specifically associated with the memory formation, it should not occur if the memory is blocked. I describe experiments making use of this criterion in the following chapter. However, there is a further problem with meeting this criterion that I failed to recognise in those more simplistic days. As I've pointed out, an animal can only tell an experimenter that it has learned and remembers a task by some change in performance. But suppose that the treatment in some way impedes not the acquisition of the memory but its subsequent expression in the animal's performance of the task, then it would appear that the memory had not been formed or was lost entirely. This can be controlled for by carefully designed experiments and I will describe some later, but it is yet another warning that interpretation of experimental results is not always unambiguous!

Criterion five – Removal of the anatomical site at which the biochemical, cellular and physiological changes occur should interfere with the processes of memory formation and/or recall, depending on when, in relation to the training, the region is removed.

This may also seem obvious; a logical analogue of the preceding two criteria. If the changes in connectivity which form the memory are localised to a particular small set of cells and their connections within the brain, rather than being widely diffused, then removal of the set of cells should also remove the memory – or prevent it from being formed. Granted that it is experimentally quite simple to make small localised holes in the brain without causing widespread damage, it should be easy to test the claims for any such region of being 'the' site of memory for any particular piece of behaviour. If the result of making such a lesion is amnesia or failure to learn, it would support the claims for having found such a memory site. However, this does imply a rather static and mechanical view of the way in which memories may be fixed within the brain. If the process of storage is more dynamic, perhaps with multiple sites being involved, then the experiment won't work. And, just as with the inhibitor experiments, the lesion may prevent not the formation but the expression of the memory. Also it ignores the possibility of the brain's own plasticity – that is, if one site is removed, another may become available to take over its tasks – issues with which the experiments of chapter 11 will have to try to come to terms. Finally one must never forget the fundamental ambiguity of all lesion experiments – remember Richard Gregory's radio and its howl-suppressing transistor.

Criterion six – Neurophysiological recording from the sites of cellular change should detect altered electrical responses from the neurons during and/or as a consequence of memory formation.

If a Hebb-type hypothesis about memory being stored in the form of altered synaptic strengths is valid, then these altered connections should be associated with changed electrical behaviour in the cells the synapses connect; that is the firing patterns of the neurons should change as a result of training. The way I have phrased this criterion implies that we should begin by looking for the biochemical and cellular changes and then on this basis seek the neurophysiological ones, and that in some way the neurophysiology is a mere incidental product of the biochemical and structural changes. Of course, this way of looking at things reflects my own bias as only a partially reconstructed neurochemist; in practice the neurophysiology may well lead – indeed, in the important cases of Aplysia and long-term potentiation discussed next, has led – the biochemistry and cell biology, pointing the way towards cells whose electrical properties and therefore their biochemical properties change during memory formation. I do not mean to imply that the biochemistry is primary, or any more fundamental in the reductionist sense than the physiology; what I am saying is that changed biochemistry translates into changed physiology just as it does into changed behaviour.

These six criteria, then, have shaped my own research from the start of the 1980s. I have tried to identify biochemical, morphological and physiological changes occurring in specific regions of the chick brain in the minutes to hours following training on a simple task, to show that the changes are not the results of other aspects of training than memory, to show that blocking the changes prevents the memory, and vice versa, and, finally, to examine the consequences of removing the brain sites of change, either before or after the chick has been trained. What I have found using these criteria will be the subject of the following chapters, of which the chick is the sole and proud subject, but the rest of this chapter will be concerned, not with my own experiments, but those of the other hunters for God's organism whose work has dominated the last decades.

All the sharp criteria and clear theoretical thinking in the world are of no help without good experimental model systems in which they can be explored. As the 1980s rolled on and the earlier uncertainties about choosing the right task and the right organism receded, consensus began to develop around a small number of such models, with several different groups of researchers each arguing the case for their own new versions of God's organism. True, God seemed to have chosen very diverse tasks and species, which bore little relation to experimental tasks that an earlier generation of psychologists would have recognised as relevant. But they seemed to work for those whose single-minded concern was to delineate the biochemistry, neurophysiology and cell biology of learning and memory.

The new models: flies

Some I have already referred to in passing. For instance, Tim Tully, at Cold Spring Harbor on Long Island, among others, has exploited the behavioural and biochemical possibilities opened up by mutations amongst fruitflies; for him and some other neurobiologists Drosophila has become as popular as it has been for most of this century for geneticists. Their argument has been based on the fact that in general any specific single mutation will result in the alteration or absence of a single protein in the organism as it develops. Such proteins may be enzymes, or membrane components, for example. Thus if a learning or memory-deficient mutant is produced, the deficiency must result from the lack or malfunction of the specific protein. If one can discover which protein is missing, then one has a clue to its necessary role in memory formation. In the early phase of these experiments, many fly mutants were made and screened behaviourally. Those that showed learning or memory deficits were then explored for the biochemical processes that the mutations had generated. Nowadays, though, it is possible to do the reverse experiment: to specifically delete or modify a biochemical pathway by targeted genetic manipulation and see what effect this has on the behaviour. Pioneered by Seymour Benzer, Yadin Dudai and others, the main advocate of this approach is now Tully.* He began by showing that longer- but not shorter-term memory was blocked by protein synthesis inhibitors, but more recently has focused on the involvement of one particular portion of the molecular biological cascade that results in the switching on of particular genes and hence

---

*Dudai has moved on to work with rats, and to write two of the most significant books in the field.[3]

protein synthesis. This involves a molecule with the acronym CREB, and Tully has vigorously promoted the idea that because CREB may be centrally involved in memory formation, its manipulation may provide a therapeutic treatment for memory disorders.[4] Like so many in the aftermath of the biotechnology revolution, he has even formed a private company to develop these possibilities, though the clinical relevance of CREB remains to be established.

Back, however, in the world of science rather than shareholding, studying behavioural mutations in Drosophila – or mice – is a bit like using inhibitors to block particular metabolic processes, and has both the strengths and weaknesses of such methods, discussed in the previous chapter and in Criterion 4. The Drosophila studies have not solved the memory problem, but they have certainly supplemented our understanding of its biochemical mechanisms. One of the most important results to have come out from the work is the demonstration that similar molecular processes seem to underlie memory formation in fruitflies as in other larger and more conventionally studied organisms – including CREB. This strengthens the claim that there are real universal biochemical principles involved in such mechanisms of neural plasticity.

The new models, Aplysia or – 'learning in a dish'

Another very popular focus for the new studies of memory has been the molluscs, for the reasons discussed in chapter 7, and in particular because of their large neurons and accessible nervous systems. Although there are several interesting species of land snails, the best known molluscs amongst neurobiologists, if not gastronomes, have been the giant sea slugs, while for those choosing to work with vertebrates, the tendency has been to abandon the mazes and Skinner boxes of earlier generations of psychologists in favour of classical conditioning of very simple reflexes, such as heart-rate or eye-blink, in rabbits, where the neural circuitry can be reasonably clearly mapped.

Ask any graduating neuroscience student which organism had

been most extensively used to study the cell biology of learning and the answer would probably be unanimous: Aplysia. Furthermore, asked to identify the researcher central to the project, it is odds on that the name they would come up with would be that of Eric Kandel, charismatic Howard Hughes professor in the College of Physicians and Surgeons in New York, author of one of the key neuroscience textbooks,[5] and prolific contributor to others, tireless and brilliant proselytiser for Aplysia as God's organism for the study of memory, and for reductionism as the methodological and philosophical route to its understanding. (So committed has Kandel been to reductionism as philosophy as well as methodology that he once gave a talk to an audience of psychiatrists under the title 'Psychotherapy and the single synapse'.[6])

Kandel, trained as a psychiatrist, spent a period working with Aplysia with Ladislav Tauc in Paris in the 1960s, saw the potential of the organism, initially for the study of short-term processes such as habituation, and over almost forty years in New York has made its study peculiarly his own and that of the generations of researchers who have cut their teeth in his Columbia laboratory. There is no doubt of the major contribution that Kandel and his school have made to the study of the neurobiology of short-term processes in memory formation, both in terms of experimental insights and in bringing physiological respectability to a research field which many, in the aftermath of some of the debacles of the 1960s, felt chary of entering. There was general recognition therefore that his efforts had been appropriately rewarded with a share of the Nobel in 2000 even though personal relations among researchers working on Aplysia, and between the Aplysia group in general and those working with other molluscans, have not always been easy, and even sometimes abrasive, to the extent that they attracted science writer Susan Allport to devote an entire book to describing them.[7] Nonetheless, the main thrust of Kandel's findings and the theoretical framework within which he set them during the 1970s and 1980s has until recently scarcely been challenged. More and more over the last few years, however, findings both from his own lab and others have tended to enrich the somewhat simplistic reductionist framework within which the Aplysia findings had earlier been set. To appreciate the critique, however, it is important first to present Kandel's reductionist case in its strongest form.

Chapter 7 described some of the reasons which made Aplysia a strategic choice for researching the neurobiology of certain basic forms of memory formation. It has a seemingly simple and limited behavioural repertoire, including various forms of learning, while its relatively easily mapped central nervous system contains only a small number of cells – no more than 20,000 neurons in all, arranged in a system of distributed ganglia and including amongst them a population of very large cells which can be recognised easily and reproducibly from animal to animal. The key to Kandel's approach has been the study of a simple piece of behaviour which can be studied in the intact animal. The behaviour is a simple reflex, the gill and siphon withdrawal reflex and its habituation and sensitisation. This behaviour, or its neural analogue, can, it is argued, be 'isolated' within the animal by progressively reduced cell populations. The culminating step in this reduction is the interaction of two specific microdissected neurons which can be induced to make synaptic contact while preserved in isolation in a dish. Kandel argues that the interactions and responses of these neurons to artificially administered neurotransmitters represent, in ultimately reduced form, the memory for the reflex itself. What is the evidence in support of this claim?

First, the reflex itself. Aplysia breathes through its gill, which is located in a cavity on the top (dorsal) side of the animal; the rear end of the cavity forms a fleshy spout or siphon. If the area around the siphon or gill is touched, both retract, a form of simple protective reflex. The neural mechanisms for this reflex include a small number of sensory neurons (some fifty in all) which respond to touch sensations on the skin in the region of siphon and gill; these sensory neurons connect to some twenty different motor neurons, both directly and by way of intermediate neurons (interneurons). The motor neurons, which are located in the animal's abdominal ganglion, in their turn make synapses onto the muscles that produce the withdrawal behaviour. A schematic version of this relatively simple circuit is shown in figure 9.2.

While a single stimulation of the body surface close to the gill or siphon produces a reflex withdrawal, on repeated stimulation the response habituates; that is, the response to the repeated stimuli steadily diminishes and finally disappears completely for a while. The habituated response can be dishabituated or sensitised by strong stimuli to another part of the animal, say the tail, in which case the response reappears in all its original strength. Because these are short-term and rather non-specific behavioural changes, they must be regarded as forms of non-associative learning, but important to Kandel's argument is that classical conditioning is also possible; in this the unconditioned stimulus is a shock to the tail and the conditioning stimulus a mild tactile stimulus to the siphon. This mild stimulus normally produces only a weak

Fig. 9.2a
The gill and siphon withdrawal reflex in Aplysia
The reflex in the intact animal. In the sequence from the left, a squirt of water is applied to the siphon via a water-pick; the siphon and gill retract.
Fig. 9.2b
A wiring diagram for the 'reduced' preparation.

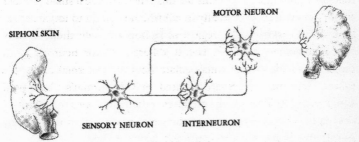

MOTOR NEURON

SIPHON SKIN

SENSORY NEURON   INTERNEURON

withdrawal; following conditioning a strong withdrawal is produced by the weak stimulus to the siphon as well. This effect persists for a relatively long time, and as there is a specific relationship between the stimuli and the responses, it is regarded as a genuine form of associative learning.

Kandel and his colleagues began by asking what was the neural circuit which underlies the gill and siphon withdrawal response. This turned out to be a straightforward problem to address by classical neurophysiological methods, and from then on the research strategy involved a series of reductive steps. To achieve more precise control over the response and quantify it, the reseachers immobilised the slug by pinning it to a stage and standardising the tactile stimulus by using a jet of water delivered with a water-pick. The contractions of the gill could also be directly quantified with a photocell. Granted that the circuitry for the reflex was known, the researchers could then ask the question: when habituation occurs, which part of the circuitry is involved? Do any specific cells or synapses show changes which correspond to the behavioural adaptation? – a question, which, of course, relates to the first of my criteria above. Being neurophysiologists rather than biochemists, the research group's efforts to answer this question began with the electrical properties of the cells – that is, my sixth criterion.

The technology required to offer an answer to this question, however, required a further reductive step, by which the active, alive Aplysia was transformed into an inactive, manipulable 'preparation'. It is possible to dissect open the animal's body so as to expose the abdominal ganglion and its cells, or even to isolate completely the ganglion and the nerves connecting it to pieces of attached skin and gill. With this degree of isolation, Kandel could ignore any other sources of inputs to the system being studied – other peripheral nerves, circulating neuromodulators and so forth. The large cell bodies of the motor neurons could be found, and, as explained in chapter 7, the 'same' cell repeatedly identified in animal after animal (see figure 9.2). The living animal has by this process been transformed into something approximating a circuit board in a computer, and the researchers can go about exploring its properties rather as if they were electrical engineers, presented with a novel piece of equipment and trying to understand from scratch how its circuits function. In this system it is possible to replace the tactile, behavioural stimulus by its neurophysiological analogue, that is, by direct electrical stimulation of the sensory nerve inputs. Similarly, the muscular output – the withdrawal response – can be generated by direct stimulation of the output nerves from motor neurons to the gill muscles.

This isolated and reduced preparation could then be used to ask where habituation occurred – that is, which bits of the circuit showed reduced outputs in response to repeated stimulation. By the mid-1970s, it was clear that neither sensory inputs nor motor outputs had properties which corresponded to the behavioural habituation, as neither showed such decrements in electrical response. It followed that the cells responsible for the habituation must lie centrally, within the sensory-motor interconnections in the abdominal ganglion. And in accord with this prediction, when recordings were made from the motor neurons within the abdominal ganglion during habituation, it was found that there was indeed a progressive decline in the firing rate of the cells as habituation occurred. The conclusion was that the 'site' of habituation must lie between the sensory input and the motor neuron.[8]

There is still, even in this highly simplified preparation, quite a lot of circuitry and many thousands of cells. In particular, the sensory neurons make both direct and indirect connections with the motor neurons; the direct connections involve synapses between an axon of a sensory neuron and a dendrite or cell body of the motor neuron (this is known as a monosynaptic pathway); in the indirect, polysynaptic pathway the sensory neuron first makes synaptic contact with an interneuron, which itself then synapses with the motor neuron. (Two further related terms should be introduced here; when the effect of one cell on another is directly by way of a modification of the synapse that the first cell makes on the second, this is known as a *homosynaptic* effect; when the effect of the first cell on the second is modulated by the behaviour of a third cell synapsing on either of the other two, this is called a *heterosynaptic* effect.)

Analysis of the recordings made from Aplysia motor neurons following sensory stimulation shows that they are responding both directly, monosynaptically, and polysynaptically, by way of interneurons. Asked to predict the most likely site of synaptic plasticity, theoreticians would probably have opted for the interneurons as these can clearly receive and modulate signals from many different inputs before dispatching them to varied outputs. The simple learning model proposed by Hebb and described in figure 6.1 requires the participation of three neurons, that is, it is, if it occurs, a heterosynaptic phenomenon. However, to many people's surprise, by the early 1980s the Kandel group had shown that 'the' locus of habituation was extremely simple: the direct synaptic connection between sensory and motor neurons and in particular, the synapse between a sensory neuron and one particular large motor neuron; the contact was monosynaptic and the modulation was homosynaptic.

Having steadily reduced the preparation from organism to circuit, the stage was set for the final reduction; Kandel's colleague Samuel Schacher dissected out the specific sensory and motor neurons and incubated them together in a dish (a procedure known as tissue culture). It has been known for many years that neurons, like other cells, can be maintained alive and well under such conditions for periods of many days or even weeks, provided they are kept warm, aerated and well fed with glucose and other essential molecules. Many types of cell will divide in such cultures; although neurons will not do so, they can grow, put out axons and dendrites and even make synaptic connections. In Schacher's cultures, the sensory neurons form synapses onto the motor neurons, and electrical stimulation of the sensory neuron results in the motor neuron making an electrical response in turn. Repetitive stimulation of the sensory nerve resulted in a steady decrement in the motor neuron response; it was, in effect, habituating,[9] and the Kandel lab had produced what he was to describe as a single synapse which showed 'learning in a dish'. This was indeed a dramatic triumph for the reductionist strategy the group had pursued, and seemed set to vindicate Kandel's claim that the goal of his research was to discover the 'cellular alphabet' of learning. The sensory-motor synapse would certainly seem to be one letter of such an alphabet.

All that has been described so far is essentially the province of neurophysiology. What can be said about the biochemical mechanisms involved in the response, at any of the levels of cellular organisation Kandel has studied? If habituation occurs by reduction of the post-synaptic response at a single synapse, it could logically be a consequence of either pre- or post-synaptic processes, or of course a combination of both. For instance, there could be a steady reduction in the amount of transmitter released by the pre-synaptic cell, or a modification of the receptors on the post-synaptic side to make them less responsive to a given amount of transmitter released, or both mechanisms could be operating. This question of pre- versus post-synaptic plasticity has been a major source of polemic in recent years, but with most theoreticians favouring the post-synaptic side as the main site of plasticity.

For the Aplysia group, the first neurochemical task was to identify the transmitter involved in signalling between the two cells, which turned out to be the ubiquitous substance serotonin (sometimes called 5-hydroxytryptamine or 5-HT). By the mid-1970s, they had shown that during habituation in the isolated ganglion there was a steady decrease in the amount of serotonin released from the sensory pre-synaptic terminal, without there being any change in the responsiveness of the post-synaptic serotonin receptors. The decreased release of serotonin was also associated with a change in the pre-synaptic membrane properties, in particular, with a reduction in the flow of calcium across the membrane and into the synapse – again the biochemical significance of this will become clear later. In parallel experiments the group showed that sensitisation, which is in some ways the reverse of habituation (see chapters 6 and 7) also involved pre-synaptic processes, this time requiring an *increase* in serotonin production and calcium entry into the cells. That both these processes, of habituation and sensitisation, involved pre-synaptic mechanisms came as a somewhat of a surprise for neural modellers.

A decade later, parallel experiments were made with the isolated cells in culture with similar results. Learning now it seemed no longer even required two cells in culture, but could be completely mimicked by squirting serotonin onto an isolated motor neuron. It is hard to get much more reduced than this!

I will have much more to say about the biochemistry of these events in the context of my own experiments in the next chapter, as I don't want to get into great detail here but instead to emphasise that Kandel explains the reflex and its habituation and sensitisation by a series of reductions. He first translates the complex gill and siphon withdrawal behaviour of the intact organism into a circuit which habituates as a result of interactions between just two cells. He then explains the electrical – that is, physiological – responses of the synapses in terms of a cascade of biochemical processes in the pre-synaptic neuron.

Long-term memory in Aplysia

If the mechanisms which Kandel has uncovered for the short-term processes of habituation and sensitisation can serve as a model for short-term memory, what have they to say about long-term memory? What the Aplysia group needed was some process in their favoured animal which could be unequivocally recognised as long-term memory and whose circuitry could be studied in a similar manner to that they had so effectively employed with the short-term processes; hence the attention paid in the early 1980s to finding an analogue of classical conditioning of the gill and siphon withdrawal reflex. For such conditioning to occur, the animal must learn to respond to a mild stimulus, which would not normally cause the withdrawal, as if it were a strong one, such as a shock to the tail, which does cause withdrawal. The experimental design was perfected in 1983 by Kandel's associate Tom Carew[10] who was able to mimic the pairing of conditioning and unconditioned stimulus in the reduced preparation. The unconditioned stimulus was replaced by the repeated firing of the sensory neuron, and the conditioning stimulus by squirting serotonin onto the cell.

The key feature of associative learning is that, unlike habituation or sensitisation, it is a long-lasting effect, and all the mechanisms discussed so far have been transients. If my second criterion is to be fulfilled, during associative learning in Aplysia there must be longer-term cellular changes which match the longer-term change in behaviour. Although as long ago as the early 1970s it had been shown that protein synthesis inhibitors were without effect on habituation and sensitisation, it was not until the mid-1980s that Kandel turned his attention to the longer-term cellular processes. By contrast with their failure to affect habituation, the inhibitors did produce amnesia for associative learning. Hence this type of learning could not be achieved by mere transient modulation of transmitter release; new proteins were being made, and it was necessary to discover which they were and what their cellular functions were. As a result, Kandel began to make the sort of experiments that those working on the biochemistry of memory had already been struggling with; adding radioactive precursors of protein to isolated ganglia or to cells in a dish, trying to identify the protein products, and to distinguish those made uniquely or in raised quantities during memory formation from the many others. At the same time it was necessary to speculate as to how transient changes in neurotransmitters such as serotonin, or the flow of calcium ions across the synaptic membrane might in turn trigger the specific synthesis of the new proteins that long-term memory demanded.[11]

Cellular alphabets or neural systems?

Because facing such questions has brought the work on long-term memory in Aplysia into the same biochemical arena as my own in the chick, I want to postpone considering them for the present and instead look at some of the problems which, in its singlemindedly reductionist approach, Aplysian orthodoxy – at least the orthodoxy of the mid-1980s, as I suspect that the position is now becoming much more flexible – has ignored. It is important that the purpose of this criticism should not be misunderstood; the theoretical and experimental contributions that the Aplysia group have made to the cellular study of memory over the past decades have been substantial, but the very

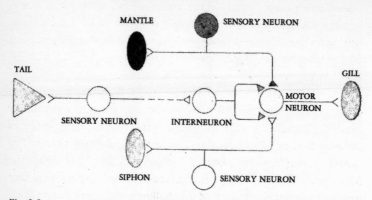

Fig. 9.3
Circuit for classical conditioning in Aplysia
The unconditioned stimulus is a shock to the tail, which excites facilitating interneu-rons that synapse onto the terminals from two other pathways, from the mantle and the siphon. Pairing of a weak shock to the mantle (conditioning stimulus) with the strong shock to the tail strengthens the pathway from the mantle, so that a weak mantle shock now elicits the gill withdrawal, whereas weak shocks to the siphon are unstrength-ened and still do not elicit withdrawal.

intellectual certainty of the group and the charisma of its leader have tended to suppress some of its problems and sideline those who have articulated them. I don't want here to get mired in issues of person-ality and priority, some of which have been brought into the public domain by Susan Allport in her book *Explorers of the Black Box*, but to concentrate instead on some of the more problematic theoretical issues.

Some of those unhappy about the strong claims of Kandel's cellular alphabet metaphor have pinned their arguments to the assumed differences between Aplysia, as an invertebrate, and verte-brate learning. The Aplysia nervous system contains relatively few nerve cells, but among them are some which are rather large. This can be contrasted with the situation in the vertebrate brain, which contains many but small neurons with a multitude of rich interconnections. Thus it has been suggested that a key difference between invertebrate and vertebrate brains is that in the former a great deal of power and responsibility could be invested in a single cell or even synapse which in the vertebrate nervous system would be more widely distributed. While this might be true, for some years now invertebrate neurophysiologists (by which is meant those who study invertebrates, not a special group of researchers without backbones!), who used to speak of their pet organisms as having simple nervous systems, have rephrased their claim, and refer instead to them as having 'simple' nervous systems, the quotation marks being deliberately inserted as a recognition that the complexity of these systems is still many orders of magnitude higher than in the genuinely simple wiring that one might expect of a mere computer. Many invertebrates, such as insects, have nervous systems packed with tiny nerve cells, as indeed do molluscs with large brains such as octopus or squid. *Aplysia* may be a special case because it is easy to study, but it would be straining credulity to believe that it organised its learning behaviour along fundamentally different principles from those of other invertebrates, or indeed vertebrates with reasonably sized nervous systems. There is ample room within the synaptic interactions of even 20,000 neurons for their properties to be those of the system rather than of its individual cells, and claims which were once popular that within insect and crustacean nervous systems one could find key 'command' neurons have gone the same way as, in eastern Europe, parallel enthusiasm for 'command economies' – that is, they turn out to be not a good way to organise individual behaviour any more than to run a country.

Another line of attack has been that of some psychologists who have concentrated their fire on the question of whether the type of experimental procedure designed to produce associative learning in Aplysia can 'really' be said to fulfil the conditions required for classical conditioning.[12] Such infights about terminology, however, concern me less here than some other matters. Let me phrase these in terms of my criteria of necessity, sufficiency and specificity. Despite the remarkable analogy between habituation and sensitisation in the intact Aplysia and the responses of its isolated sensory-motor synapse, which certainly fulfil some of my criteria, there is a conspicuous gap in the logic. Although changes at the sensory-motor synapse might occur during the habituation of the gill and siphon withdrawal reflex, they have not yet been formally shown to be either *necessary* or *sufficient* for that behaviour. I have already hinted, in my account of the reductive steps the group employed, that a variety of experimentally or theoretically inconvenient processes that also occurred during the behaviour, such as a contribution of the peripheral nervous system, and some of the polysynaptic inputs onto the motor neuron, were dissected away and no longer taken into consideration. Can habituation, sensitisation, or associative learning of the gill and siphon withdrawal reflex occur in Aplysia if the key sensory-motor synapses are lesioned (Criterion 5)? And are changes at these particular sensory-motor synapses the *only* ones that occur during short- or long-term learning?

One of the more persistent of Kandel's critics has been the Calgary-based neurophysiologist Ken Lukowiak. He points out that the 'causal' translation between neural and behavioural response implied by Kandel has never been tested directly in the intact animal. For example, if the entire coding for the strength of the response depended on Kandel's single synapse, there should be in the intact animal a direct correlation between the frequency or amount of firing of that specific motor neuron and the strength of the withdrawal reflex. Yet when Lukowiak looked for such a correlation he could not find it; it seemed as if control of the strength of the reflex was not vested in any single cell of the abdominal ganglion, but was instead a property of the interactions between the ensemble of cells as a system.[13]

Evidence which also points in the same direction has come not from critics but from within the Kandel group itself. For instance, the morphologists Mary Chen and Craig Bailey have spent several years studying and measuring the synapses of the Aplysia abdominal ganglion. They find that, when associative learning occurs, there are also characteristic changes in the appearance and number of these synapses (changes which, as it happens, are rather analogous to those we find in the chick and which I will describe in greater detail in the next chapter). Some of these changes are transient, perhaps corresponding to short-term processes, and some, especially in the actual number of synapses, seem more permanent. If this is the case, and long-term memory for the simple association is reflected in a widespread increase in numbers of synapses, it is difficult to argue that the memory is 'represented' by but a single set of synapses at a particular motor neuron; thousands must be involved, distributed across many cells.[14] The final piece of evidence in this context comes from the work of Tom Carew, while working at Yale (he is now at Irvine, in California). He studied the development of Aplysia, from its tiny, free-swimming larval form through a series of intermediate stages to its adulthood, and in particular has mapped the development of the animal's nervous system and of its behaviour. The capacity to show habituation, he observed, occurs relatively early on in the development of the baby Aplysia, while sensitisation does not appear until a relatively late stage. The very young Aplysia has a nervous system consisting of relatively few neurons, while the period of onset of the capacity to show sensitisation as a behavioural phenomenon matches that of a great increase in neuronal number; yet if all that sensitisation required was the facilitatory response in a set of three neurons of the network described earlier, it is hard to see why, by contrast with habituation, it should be dependent on such an increase in neuronal number.[15]

I have devoted some time to the Aplysia story here, not only because of the significance of Eric Kandel's achievement in terms of its wealth of experimental data and theoretical model building, but perhaps above all because both of the place it has come to occupy, not only in the textbooks in making memory research neurophysiologically respectable, but in the framing of the research field. Kandel also has the merit of offering a clear theoretical perspective which has led to an explicit but, I believe, ultimately flawed reductive philosophy and strategy in the search for the mechanisms of memory. Of course, no creative scientist holds rigidly to a fixed position in the light of new evidence, and Kandel, although still committed to a campaigning reductionism, has moved on dramatically in experimental terms. Although Aplysia work still goes on in his lab, he has more and more turned to work with a new, gifted group of collaborators – including for a period my own one-time colleague Rusiko Bourtchouladze[16] – exploiting the latest in molecular genetic techniques in the study of new models of mouse learning. This has led him too to the study of CREB, and, like Tully, to forming a private company to exploit its potential. But now it is time to turn to a consideration of what has become in the last three decades, and is still as I write, perhaps the single most popular learning model in the trade today.

The new models: long-term potentiation and the hippocampus

During the 1960s, sporadic reports appeared in the neurophysiological literature to the effect that, if the neural pathways to certain regions of the cortex were stimulated repetitively at relatively high frequency, there were long-lasting increases in the spontaneous electrical activity of those regions. This effect, essentially an increase in the efficacy of transmission between pre- and post-synaptic cells, was termed potentiation. Could such potentiation be a form of neurophysiological memory? In 1973, in what has become one of the most frequently quoted papers in the literature of memory research Tim Bliss, from the National Institute for Medical Research, in London, and Terje Lømo, working together in Per Andersen's laboratory in Oslo, anaesthetised a rabbit, then exposed its hippocampus and the nerves leading to it. They placed stimulating electrodes onto one of the nerves, known as the perforant pathway, and recording electrodes within a hippocampal region at which the nerves of the perforant pathway made synapses, the dentate gyrus (figure 9.4). When they then stimulated the perforant pathway with a train of electrical impulses, at the rate of 10-100 per second for up to 10 seconds, they found an extraordinarily long-lasting increase in the firing of the hippocampal neurons of the dentate gyrus, persisting for up to ten hours.[17] They called the phenomenon long-term potentiation, soon abbreviated to LTP. LTP, however, lasts much longer than a mere ten hours; the effect can also be found in unanaesthetised animals implanted with permanent electrodes, and in such animals the potentiation has been observed as much as sixteen weeks after the initial brief burst of stimulation. The brief period of electrical stimulation of the hippocampal cells had seemingly permanently altered their electrical properties.

Bliss, Lømo, Andersen and many others in the neuroscience community were immediately intrigued by the phenomenon (many in the field were surprised that Bliss did not share the Nobel with Kandel for this discovery). It was a large effect, specific, reproducible, and above all, very amenable to physiological – and later biochemical, pharmacological and morphological – investigation. The mammalian hippocampus was already a very well understood structure, its neural connections, input and output pathways were clearly mapped and easily identifiable from preparation to preparation even if its individual neurons were not as directly recognisable as are those of Aplysia.

A lasting cellular change in output in response to a defined input is at the least a dramatic example of neural plasticity, but even more than this, the very specific form that the response takes could be regarded as a form of memory. The hippocampus was already well known to be a structure which, in humans and non-human mammals alike, was in some way involved with memory. So might LTP be a mechanism by which memories were formed? Could it not at the least be studied by physiologists as a model for memory? Although Bliss and Lømo nodded vigorously in this direction by referring to their stimulating procedure as a 'conditioning train' of pulses, they concluded their paper with enigmatic caution:

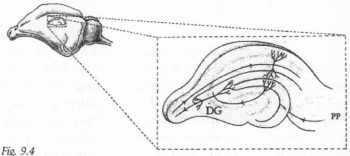

Fig. 9.4
The rat hippocampus
Enlargement shows the hippocampal slice with the dentate gyrus (DG) and perforant pathway (PP) marked. Other regions which can show LTP include the CA1 and CA3.

Whether or not the intact animal makes use in real
life of a property which has been revealed by synchronous, repetitive volleys to a population of fibres, the
normal pattern of activity along which is unknown, is
another matter.'[17]

LTP is a phenomenon that is easily produced and manipulated
by classical neurophysiological techniques, so its popularity as a
potential memory-model is scarcely surprising. In the years that
followed their initial observation, Bliss in London, Andersen in Oslo
and an increasing number of labs began to investigate in immense
detail the neurophysiology of LTP. It was shown to occur not merely
in anaesthetised and unanaesthetised rabbits, rats and other
laboratory species, but also in *in vitro* preparations.

The hippocampus is a structure which can readily be dissected
out from the brain together with its input pathways such as the
perforant pathway. Its three-dimensional organisation is such that
thin slices can be cut, as shown in figure 9.4, leaving the inputs to
the cells of the slice intact. The slice can therefore be maintained
and its electrical properties studied in isolation; the lost McIlwainian
techniques of the 1950s thus became restored to neurobiological
fashion in the late 1970s. In such a slice, appropriate stimulation of
the input pathways will also lead to LTP which persists for as long
as the slice can be maintained alive.

Whether in slices or in the intact brain, LTP turned out to have
a similar range of properties. First, the effect is pathway-specific;
that is, it only occurs in the cells to which the conditioning train
is delivered, rather than spread across to others – it is thus the result
of the functioning of a network of specific connections rather than
a wave of diffuse activity; because there are several distinct input
pathways to separate areas even within a single hippocampal slice,
this specificity can be elegantly demonstrated. Second, to trigger it
requires a repetitive train of reasonably high-frequency pulses; the
same number of pulses delivered more slowly are ineffective, so
there is a threshold below which LTP cannot be induced; above
this threshold, it can be developed gradually or in an all-or-none

fashion depending on the pattern, intensity and frequency of the conditioning train. The development of LTP appears to proceed through at least two (possibly three) phases, in which a brief initiation period is followed by a longer-term maintenance phase, which have been seen as analogous to the transitions between short- and long-term memory.

Third, and perhaps most interesting from the point of view of cellular analogies to memory, by the early 1980s it had been shown that a form of associative LTP is possible. In this a weak input which cannot sustain LTP in its own right may be encouraged to do so if combined with a strong stimulus arriving from a second pathway.[18] The two inputs have to be combined or associated in time in the same sort of way that conditioning and unconditioned stimuli have to be combined for association learning to occur. Indeed it is even possible to produce a form of associative learning in which behavioural and neurophysiological inputs are mixed. A rat can learn to cross a barrier between one side of a box and the other in response to a signal which consists merely of the train of impulses to the hippocampus as the unconditioned stimulus.[19] Taken together, all these properties of LTP would seem to make a powerful case for its study as, at the very least, an intriguing model for memory.

The hippocampus as a cognitive map

What probably clinched the appeal of hippocampal LTP as the memory model of the 1980s was the increasing body of evidence coming from psychologists concerning the role of the hippocampus in animal learning. While the human studies had suggested a role for the hippocampus in the transition between short- and long-term declarative memory – a role supported by study of hippocampal lesions in monkeys – another facet of the hippocampus's role in memory was being uncovered in rats. Here, one of the striking effects of lesions is to affect the animal's capacity to learn spatial tasks – for instance to run mazes. Although shown previously with more traditional maze-learning tasks, the best demonstration of this effect comes in a test devised by Richard Morris, then of St Andrews, now at Edinburgh. The equipment consists of a circular, high-sided tank, a couple of metres in diameter, filled with warm water which is made cloudy by adding some milk. The tank is located in a room whose walls contain recognisable orienting cues; thus on the north wall there may be a clock, on the south a source of light, on the east an animal cage and so forth. At one point in the tank there is a platform just below the water level, but invisible because of its cloudiness. A rat, put in the tank, swims at random until it locates the platform, more or less accidentally and climbs onto it. A video camera mounted above the tank can track the route taken by the swimming animal. After a few trials, the rat will swim more or less directly to the shelf, locating it by cues in the environment such as the clock, light and cage. The effects of drugs, lesions and other manipulations can readily be tested in this type of equipment by a study of their effect on the speed and directness with which the rat can find the hidden platform. Indeed, so popular has it become that its designer has achieved the ultimate scientific accolade of eponymy – having a phenomenon, method or piece of equipment named after oneself – for the apparatus is known as the 'Morris water maze' (figure 9.5) and for more than a decade it replaced Skinner boxes as the standard system for memory research. However, fashions change and over the last few years it too has been upstaged by simpler memory tasks in which the animal learns to associate a specific environment (a particular cage) or a particular sound or light signal with a footshock (so called conditioned fear response). This model, although we are back in the world of one-trial learning and responding to pain rather than increasing navigational skills, is interesting because it appears that different brain regions are involved in learning the context – cage – from learning the cue – sound or light.

But back to the water maze and its uses. How does the rat learn to locate the invisible platform? Does it measure the distance it has swum from the start point, for instance, or does it orient by use of the environmental cues given by the objects visible on the walls surrounding the tank? Such possibilities are easy to test. Altering the location at which the rat is put into the tank is almost without effect on its capacity to find the platform. On the other hand, if the room cues are rotated so that, for instance the clock now appears in the south instead of the north, the rat will become confused, swimming to the region of the tank at which the platform would have been relative to the clock if the latter had not been shifted. Thus the animal locates itself in space by use of environmental reference points (this is of course more-or-less what a psychologically untutored lay person might have guessed would happen, but it was not what psychologists brought up on a diet of Skinner would have theorised). Lesioning the hippocampus, however, profoundly damages the rat's capacity to learn or remember the spatial cues and thus dramatically impairs its capacity to work out an effective escape route in the tank.

The water maze offers a number of advantages for the study of spatial learning in that within the tank the animal is quite unconstrained as to the direction it may take, though this must be balanced against the fact that the swimming task is somewhat stressful and the animal is learning how to reach a relatively precarious goal. Before Morris introduced this type of maze, it had become common practice to study spatial learning in a more conventional version of the same task, in which rats are placed in various forms of radial mazes with four, six or eight arms, and must learn to run to a goal box containing food or water at the end of one of the arms. Again, there are cues both internal to the maze and on the walls surrounding it, and the maze can be rotated relative to these external cues. This type of maze allowed David Olton, in Baltimore, and John O'Keefe and Lynn Nadel (both expatriate Americans, then working in University College, in London, though Nadel later returned to the United States, to Tucson, Arizona) to distinguish between working and reference memory cues in learning the task. The rat can use the cues within the maze itself – for instance, 'to turn second right at this point'. This is a form of working memory, as the cue is only meaningful if the animal remembers where it has just come from. But the rat can also refer to cues offered by the external environment – for instance,

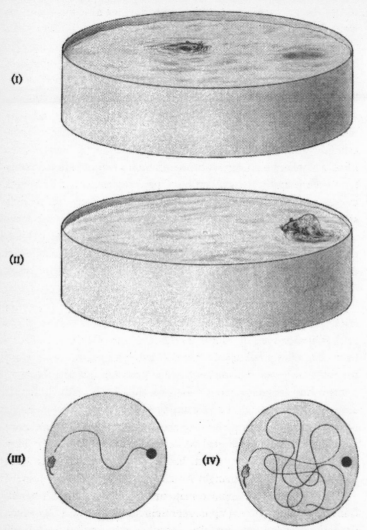

Fig. 9.5
The Morris maze
In this task the rat is placed in a tank full of milky water (i) and learns to swim to safety on a submerged platform (ii). After several training sessions, the animal learns to swim directly to the platform (iii). However, animals with hippocampal lesions, or whose memory for the task has been disrupted by drugs, swim erratically and fail to find the platform except by chance, as if they had never learned it (iv).

to use the rule 'turn left in relation to the clock on the wall' – as fixed or reference memory cues.

O'Keefe and Nadel implanted recording electrodes into the rat's hippocampus and studied the electrical activity of hippocampal cells during the learning of such spatial mazes. A fair proportion of the cells they recorded from gave rhythmic bursts of high frequency firing, at the rate of some 4–12 per second, more or less irrespective of what the animal was doing; this rhythmic activity is interesting because it corresponds with the so called theta rhythm of the EEG, and may be an aspect of the attentional processes necessary for the learning or remembering of particular activities. However, even more interesting were the large number of cells which only seemed active when the rat visited a particular place in the maze and/or carried out particular behavioural acts (food searching, drinking or whatever) in that place. O'Keefe and Nadel called such cells 'place cells' and the parts of the environment in which they are active 'place fields'.[20]

From such observations they generalised to a theory of *The Hippocampus as a Cognitive Map*, the title they gave to their 1978 book. The title not only confirmed the centrality of the hippocampus to studies of animal learning, but was also symbolic of the conceptual shift amongst psychologists away from the crudities of behaviourism and simple associationism towards an understanding of animals, like humans, as cognitive organisms. Cognitive behaviour is not reducible to simple sequences of contingencies of reinforcement but instead reflects goal-seeking activities, hypothesis making and many other features which had hitherto been dismissed from consideration within the Anglo-American tradition in psychology.[21]*

The concept of a cognitive map, in O'Keefe and Nadel's hands, is more than just a topographic representation of the space in which the animal is located; but also describes the distribution of cell

---

*In passing, I would draw attention to the qualifier 'Anglo-American' in the preceding paragraph. In chapters 5 and 6 I contrasted the arid, reductionist abstraction which dominated the Anglo-American behavioural tradition from the 1920s until at least the 1950s and 1960s, with the much richer perspectives offered by other European traditions, but perhaps above all that developed – often in the face of much ideological persecution – by some of Pavlov's pupils and followers in the Soviet Union. In particular,

systems concerned with the analysis and integration of spatial cues
within a framework of behavioural meaning for the animal. Although
in the O'Keefe model there are indeed specific place cells (and during
the 1980s and 1990s other researchers were able to identify, in
monkeys, cells which fired in response to even more precise inputs,
such as photographs of particular faces) the concept of a cognitive
map is in many ways the precise antithesis of the 'cellular alphabet'
model of behaviour offered by Kandel. No way would it be possible
to dissect out one of O'Keefe's place cells and show it 'learning in
a dish'; the cell's responses are only meaningful in the context of

the neurophysiologist Peter Kuzmich Anokhin (whose research career spanned the whole
period from the Bolshevik revolution until his death in 1974) had persistently emphasised the need to view the workings of the brain as a functional system in continuous
integrated interaction with the external environment,[22] and Anokhin's pupils had early
on developed techniques for recording from neurons distributed across the brain of
animals – generally rabbits – going about their day-to-day business with as much freedom
as possible within the confines of a highly restrictive laboratory environment. At least
as early as O'Keefe and Nadel, they were reporting the detection of cells in many brain
regions which had the property of firing when, and only when, their rabbit was in a
particular location and performing a particular act. Even LTP was observed, and its
implications noted, by the neurophysiologist Olga Vinogradova – one of Anokhin's
pupils.[23]

Such work was – and still largely is – ignored in the West. The reasons for this are
instructive. First, it has mainly been published in Russian, where the practice and standard of scientific publication not only doesn't often correspond to Western norms, but
is hard to access for a scientific community which, thanks to the cultural and technical
domination first of British and then of United States science since the 1930s, is less
and less able to read any language other than English. Second, and particularly characteristic of the development of United States science in the post-1945 period, has been
what has been called the 'NIH syndrome'. Not Invented Here symbolises a sort of scientific chauvinism which tends to ignore or discount anything not done in a United States
laboratory; if you are not a United States based researcher, the best you can hope for
is to work in the tiny number of European, Australian or Japanese institutions which
are regarded by your United States colleagues as sort of honorary American. Third, the
Soviet research was often done with equipment considerably less sophisticated than that
in Western labs and therefore doesn't seem to be 'state of the art'. And fourth, and
perhaps most importantly, until the last few post-Marxist and then post-communist
years, Soviet and Russian psychology and neurophysiology was set within a specific
philosophical tradition which explicitly counterposed a dialectical understanding of
mind–brain relations to the mechanistic reductionism which dominates Anglo-American
science.[24] Within a continuing climate of cultural cold war suspicions, and a naïve belief
that to be a reductionist was to be ideologically free,[25] the research was therefore instantly
discounted as 'biased'. As, at least in neuroscience, the theoretical limitations of naïve
reductionism become increasingly apparent, and cold war suspicions recede into history,
the time is ripe for a reassimilation of the autonomous Soviet tradition in neurophysiology and psychology into a more integrated and universalistic neuroscience.

the entire nervous system and the behaving organism in which it is embedded.

The combination of the reproducibility and reliability of long-term potentiation as a physiological phenomenon, the evidence of the central role played by the hippocampus in mammalian memory and the renewed enthusiasm about the prospects for productive research into the cellular processes of memory combined, in the early 1980s, to produce an extraordinary bandwagon in hippocampal studies. Labs which for years had worked on more classical memory tasks found themselves funded to purchase Morris mazes and set up LTP facilities. It helped that the hippocampus is very easy to work on. It is a large structure present in standard laboratory animals such as rats and rabbits that psychology and neurophysiology labs were familiar with, and did not require more exotic facilities like seawater tanks for Aplysia or a knowledge of a novel neuroanatomy like the chick; and the techniques for investigation such as recording electrodes and drugs were all to hand. The phenomena of LTP can be studied at levels from the more-or-less intact organism to the tissue slice. By the 1990s more papers were being published on the hippocampus than virtually any other brain structure, and it even warranted a research journal entirely devoted to it. Even previously committed invertebrate labs like Kandel's group were making the switch.

The biochemical mechanics of LTP

The neurophysiological parameters of LTP having been mapped with exquisite precision, the question of interest became its cellular mechanism. As a completely physiologically induced, and in that sense artificial phenomenon, some of the criteria relevant to memory summarised at the beginning of this chapter are not relevant at this stage. What becomes of interest are the cellular processes which initiate and maintain LTP and what happens if these processes are inhibited. Also as more and more brain regions were found to show LTP-like effects, it becomes important to know whether LTP is one or many phenomena – that is, whether the mechanism whereby it is initiated and maintained in one region is the same as that in others.

Because LTP is a post-synaptic effect – that is, it occurs in a neuron as a result of incoming stimuli along a pathway which synapses on it – one of the first questions was to identify the neurotransmitter involved in this signalling. It was soon apparent that the vital molecule was the transmitter amino acid, glutamate, well known as one of the commonest of the excitatory neurotransmitters of the brain and present in high concentration within neurons. Like all transmitters, glutamate is released from a pre-synaptic terminal when the nerve axon running to that terminal fires. Annette Dolphin, working with Tim Bliss, showed that, when the perforant pathway is stimulated *in vivo*, there is an increased release of glutamate in the hippocampus, and the biochemical mechanisms of this release were mapped in some detail by Marina Lynch. The glutamate is released from the pre-synaptic side of the synapse between the incoming perforant nerve and the hippocampal neuron. On this basis, Lynch and Bliss were to argue, rather as Kandel had done earlier for serotonin in Aplysia, that it was pre-synaptic plasticity that was important for the initiation of LTP, and the post-synaptic cell was simply doing what it had to as a result of the increase in the strength of the glutamate signal it was receiving.

Nothing in biology turns out to be simple, however. Although glutamate is one amongst many dozens of transmitters, it itself interacts with post-synaptic cells in several different ways; there are at least three different types of post-synaptic glutamate receptor, each differently distributed amongst cells responsive to glutamate, each with rather different pharmacological properties and each producing rather different types of post-synaptic responses. Thus although each receptor type responds to glutamate, some will respond to chemically similar molecules as well, others show different forms of specificity. One class of glutamate receptor is known as the NMDA receptor, because the effects of glutamate can be mimicked by injection of the chemically similar substance N-methyl-D-aspartic acid. Injection of drugs which can specifically bind to and poison NMDA receptors will prevent LTP, though not its maintenance if already established. Drugs which interact with the other types of glutamate receptor are without effect. Thus it can be concluded that the NMDA type of glutamate receptor is essential for the initiation of LTP, and in contrast to Bliss's group, other labs reported that there was an increase in the number of these receptors in hippocampal neurons following induction of LTP, thus moving the focus of interest about mechanism from the pre- to the post-synaptic side.

How can an increase in glutamate release, or in the receptors responding to it, result in further pre- or post-synaptic changes? As with Aplysia, a key player in this process appears to be calcium ions. If the calcium concentration is increased during incubation of hippocampal slices, then it becomes easier to induce LTP, while if it is removed from the medium in which the slices are bathed, then LTP cannot develop. If molecules which bind to calcium ions and remove them from solution are injected into the post-synaptic cell, then once again LTP is blocked. This led Gary Lynch,* working in Irvine, California, to propose that LTP was initiated in a process involving enhanced calcium uptake into the post-synaptic cell. Although as more results have come in, the details of Lynch's model have become enriched, in an early version, constructed with his long-term collaborator Michel Baudry,[26] the effect of the calcium was supposed to activate an enzyme present in the post-synaptic site which breaks down proteins. The activated enzyme was then supposed to eat away at the synaptic membrane so as to expose more NMDA receptor sites which, until thus exposed, remain buried in the membrane surface and hence inactive. More NMDA sites would mean a post-synaptic cell more responsive to glutamate and hence more likely to fire.

However, it is clear that calcium has a multiplicity of effects within the cell, and there are other ways in which it can affect the synaptic membrane. Amongst the key molecular components of the

---

*The two Lynches, Marina and Gary, are, so far as either of them know, unrelated; apart from gender and nationality – she is Irish, he Californian – they are to be distinguished by the former being the pre-synaptic and the latter the post-synaptic Lynch.

membrane are a number of proteins which are capable of forming reversible chemical links to phosphate ions. When a phosphate ion binds to such a protein (this process is called phosphorylation) the protein changes its shape, curling up or stretching out within the membrane, so as to open or close channels which run across the width of membrane from the outside to the inside of the cell. These channels make the membrane permeable to ions or molecules, which can then enter the cell and act as signals for the initiation of the biochemical cascades which ultimately lead, in ways that I will describe in the next chapter, to the synthesis of new synaptic membrane components and hence to synaptic remodelling. There are a number of membrane proteins which can be phosphorylated in this way, some post-synaptic, some pre-synaptic, and the enzymes responsible for catalysing the phosphorylation are known collectively as protein kinases. One of these protein kinases is specifically activated by calcium (and is therefore, in another of the acronyms beloved of biochemists, known universally as PKC). In the late 1980s it became clear from the work of several labs that drugs which inhibited this enzyme could block LTP. As a result the models for the mechanism of LTP had to be revised to include an effect mediated through the long-lasting phosphorylation of specific pre- and post-synaptic membrane proteins, conformational changes which it has been hypothesised, are a way of holding even long-term memory.[27]

Hippocampal long-term potentiation is clearly a model system of immense significance – and indeed one which has already yielded much information – about the ways in which neurophysiological changes can be translated into biochemical and structural mechanisms. However, I am not convinced that the most relevant biochemical questions have yet been asked, at least in part because so much attention has been devoted to the intimate synaptic processes involved in the initiation of LTP that there has been surpringly little directed towards what seems to me most interesting about it – the very long-term nature of the phenomenon. Changes in the entry of calcium ions, or the phosphorylation of membrane constituents, or the activation of NMDA receptors, all seem plausible

ways of bringing about a temporary change in the electrical properties of a cell, but what makes the change persist – what puts the L
into LTP – should be the important question if LTP is really to serve
as a model for long-term memory.

The particular power of LTP as a model, though, apart from the
possibility of moving readily between levels of analysis even more
striking than is the case with Aplysia, from intact organism to slice,
lies in its geometry. The cellular and biochemical changes which
must be translated into behavioural processes such as memory
formation must be precisely located in space and time, as the criteria
with which I started this chapter have emphasised. The hippocampus
is one of the regions of the mammalian brain whose structure,
connectivity and geometry are well understood and which should
therefore in principle make such a mapping possible. For memory
modellers it is therefore particularly rich in offering the possibility
of playing with Hebb-type learning rules in synapses whose
connections are genuinely understood rather than merely guessed
at.[28]

But to return to the question with which I began the discussion
of LTP, is it really a model for long-term changes in the nervous
system, or is it something more, a mechanism for memory itself?
The case in favour of it being a mechanism by which real memory
is stored in the brain derives primarily, as I have implied, from the
known role of the hippocampus in various forms of memory
processes, and the fact that forms of associative LTP can be shown
to occur. But beyond this point, the arguments become inferential.
For instance LTP is increased in rats trained to find food in an
operant task, while drugs which block LTP also prevent learning in
tasks such as the water maze. Aged rats lose both their capacity to
learn new tasks, and to show potentiation. Match these arguments
against the criteria with which this chapter began, however, and it
will be seen that they are far from conclusive.

One of the most intriguing sets of experiments in this context
has come from Richard Morris's lab, working with his water maze.
Relatively early on, he was able to show that just as blocking the
NMDA receptor with specific drugs also prevented LTP, so, in

parallel, it prevented rats learning their way round the water maze. However, once they had acquired the memory for the maze, NMDA inhibitors no longer affected memory or performance. This seems powerful evidence in favour of the LTP = memory hypothesis.[29] However, Morris went on to do another experiment. Rats were trained on one water maze as before, and then retrained on a second maze on a different floor of his lab. NMDA blockers did not prevent the rats learning the new maze. So maybe NMDA (and LTP?) are only involved in assimilating some entirely novel situation and experience, not in solving a further example of the same set of problems?[30]

Thirty years after Bliss and Lømo's discovery, in May 2003, the Royal Society in London hosted a birthday party for the phenomenon, and its discovery. The principals reviewed all the exotic new techniques, which the intervening period had brought to bear on elucidating LTP's mechanism and function. And the conclusion? We still don't know just how far the physiological phenomenon really mimics what goes on when a rat learns to run a maze, still less whether it is relevant to the complexities of human autobiographical memory.

Chapter 10

Nobody here but us chickens

AT LAST; SIX DETOUR CHAPTERS, AND I CAN GET BACK TO MY CHICKS. I'm sorry it's been such a long journey, but I couldn't find a quicker route. A few more paragraphs, and we'll really have arrived. In 1977, I took my first ever sabbatical, and spent a couple of months at the Australian National University in Canberra. I chose Canberra because there was quite a nest of chick workers there, including one of my own past students. By this time I was becoming a little dissatisfied with the imprinting set-up that had become the stock in trade of my lab over the past few years, and I planned to try using an alternative form of learning in the chick. The Canberra group had begun to study the effect of a variety of drugs on memory formation in the chick, using a novel learning task that they had adapted from work done in Los Angeles by the veteran neurobiologist Art Cherkin.

Cherkin's own description of how that task was developed during the mid-1960s was characteristic of the man. He had been watching young chicks, he said, and noted how they explored their environment by pecking at crumbs or other small objects, including their own droppings, but quickly learned to distinguish edible from inedible items. 'Well', he said, 'I was damned if I was going to have people going around talking about my system as the Cherkin shit experiment,' so he and Elaine Lee-Teng hit upon the device of offering the chick a small coloured bead to peck. If they made the bead taste bitter, by dipping it in alcohol, or quinine, or the pungent methyl-anthranilate, then the chick would peck once, show disgust by shaking its head vigorously and wiping its beak on the floor of its pen, and then back away, refusing to peck at a similar, but dry bead when offered any time from a few seconds to a few days subsequently. This is the basic one-trial passive avoidance learning model that had attracted me.

It is one-trial, because it requires only a single peck for the bird to learn; it is avoidance, because the result of the learning is for the bird to stop doing something it otherwise would have done; and it is passive because the bird is not required actively to avoid, as it would if it had to escape from some unpleasant condition, instead needing merely to refrain from pecking.

Dramatic as imprinting is as a form of learning, it suffered from my point of view from the problem that for a bird to become imprinted requires exposing it to the stimulus, the flashing light or whatever, for a couple of hours; memory builds up slowly over that time, and so the cellular changes that are going on during the period inevitably intermingle the effects of learning and of visual stimulation with those of memory formation. Pat, Gabriel and I had spent some years unpicking these variables, but if I wanted to study the cellular events occurring in the minutes after the behavioural stimulus had ceased and representing distinct stages in memory formation, then imprinting wouldn't be the model of choice.*

Passive avoidance learning, just because it was a precisely timed,

---

*There is one further point about imprinting. Whenever we presented our work to an audience which included psychologists, someone would ask the question whether the findings had any relevance to other forms of learning because 'imprinting is unique'. Pat Bateson, as an ethologist, has exhaustively answered that criticism,[1] and we saw no reason to believe that imprinting is so special that its cellular mechanisms are likely to be very different from any other form of vertebrate learning. Passive avoidance learning, however, is a form of learning which psychologists feel happier about, even though, as I was later to discover, the fact that learning occurs within a single trial, which I regard as a strength of the system, makes some psychologists feel uneasy because they can't quantify it in the same way as they can tasks which might involve tens or hundreds of trials for an animal to learn. But there is no pleasing some people.

one-trial task, seemed to offer that prospect. Further, there was a practical point. The apparatus required for imprinting, and then for measuring the efficacy of the imprinting response, was large and elaborate; it was impossible to train more than a few birds at a time. This didn't matter so much for physiological or anatomical studies, where one could only work with small numbers of animals anyhow, but for biochemistry, when larger numbers were needed, it made progress very slow. All that the passive avoidance training required was a set of simple, small 20 x 25 cm pens into which a couple of chicks could be placed (in the United States, they used quart-sized milk cartons) – the setup I described way back in chapter 2. And testing meant simply showing each bird the bead again for a fixed period of time – perhaps ten seconds – and noting its response.

By the time I arrived in Canberra the person who had set up the passive avoidance work at Canberra, Marie Gibbs, had moved to La Trobe, a campus in Melbourne, several hundred kilometres distant. Her main interest was in the time-course of memory formation, and she had been using a variety of drugs, including agents which disrupt entry of ions such as potassium into the cell, and also protein synthesis inhibitors, to dissect out a series of phases, which she described as short, intermediate and long-term memory. Each phase was, she argued, sensitive to a different class of drugs. While the earliest phase lasted only a few minutes after the training trial, and the intermediate ones declined within the hour, long-term memory seemed to build up slowly over the first hour after training, and protein synthesis inhibitors would no longer disrupt it if administered more than an hour after the training. (See figure 10.1.)[2]

My interest, however, was in the biochemistry, rather than the pharmacology, of the memory formation. Marie and I agreed that she would train the chicks at La Trobe, using exactly the protocol that she herself had modified from Cherkin, which is essentially how we still do it even today, and as I described it back in chapter 2. She would then freeze and code the brain samples and send them to me – by air, as there were no other practical links – in Canberra. There, I would dissect the brains into the same crudely defined regions we had adopted for

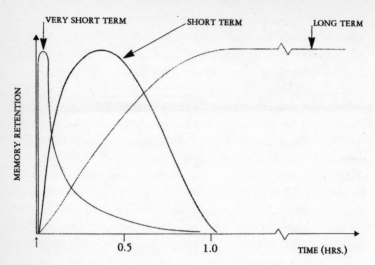

Fig. 10.1
Time-course of memory in the chick
The three phases of memory formation in the chick, as proposed by Marie Gibbs. (She subsequently included a fourth, intermediate-term memory phase, not shown on this graph.)

the imprinting studies and do the biochemistry. At the time, I was particularly enthusiastic about the possibility of the involvement of one of the major neurotransmitters, acetylcholine, in memory formation, and had set up a simple, fast assay for the brain's acetylcholine receptor (called the muscarinic receptor, to distinguish it from other types of acetylcholine receptor*). I already knew that the forebrain roof of the chick was particularly rich in the receptor, and the assay method was fast enough for me to measure it in a couple of hundred brain samples during a single twelve-hour day (although, inevitably, it took

---

*The assay involves using a radioactive drug which binds quantitatively to the acetylcholine receptor, the amount of radioactivity bound being proportional to the amount of receptor present. The drug is code-named QNB, and although it was easy enough to obtain from other labs, my scientific colleagues were quite chary of telling me where they had got it from. Only some time later did I learn that this was because it had been developed as a potential chemical warfare agent and came via Porton Down (it was originally called BZ, and much touted by the United States Chemical Defense Corps during the Vietnam war period as a disorienting agent which, if sprayed on enemy troops, would cause them to lose their will or ability to fight). Granted my very public opposition to

several subsequent days to analyse and calculate all the results).

The plan was for Marie to train groups of chicks to avoid the bitter, methylathranilate-coated bead. A matched group of birds would be 'trained' by being given a water-coated bead to peck instead. These birds would later peck a dry bead when offered it on test and therefore serve as controls for the methylanthranilate-trained group, for they had not learned to avoid the bead. (If you've followed the logic of the earlier chapters you will at once see that this isn't a perfect control, as the water-trained birds may be learning something else about the bead – but I'll come to that later. For the moment it seemed a simple enough experiment to do.) I could then measure the amount of the muscarinic receptor in brain regions from the methylanthanilate-trained and the water-control birds at various times after they had pecked the bead, to test whether there were any transient or longer-lasting changes in the amount of the receptor.

There should have been plenty of time for all the work we planned, but what with all the delays of getting the assay going in an unfamiliar lab, as well as making a quick canter round a dozen or so Australian campuses to give seminars, it wasn't until almost the last few days of my visit, during a long car journey through the outback to attend a biochemistry congress at Brisbane, that I managed to decode and assemble all the data. To my delight, thirty minutes after training on the bitter bead there was a substantial increase in the amount of the receptor in the same region of the brain in which we had found changes during imprinting. The increase was transient, though, for it had disappeared again by three hours after training. Elevated muscarinic receptor seemed therefore to be associated with the early phases of memory formation.

chemical and biological warfare research, then and now, I am not surprised at my colleagues' reticence, though I had always viewed the claims for the military utility of such an agent rather sceptically. The question arose again in 1991, in the aftermath of the Gulf War, in which Iraq's chemical weapons capacity became a matter of much public concern. It turned out that for several years British companies had been supplying the Iraqis with vast quantities of QNB, ostensibly for use in relieving digestive disorders in the Iraqi army. The question arose once more after the disastrous failed attempt to spray gas to rescue hostages held in a Moscow theatre in 2002.

Of course, the experiment was not conclusive – as any inspection of my criteria in the last chapter would make clear – for instance it could have been that the increase was due to the taste of the methylanthranilate itself rather than the learned association of pecking and tasting. After I had gone home, Marie checked this by the somewhat crude device of blindfolding the chicks and putting a cotton bud dipped in the methylanthranilate in their bills. There was no increase in the muscarinic binding under these conditions. However, I was still a little cautious, and waited until, a few months later, she paid a return visit to England. By that time I had copied her training setup in my own lab, and we were able to repeat the entire experiment as before. The effect held up, although this time it was somewhat smaller, and we published the first paper describing the results in 1980.[3]

With the success of the muscarinic experiment, I felt committed to the passive avoidance model, to the exclusion of almost everything else. We dismantled the imprinting equipment to make way for the passive avoidance pens and I set about raising the grant money to let us move into full swing. In the years since 1980, the lab has worked almost exclusively with the task.

There are several sorts of truth I could tell about these decades of work. One would be the version to be found in the published research papers, those strange constrained pieces of writing whose conventions are as rigid as a sonnet, whose scientific account is always a set of clearly designed, unambiguously conducted and conclusive observations, building from earlier data and, in their final paragraphs, pointing the way to future experiments ('more research is needed to . . .'). Or I could tell the story behind the papers, a chronology in which I describe the sequence of experiments as we actually conducted them. This would reveal a sort of untidy to-and-fro between problem areas and techniques, from biochemistry to behaviour. (Our lab is much more magpie-like in picking up techniques than most others I know; for us, anything goes provided (a) we can afford it, (b) it moves us forward on our central question and (c) we can find out how to do it – or hire someone who can.) Some obvious and important matters were left to one side for years because I couldn't see a way forward, or had no time to do the experiments – or couldn't find the funds to buy the equipment or chemicals needed. Others were picked up opportunistically because a visitor or a student arrived with just the right skills or interests to move forward on a front I might otherwise have neglected. Still others were suggested by a casual reading of someone else's research paper in the train home one evening, or from a talk heard almost by chance at a conference. In some cases an experiment begun at one time was transformed in design and intention by a result coming from elsewhere in the lab. As Peter Medawar pointed out many years ago in his classic essay 'Is the scientific paper a fraud?'[4] these essential elements in how research is done get refined out from the account as it appears in the finally published papers or scientific reviews, just as they have largely, though not entirely, been filtered from the discussion of Aplysia and LTP in the last chapter. I wasn't proposing to do it that way for anyone's work but my own!*

In this and the next chapter I am going to tell two different stories. These are the stories in the first edition of this book and I have left them virtually as I wrote them then; a decade more work, culminating in our identification of a potential therapy for Alzheimer's disease, now forms an entirely new chapter 12. The story in this chapter, however, is not chronological but logical; that is, it will set out how, having invented the six criteria with which I began chapter 8, I have tried to meet them using the passive avoidance task. This story will, I hope, both be and sound convincing, for it is the way I have tried to tell it to my neuroscientific colleagues, but it will be in one important sense economical with the truth. Not that I am

---

*Yet another story would be more human than the chronological one; it would talk of the individuals and their interaction within the research group, who we are and what we look like; what we wear, who is having or has had an affair with whom. The psycho-dynamics of small research groups bound together by common problems and working in a campus university rather far from any major conurbation has its own fascination; the internal and external rivalries, endless hassles over grants and space, the trade-offs between teaching and research ... plenty of stuff, if one wanted, for a novel here, granted a group of twenty-five or so, mainly transient, youngish people, staying for anything from a month to three years and from half-a-dozen different countries (C. P. Snow tried it years ago, and might have done better if he hadn't written such constipated stuff, for he knew what he was talking about). But I'm not ready to hazard that yet.

deliberately distorting or mis-speaking our findings, but because I have selected and imposed an order on the research which suits my theoretical and creative purposes and which therefore paints nature in the colours in which I wish to view it. The second story, in chapter 11 is different; it reaches no tidy conclusions, and seems to turn my imposed order into something more chaotic once more. But perhaps it reveals more of the inherent uncertainty of experimental research than the tidiness which precedes it. To discover whether chapter 12 recreates order must wait until we get there.

But enough of prologues and manifestos; it is time for the first story.

The first story – order out of chaos

Criterion one – Something, somewhere, has to change

Begin at the logical beginning. If memory storage requires alterations in the biochemistry and structure of particular cells, then when memories are formed, something, somewhere must be changing within the brain, but we don't know exactly what or where. What's worse, although one might start with some hunches in a mammalian brain, the anatomy of the chick brain is very different from that of mammals, and even now not well mapped, so I couldn't afford inspired guesses derived from mammalian expectations – chickens hardly have anything worth calling a hippocampus, for example. So to begin with I needed a method that was agnostic about location and mechanism. Virtually any biochemical process, certainly anything that means that neurons are becoming more active or are synthesising macromolecules, is going to demand energy. Energy in the brain comes from burning glucose, so if we could find out if, where and when more glucose was being used in the minutes after training, we would have a clue as to which areas of the brain were relevant to the memory storage process. Fortunately, there is a relatively straightforward technique for discovering this. It makes use of the existence of a synthetic chemical closely related to glucose, 2-deoxyglucose, or 2-DG. If 2-DG is injected into the bloodstream, it fools neurons (along with all other body cells) into taking it up as if it were glucose. Inside the cell, the first of the series of enzymes that normally breaks glucose down also thinks the 2-DG is glucose, and therefore converts it into the molecule 2-deoxyglucose 6 phosphate, or 2-DG6P, normally the first step on the pathway of glucose breakdown. However, the next enzyme in the sequence, which would have started work on glucose 6 phosphate, is smarter, and won't have anything to do with 2-DG6P. So the substance accumulates in the cell, and the amount that is there serves as a measure for how much glucose the cell is using. If the 2-DG injected into the bloodstream is radioactive, radioactive 2-DG6P accumulates, and all one then has to do is measure the radioactivity in the cell.

So the experiment involves injecting 2-DG into methylanthranilate-trained and control chicks, waiting half an hour or so for the 2-DG6P to accumulate, killing the chicks, removing and freezing their brains, and subsequently counting the radioactivity present. But the aim is to discover not merely whether there is more radioactivity in the brains of the trained compared with the control chicks but just where in the brain it is located. This is where the 2-DG technique is so neat. The frozen brain is mounted in the laboratory equivalent of a tiny meat slicer, called a cryostat, and sliced sequentially into very thin sections. The sections are put onto microscope slides pressed against a sheet of X-ray photographic film, wrapped in light-tight black paper and placed in a dark-room. Then one must wait anything from days to months – just how long depends on how much radioactivity is present – before eventually the film, now called an autoradiogram, can be developed (the process is called autoradiography).

Each section will have left an image on the film; the more radioactivity present, the blacker the image will be. Just how dark each region is can be measured in an automatic scanner which passes a tiny beam of light through it and records how much is transmitted. The black and white images can be converted to false colour by computer, which looks much prettier, and is easier to interpret by eye, although it doesn't really give more information. One can then compare the amount of radioactivity in control and trained chicks region by region and look for differences. I did this experiment in four frenzied weeks in 1984 with a fanatically hard-working, Warsaw-based autoradiographer, Margaret Kossut, and repeated them in more detail the following year with a neuroanatomist from Budapest, Andras Csillag, who helped identify the anatomical structures in which Margaret and I had found the changes.

The results were clear. One region, with the dog-latin anatomical name Intermediate Medial Hyperstriatum Ventrale (henceforward, IMHV) and another, the Lobus Parolfactorius (henceforth LPO) 'lit up' in the trained compared with the control animals. What is more, just after training the increase in activity was greatest in the left IMHV and left LPO. That is, although the chick brain, like the mammalian brain, is bilaterally symmetrical, being composed of two apparently identical hemispheres, the effects of training are asymmetric – when it comes to learning, chicks are left-hemisphere creatures.[5]*

These results were important for us in several ways. First, it was particularly interesting to find changes in the IMHV after passive avoidance training because Gabriel Horn had already been able to identify this as a key brain region for imprinting. Thus results from passive avoidance and imprinting might begin to converge, which should be good news for both labs. However, neither Gabriel's nor our lab had much idea of how, if at all, the IMHV and the LPO might be functionally connected, or what part each region played in the general economy of the brain. So far as we know, the IMHV in

---

*There are lots of assiduously propagated myths about the almost mystical implications of lateralisation in the human brain, ranging from radical feminist and biological determinist views on left-brain cognitive masculinity versus right-brain affective femininity to the catholic Nobel-prizewinning neurophysiologist Sir John Eccles, who claimed that only humans show such functional lateralisation, and that the left hemisphere is the seat of the soul.[6] Although these days couched in the more sophisticated language of modern neurobiology, the origin of these obsessions with the significance and uniqueness of human lateralisation date from the second half of the nineteenth century, when, on the basis of autopsies of people whom strokes or other brain damage had left without the power of speech (aphasia) the French neuroanatomist Paul Broca located a 'speech centre' in the left frontal lobe. From this Broca – and after him many others – developed an entire speculative apparatus about how functional brain asymmetry was a uniquely human characteristic, and how adults, males and whites showed much greater such asymmetry than children, females and blacks. Then and now, such stories are generally little more than ideological fantasies.[7] But if Eccles did turn out to be right, and functional lateralisation is the key to possession of a soul, then any of my chicks would have as good claims as Sir John to possessing one.

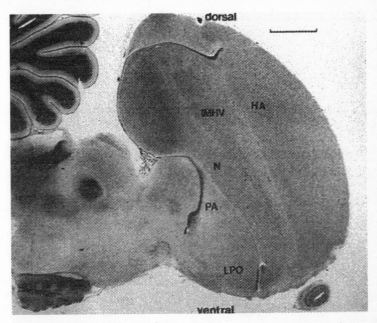

Fig. 10.2
Section through the chick forebrain
The microscope picture of the brain (the scale bar is 1.5mm) shows the location of the IMHV and LPO; other regions marked are the HA (hyperstriatum accessorium), N (neostriatum) and PA (paleostriatum augmentatum). The darkly staining tree-like structure at left is the cerebellum. (Photo courtesy Mike Stewart.)

the chick is a bit like the 'association cortex' in mammals – a region of the brain where inputs from many different sense systems converge and presumably become integrated. As for the LPO, no one was quite clear; some researchers thought it was primarily an 'output' region, coordinating motor responses such as pecking; others saw it as more to do with the bird's emotional responses, which would certainly include fear and distaste (see figure 10.2).

Second, the results proved something we had already begun to suspect, that there are important functional differences between left and right sides of the chick brain. There was a lot of evidence accumulating at the time about lateralisation of function in bird brains – for instance, it appears that chicks respond behaviourally in different ways when they view things with left and right eyes,[8] while in song birds like canaries and zebra finches, the 'song centre' is located in a left-hemisphere region, rather close to our IMHV.[9] What these differences between the two halves of the brain might be telling us, we had no idea at the time – but some clues will begin to appear by the end of the next chapter.

Third, and of more practical importance, we now knew where to look for any further changes; by being able to concentrate on IMHV and LPO and discard 'irrelevant' tissue we might hope to magnify any effect we were studying by diminishing background noise. The two brain regions are quite small – dissected out each weighs no more than a couple of milligrams – and Andras invented a special plastic mould into which we could drop the brain, slice slabs out with a razor blade and then use a fine scalpel to cut round the regions, guiding the dissection under a microscope. We were set to move on.

Criterion two – The time-course: biochemistry

If Marie Gibbs' time-course was right, I should expect to find a sequence of cellular changes in left and perhaps right IMHV and/or LPO, associated with the several phases of memory formation, in the minutes to hours following the bird's pecking at the bitter bead. As the next few paragraphs are going to get quite biochemical, and I see no easy way round them, figure 10.3 summarises the whole sequence for anyone who really can't bear the details (you could then skip to page 301). But they are after all my bread and butter, so I would rather hope that they are worth at least a quick read.

We had of course already found a transient increase in the muscarinic acetylcholine receptor. If I had been working properly systematically I should have gone back and looked at this – and other receptors – in detail in IMHV. But it wasn't until some years later that I came back to the question of the receptors and showed that the most dramatic effects involved the NMDA glutamate receptor I mentioned in the last chapter, but won't discuss further here. Instead, my attention was caught by the evidence coming from the hippocampal work, also discussed in the last chapter, about the role

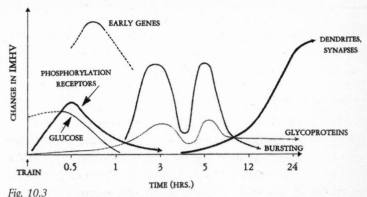

Fig. 10.3
The molecular cascade of memory
The graphs show, schematically, the sequence of molecular and physiological changes found in the chick IMHV at various times after training on the bitter bead.

of the phosphorylated proteins of the synaptic membrane. Perhaps because my own PhD, many years before, had been taken up with working on protein phosphorylation without fully realising its significance (chapter 3) the temptation to explore it in the chick proved irresistible.

The pre- and post-synaptic membranes can be separated out from IMHV and studied in isolation by centrifugation, rather like the method I described in chapter 3. The membranes of course contain both the proteins and the enzyme that phosphorylates them, protein kinase C. If radioactive ATP is added to a tiny sample of the membranes, and incubated together for a few seconds in a miniature test-tube, the membrane proteins become both phosphorylated and radioactive. Another simple but ingenious technique enables one to separate the individual proteins and measure the amount of radioactivity in each. The method makes use of the fact that the many hundreds of different proteins in the membrane all differ from one another in molecular weight and electrical properties, each carrying a specific array of positive and negative ions. To separate the proteins, one makes a small rectangular slab of inert jelly (called a gel), from starch or acrylamide, puts a drop of a solution containing the protein mix at one end, then passes an electric current across the gel. The proteins move in the electric current at a speed which depends on their electric charge and molecular weights, and so within a few hours they have become distributed along the length of the gel – the procedure is called gel electrophoresis. The gel is then soaked in a dye which stains the proteins, which then appear as a series of bright blue bands, like ink lines, on the gel. The band of gel containing each protein can either be cut out with a razorblade and the radioactivity in it counted or the whole gel can be placed against X-ray film and an autoradiogram made, just as with the 2-DG experiment.*

The resulting picture looks like figure 10.4.

We measured the phosphorylation of the proteins of synaptic membranes prepared from brains dissected at various times after training, and, sure enough, the phosphorylation of one key pre-synaptic protein was affected thirty minutes after the birds had pecked the bitter bead. The change was transient; by three hours after training it had vanished. And when we measured the activity of protein kinase C in the membrane we found that it too increased in activity in the left IMHV thirty minutes after training.[10]

So training produces a transient change in the phosphorylation state of a specific pre-synaptic membrane protein, regulated by a specific protein kinase enzyme. But it is only a transient change, so although it may be necessary if long-term memory is to occur, it cannot be the biochemical representation of that memory. Something more permanent is required, something that will in some way produce some lasting remodelling of synapses. It is this remodelling which must require the synthesis of new proteins.

---

* The first book I ever wrote, many years ago now, was about biochemistry. In trying to describe what doing a biochemistry experiment was like, I suggested there were strong similarities between a lab and a kitchen; to me almost the single most extraordinary thing about how we spend our experimental time is still this strange mix between using machines which are capable of great power, like centrifuges which can generate gravitational fields of half a million and more, chemicals of great hazard, like radioactive isotopes and powerful toxins, measurement of quantities almost inconceivably tiny, thousandths of millionths of grams; and yet the principles of separation and handling of materials we use are instantly familiar to any cook experienced at making a sauce or baking a cake.

Fig. 10.4
Proteins from the synaptic membrane
The drawing shows two gels. Samples of synaptic membrance proteins have been placed at the top and electrophoresed for several hours. The several thousand different proteins migrate down the gel at various speeds. The gel at the left has been stained with a dye that colours proteins. Note the many different bands, each representing one or more of the migrating proteins. The gel at the right is an autoradiogram pattern after phosphorylation, derived from the one on the left. Of all the many protein bands, only some four have been phosphorylated. The heavily staining band in the middle of the gel is a protein called B50, with a molecular weight of about 50,000. It is a specifically presynaptic protein, and it is this band which is affected by the training.

Now proteins are synthesised on the basis of information provided by the DNA, that is, the genes present in the cell's nucleus. If new proteins are to be made, the DNA must be activated in some way, so as to switch on the relevant genes. So the changed phosphorylation of the synaptic membrane, which probably results in calcium entering the cell, must act as some sort of a signal to the DNA in the nucleus. At the time of the first edition of this book I didn't know how this worked, but recently we were able to show that in the minutes after the chick pecked the bitter bead there is a surge of calcium ions into the cell. Once inside in a sort of chain reaction they mobilise more calcium release from so-called 'intracellular stores'.[11] What happens next is that the calcium activates a complicated enzyme jungle the study of which has kept molecular biologists happily engaged for many years. A key molecule in this cascade, but only one of several, is the CREB that I mentioned in the last chapter as a 'memory molecule' of interest to Kandel and Tully. The complexity of this molecular jungle is such, however, that there are many paths through it. The important point is that once that signal does get to the nucleus, what happens next in any process of cell plasticity and growth – a step first detected in rapidly dividing cancerous cells but soon recognised to be a rather universal mechanism – is the activation of a group of 'immediate early genes'. These genes are the mechanism by which information arriving at the cell nucleus is translated into instructions for the later synthesis of key structural proteins – that is, proteins which will eventually be inserted into the synaptic membrane so as to change its structure and shape. These structural proteins are coded for by more orthodox 'late genes', all that the early genes themselves do is ensure the synthesis of a group of intermediate signal proteins, rejoicing in even more than usually barbaric names (c-fos and c-jun). In turn c-fos, c-jun and their relatives act as further signals to the nuclear DNA, switching on the relevant 'late genes'. This complex cascade of signals is shown diagrammatically in figure 10.5.

It is the structural proteins which are of real interest, as they go about the business of actually modifying cells; the immediate early gene mechanism is a piece of molecular biological housekeeping, which probably seems arcane not merely to most non-biochemists but to biochemists as well. C-fos and c-jun are of interest though not merely because they provide a key mechanistic link between early events at the cell membrane and nuclear protein synthesis, but because they only become active in cells showing plastic changes, and they can be measured and localised with exquisite sensitivity by variants of the autoradiographic techniques I have already described. When, in 1989, we started to explore the involvement of this mechanism in passive avoidance learning there had already been a lot of speculation in the molecular neurobiology literature about whether it would be possible to show that c-fos and c-jun were specifically activated during memory formation, but no one had yet done the key, unequivocal experiment.

I am no molecular biologist, and wouldn't have dreamed of learning the techniques required to detect the immediate early genes,

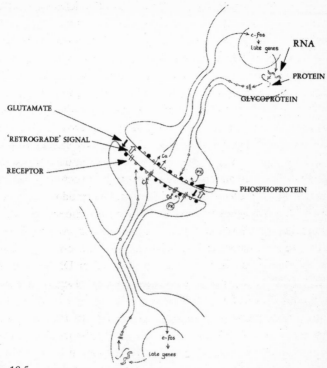

Fig. 10.5
Signals between synapse and nucleus
The drawing shows (not to scale of course) a synapse onto a dendritic spine, and the pre- and post-synaptic cell bodies to which they are connected. During memory forma-tion, transmitter (dark arrow: glutamate) released from the pre-synaptic side interacts with receptor (11) on the post-synaptic side, resulting in phosphorylation of membrane proteins (•) by protein kinase C (PK) and entry of calcium (Ca). The calcium provides a signal to the nucleus for the synthesis of early (c-fos) and late genes, which by way of RNA, code for the synthesis of protein and glycoprotein molecules (°) which are then transported to and inserted into the membrane, changing its size and shape. With the same time-course a retrograde signal to the pre-synaptic cell (open arrows) triggers a matching process there. In 1992 we showed that this signal is carried by the small gas molecule nitric oxide (NO).

if it hadn't been for the serendipitous arrival in the lab of a young molecular biologist from Moscow, Kostya Anokhin (grandson of the psychologist and physiologist pupil of Pavlov, Peter Anokhin, whose 'functional systems theory' I referred to in passing in chapter 9).

Kostya had access to the specific molecular biological 'probes' that the detection method required and a great appetite for laboratory work. Within a few weeks of his arrival, we had shown that, half an hour after training, that is at about the same time as the changes in membrane phosphorylation, there is a dramatic increase in the expression of c-fos and c-jun proteins in cells of the IMHV. We had found a vital step along the route from synapse to nucleus.[12]

By contrast with such complexities, the rest of the biochemistry is relatively straightforward. Almost the first experiments I had made with the passive avoidance model after completing the work with Marie, and even before we had located IMHV and LPO as the sites of change, was to look at the effects of training on protein synthesis in general, using the precursor techniques that have already been described in earlier chapters. From half an hour after training, to as long as twenty-four hours afterwards, it was possible to detect an increase in protein synthesis in the brain regions containing IMHV – a result which of course squared with the known amnestic effects of the inhibitors of protein synthesis. Because I believed that much of this increased protein synthesis was likely to be associated with the production of new synapses, or the modification of old ones, it was, however, important to look not at proteins in general but at synaptic membrane proteins in particular.

Many of the most important and prominent proteins of the synaptic membrane are of the class known as glycoproteins, which, if the description I gave in chapter 2 now seems a long way back, can best be summed up as molecules made in two parts; an amino acid chain embedded in the membrane, to which is attached a further chain made of sugar molecules such as glucose, fucose and galactose, sticking out from the membrane into the extracellular space beyond. These sugar units are 'sticky' – when they meet a matching sugar chain sticking out from the membrane of an adjacent cell, the two recognise each other and become attached. Thus glycoproteins function as cellular recognition molecules, and it seemed to me that if synapses, which are *par excellence* recognition and attachment points between cells, were going to be modified by training, then glycoproteins would be involved. The experiment I was doing all those months ago, when I began writing the first edition of this book, and which I described in chapter 2, involved using the sugar fucose as a precursor for glycoprotein.

In fact, as long ago as 1980 we had shown that, just as amino acid incorporation into proteins increases for twenty-four hours after training, so too does fucose incorporation into glycoproteins of the pre- and post-synaptic membrane. The problem is that glycoproteins are notoriously difficult molecules to analyse, and there are quite a number of different types in the synaptic membranes. We spent a good part of the 1980s in a long and often rather frustrating attempt to identify them. But the endeavour – which I will come to in chapter 12 – has finally been rewarded, and if my hunch about Alzheimer's is right, may turn out to be much more than merely scientifically exciting.

The time-course continued: biochemistry becomes structure

If the hypothesis that the glycoproteins are involved in some form of remodelling of synapses is correct, then maybe one could actually observe and measure these changes in the neurons of the IMHV? It is relatively easy to prepare brain samples to examine either under the light microscope, with its maximum useful magnification of a few thousand, or the electron microscope which can magnify by hundreds of thousands. It is much harder to move from a visual, qualitative appreciation of what can be seen under the microscope to more quantitative measures of how much or how many of any component is present, yet unless something entirely new was being synthesised as a result of training, then what we might anticipate observing would be small changes in the number, pattern or distribution of existing structures, particularly synapses. Using the light microscope, one cannot see individual synapses, but it is possible to stain individual neurons and analyse the structure of their dendrites, hence picking up possible changes. If there were changes in the terminals at the pre-synaptic side, they would have to be measured using the electron microscope, however, for they are not visible at light microscope magnifications.

Doing this sort of quantitative morphology – that is, measuring the shape, number and size of cells in the brain – is, even today, with highly sophisticated image analysis and computing systems, complex, time-consuming and, if one isn't careful, fraught with the danger of misinterpretation. How many of the hundreds of thousands of cells in each tiny brain region must one study to get a representative picture? How can one be sure that what is being seen and counted is 'really' present in the living brain rather than an artefact, an artificial pattern generated by the techniques required to fix, slice and stain the brain tissue to make it visible? How can one scale up from what can be counted in a two-dimensional section to the three-dimensions of living tissue? These were the sorts of technical problem I began discussing with my colleague Mike Stewart in the early days of the chick work, when it became clear we would want to try to make this type of measurement. Mike picked the problem up and ran with it, and succeeded in creating a really first-class lab for quantitative morphology. But of all the multitude of possibilities, what should we try to measure in the chick?

Back at the end of the nineteenth century, the Milanese anatomist Camillo Golgi had discovered almost by chance a stain, based on the use of silver salts, which has the capacity to select out, seemingly at random, a small proportion of the neurons in a section of tissue and stain each immaculately, revealing every last detail not just of its cell body but also the dendrites and even the myriad little spines which stud their surface. It is interesting that Golgi himself, who got the Nobel prize in part for this work, didn't believe that there were individual neurons within the brain, preferring to think of it as a continuous network of fibres, and persisted with this mistake despite the evidence of his own staining technique. It took the formidable Madrid neuroanatomist Santiago Ramón y Cajal to see the significance of Golgi's achievement (he too got a Nobel prize, though Golgi apparently refused to accept Cajal's interpretation – or even to speak to him). No one who sees a Golgi-stained preparation of neurons could fail to be awed by the complexity and elegance their complex branching patterns reveal. Even today, decades after I first saw my first such microscope slide, I still think them extravagantly beautiful and can easily get lost in contemplation of the cellular thicket the microscope reveals, made the more intriguing because of the curious almost three-dimensional effect the stain gives; as it brings into vision only a few of the total population of the neurons present, the cells seem to stand out like trees in a winter mist. (See figure 10.6.)

Looking at a picture of this sort, however appreciative one is of its beauty, is a very different matter from quantifying any aspect of it. What could we measure which might be different about a neuron which had changed its structure in some way as a result of learning? The surface of each of the dendrites which branch out from the neuronal cell body is studded with synapses – perhaps up to ten thousand in all – arising from the other neurons which thus make contact with them. Some of these synapses are on the shafts of the dendrites, others are attached to the tiny spines which stud the dendritic surface and which can be seen in figure 10.6. Changes in synaptic connectivity between one neuron and another as a result of learning along Hebbian lines might involve the dendrites in

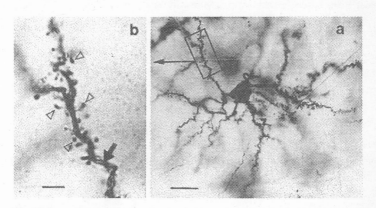

Fig. 10.6
Neurons from the IMHV
At (a) a Golgi-stained single neuron from the IMHV. Scale bar 15μ. At (b) an enlargement of one of the dendrites showing the surface studded with spines. Scale bar 4μ. (Photo courtesy Mike Stewart.)

increasing in length, or changing in branching pattern, or the numbers of spines might alter.

How effective any given synapse is at influencing the post-synaptic neuron in firing depends on a number of factors – how close it is to the cell body, whether it is on the shaft of a dendrite, or on one of the many spines, and so forth. At a synapse, transmitter released from the pre-synaptic side binds to receptor on the post-synaptic side, resulting in a change in the electrical properties of the post-synaptic membrane and a small flow of current around it. The effect that this current has on the rest of the dendrite and hence in due course the cell body depends very much on the geometry of the region around the synapse; biophysical calculations show that spine synapses are more effective than shaft synapses in spreading the current, and in any given spine, the current flow is dependent on its exact shape. Thus any change in the structure of dendrites and the location that the synapses have on them can change the neurophysiological relations of pre- and post-synaptic cell. So neuronal connectivity can be altered not merely by increasing or decreasing the actual number of synapses between two cells, but by altering the size or position of any particular synapse – for instance by shifting it from being located on the dendritic shaft to a spine.

There are good reasons to believe the branching pattern and shape of the dendrites may also be important and may well change as a result of training or other types of experience. Because to make a microscope preparation means that the tissue has to be fixed and stained, what one sees always looks as if it is a very rigid structure, but in the living organism the dendritic pattern of neurons is as mobile as the branches of a growing tree in a gentle breeze,[13] so changed branching patterns are perhaps not so hard to envisage. Such patterns aren't very easy to analyse. But amongst the easier and more obvious structures to count are the dendritic spines, and a graduate student of Mike's, Sanjay Patel, did just that in the mid-1980s. He trained chicks and twenty-four hours later took out and stained left and right IMHV with the Golgi method. He then selected a particular class of neurons, recognisable by their long axons, measured the length of each dendritic branch and counted the spines on each, which he then calculated as number of spines per μm – that is, millionth of a metre – of dendrite.

Much to our delight (and I must confess to my astonishment) Sanjay's counts showed dramatic effects. Twenty-four hours after training, there was getting on for a 60 per cent increase in the numbers of spines to be found on the dendrites in the left IMHV (but almost no effect in the right). The spines were also slightly changed in shape; it was as if the head of each had been blown up like a little balloon – just what biophysical theory would predict would need to happen if the electrical connections between the pre- and post-synaptic sides were being strengthened when the chick pecked the bitter bead; and exactly the sort of change which might be predicted as a consequence of increased glycoprotein synthesis.[14]

So here was unequivocal evidence for quite major post-synaptic changes in structure as a result of the training experience. Mike Stewart's morphological methods, however, could go beyond those of light microscopy to that of the electron microscope. At the levels of magnification we routinely use with the electron microscope a thumbnail would be 250 metres wide, and the order that the Golgi stain seems to offer dissolves into chaos which only a disciplined imagination can control. Because one cannot simply enlarge the Golgi picture to electron microscope size and follow any individual cell and its connections from beginning to end, it is hard to relate the spine changes seen in the Golgi pictures directly to particular synapses seen in the electron microscope. Nonetheless the eye of art and experience can interpret the electron micrographic chaos to pick out individual synapses, cell bodies, axons and dendrites and measure them.

Figure 10.7 shows a synapse from the IMHV, magnified about ten thousand times and marked up to give our interpretation of what there is to be seen. The sites where pre-synaptic terminal and post-synaptic membrane are in close contact are recognisable in the photograph as characteristic dark, thick regions where the pre-synaptic and post-synaptic membranes virtually – but not quite – touch. This thickening is the area of the post-synaptic membrane which contains the receptor molecules and which traps the transmitter released from the many small vesicles visible packed into

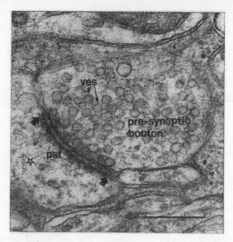

Fig. 10.7
A synapse in the IMHV
A single synapse showing vesicles. The asterisk shows synaptic contact with the dendritic spine head. Post-synaptic thickening is indicated by arrowheads. Scale bar 0.2μ. (Photo courtesy Mike Stewart.)

the pre-synaptic terminal. The number of terminals in a given volume of tissue and their average size, that is, the average volume of each such terminal, are obvious measures to make. So is the length of the synaptic thickening. With time and patience, it is even possible (if one is a graduate student or an obsessive) to count the numbers of vesicles packed within each terminal.

Mike, his students and visitors, have counted all of these parameters in the synapses of the IMHV and LPO. In the early studies we chose to look at them twenty-four hours after training on the grounds that any structural change would take time to build up; more recently he has pushed the earliest time at which changes can be found back to twelve hours after the bird pecks the bead. His findings are clear. There are increases in the numbers of synapses in the LPO, in the numbers of vesicles per synapse, and even in the length of the post-synaptic thickenings in left IMHV and LPO. All-in-all, these, along with the changes in numbers and size of dendritic spines, are what one might expect if, when the chick pecks the bead and learns the association between the pecking and the bitter taste, there is a synaptic reorganisation in IMHV and LPO to code for – to represent – this new association and the resulting change in behaviour – that is, to say 'don't peck' instead of 'do peck' when the chick sees the bead a second time.[15]

To recap so far, as figure 10.3 summarises, pecking at the bitter bead sets off a cascade of biochemical processes in two specific regions of the chick brain; these begin with transient changes in cerebral blood flow and energy use and involve brief increases in transmitter–receptor interactions, which alter the properties of the synaptic membranes such as to increase the efficacy of communication between pre- and post-synaptic side. These changes in turn provide a signal to the cell nuclei, which results in the activation, first of a number of immediate early genes, and later the genes required for the synthesis of new synaptic membrane constituents, especially glycoproteins. In the hours that follow, these glycoproteins are inserted into the synaptic membranes, increasing the numbers of dendritic spines, and the synaptic contact areas in both left IMHV and left and right LPO. Not bad, as the consequence of but a single peck!

Criterion three – Necessity and sufficiency

So far, so good, but finding a set of correlations of this sort still says nothing about whether they are necessarily part of the memory formation process unless I can find a way of showing that they are not simply the aftermath of the unpleasant experience of tasting the bitter bead; that is, I must meet my own third, reductionist criterion. Until I can do this, the entire complex cascade I have described in the previous section, and which has taken a decade of work to map, might simply turn out to be a consequence of a bad taste in the chick's mouth and nothing to do with learning and memory at all.

The rather crude experiment that Marie Gibbs made by putting the methylanthranilate in the chicks' bills when their eyes were closed and showing that this did not result in a change in receptors is one such test, but, because blindfolding the birds is itself scarcely a neutral experience for them, it can't be more than suggestive. What I wanted ideally was a situation in which I had two groups of birds, each trained on and showing the disgust response to the bitter bead, but one group then remembering and the other forgetting the association. The group which remembered should show the cascade, the group which forgot should not.

While I was puzzling over how to solve this problem, I came across a paper written by a Harvard-based neurobiologist, Larry Benowitz in the early 1970s. In it, he described how, if chicks are trained on the passive avoidance task and immediately afterwards given a mild electric shock across their heads, they seem to forget the association of taste and bead, for when offered a bead later they peck at it enthusiastically once more. However, if the shock is delayed until some ten minutes after training, memory is not disrupted, and the chicks will avoid the bead.[16] The explanation for this effect is presumably that the very early events in the cascade of memory formation involve electrical activity within the neurons and that the immediate shock disrupts this process; by the time the delayed shock is given, however, the cascade is already past this phase, and is no longer vulnerable.

This at once gave me the idea for the experimental design I wanted. I could train birds on the bitter bead, shock them, either immediately or a few minutes after training, and compare the biochemistry in the birds which had tasted the bead but forgotten the association with that in the birds which had tasted and remembered the bead. We built a simple little device which could administer a mild shock to a chick held briefly in my hand – tested on my finger, the shock is no more than a brief tingle, and the chicks seemed scarcely to notice it – and I checked out Benowitz's finding. He was absolutely right; shocked a minute after training, most chicks showed amnesia and so pecked the bead when they were tested some hours later; delay the shock to ten minutes after training and most chicks avoided it subsequently.

Two groups weren't enough for the actual experiment though; I needed six. Call them A–F. Three groups (A–C) would be trained on water, three (D–F) on the methylanthranilate, and amongst both water and methylanthranilate birds, one group (A and D) would be unshocked (or rather, to be sure, 'sham' shocked – I would go through the motions of shocking them but with the current turned off), one group shocked immediately after training (B and E) and one shocked after a delay (C and F). If we then take a biochemical marker, such as, say increased fucose incorporation in the hours after training, we expect a difference between A and D, due to the training, and between C and F, as the delayed shock group has both tasted the bead and shows the memory. The crucial comparison is between B and E, as although E has tasted the bead, it is amnesic and pecks it later. If the biochemical change is the consequence simply of the experience of tasting the bead, then the level of fucose incorporation in E should be the same as that in D and F and higher than in all the water groups A–C. If, however, the increased incorporation is associated with memory, then E should be equivalent to B and lower than either D or F. We needed, of course, to repeat the experiment with enough birds – eventually, a dozen in each group – to be sure of the results – but when we analysed the data, they were unequivocal – and I can't resist showing them in figure 10.8.[17]

The increased incorporation occurs only in the groups showing memory and not in the group which is trained but amnesic. What's more, the shock itself is without effect on this biochemical measure, as can be seen by comparing the shocked and unshocked water-trained birds. I was sufficiently pleased with this neat and simple control that we have employed it for several other of the key steps in the cascade – for instance, the increase in dendritic spines occurs only in a remembering and not in an amnesic group.[18]

There were still some lingering doubts in my mind, not because I question the logic of this experiment, but because it would be nicer to have an experiment which did not require even the distress of a mild shock to the chicks. What do I mean by nicer? The electroshock experiment is logical and elegant, but shocking the chicks is aesthetically (morally? – I am not sure) displeasing, however mild the actual experience. In 1990, Kostya Anokhin and I therefore invented an alternative procedure, one that would not involve the chicks in any type of aversive situation at all – even that of tasting a bitter bead. The chicks are placed on a surface scattered with a mix of food grains and

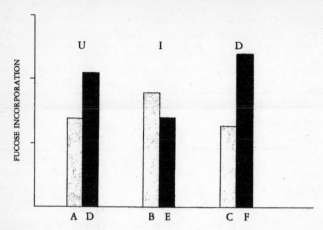

Fig. 10.8
Memory and fucose incorporation
The height of each bar represents the amount of fucose incorporated. Light-coloured bars are from water-trained animals, dark bars from methylanthranilate-trained animals. Letters on each bar show the group as described in the text. U – unshocked group, showing expected increase in methylanthranilate-trained birds. I – immediate-shocked birds, which forget, showing that the increase is abolished. D – delayed shocked birds, which remember, showing increase after methylanthranilate training. Note that there is no significant difference between the water-trained birds in any group, indicating that shock alone has no effect.

pebbles of about the same size and colour as the food. At first chicks peck at both food and pebbles about equally, but after a few minutes, they learn the difference and thereafter pick up the food and avoid the pebbles (especially if the latter are glued to the floor!). We divided the chicks into four groups and spread the experiment over two days. The design is shown in figure 10.9. Group K are birds which are 'quiet controls' on both days. On Day 1, Groups L and M are given a number of sessions on the pebble floor without food; Group N has the same number of sessions, but with food and therefore learns the distinction. On Day 2, Groups L and N repeat their experience of Day 1, while Group M experiences the pebble floor plus food for the first time. In this experiment only group M is learning on Day 2, whilst group N is repeating similar but already learned behaviour to that of group M. Immediately after the trial on Day 2 we took the birds and

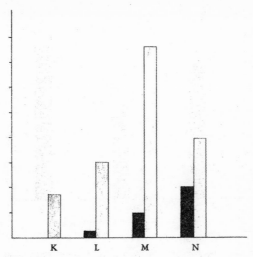

Fig. 10.9
Pebble floor experiment
The four groups of birds, K, L, M, N, were trained over two days as described in the text. Dark bars are a measure of pecking, open bars a measure of c-jun. Although the already trained birds in group N peck most, c-jun activity is much higher in birds of the learning group M.

looked for the expression of one of the immediate early genes, c-jun. Compared with the Quiet control (Group K), there was an increased c-jun expression in both the 'behaving' groups M and N, but the increase was much more marked in the learning group, M, than in the group (N), which is merely repeating an already learned behaviour, even though in fact the chicks of group N are eating the food grains even more avidly than those of group M.[19]

Criterion four – Inhibit the biochemistry, inhibit the memory?
The logic of the inhibitor approach is obvious, but for a number of years I was reluctant to get involved in doing such experiments, as it wasn't clear to me that using broad spectrum inhibitors such as those for protein synthesis would tell me anything precise about the biochemical processes which I was trying to unpick. Once we had begun to look at particular steps in the biochemical cascade in more

detail, however, I became persuaded that reasonably specific inhibitors might help cast light on relevant mechanisms. For instance, we found that, injected before training, the agents which block hippocampal LTP and spatial learning – inhibitors of the NMDA type of glutamate receptor – also produce amnesia in the chick; injected into the left hemisphere just before or just after training, inhibitors of protein kinase C also produce amnesia.[20]

However, perhaps the most interesting of the inhibitors from my point of view was introduced by Reinhard Jork, working with Hans-Jürgen Matthies in Magdeburg, in what was still then East Germany. Matthies' group, like ours, was interested in the glyco-proteins, which they had shown to increase in synthesis during various forms of more conventional training procedures in rats, and Jork had scanned the biochemical literature to find specific inhibitors of glycoprotein synthesis. He came up with a sugar called 2-deoxy-galactose (2-Dgal, which bears the same relationship to the sugar galactose as 2-DG does to glucose). 2-Dgal very specifically prevents the synthesis of those glycoproteins in which the molecule involves a link between the two sugars galactose and fucose; thus it blocks fucose incorporation into glycoproteins. In Magdeburg, he and Matthies found 2-Dgal injections produced amnesia in rats. I invited Reinhard to join me in some parallel experiments in chicks; to our delight 2-Dgal, injected either just before or up to a couple of hours after training, blocked fucose incorporation into the chick brain glycoproteins and produced amnesia in animals tested twenty-four hours later. So by both the inhibitor approach and the electroshock experiments, the synthesis of specific glycoproteins seems necessary for memory.[21]

*Criterion six – Biochemistry translates into neurophysiology**
Neurophysiology requires a set of skills – notably, apart from a dexterity in operating on small animals, a reasonable grasp of

---

*I do know that five comes before six. My reasons for postponing discussing the fifth criterion will become clear before long.

electronics – that are beyond the range of mere biochemists like me. My electrical capabilities go no further than being able to wire a plug – and even that we aren't (officially) allowed to do in the lab; the safety regulations demand that a qualified electrician carry out this skilled task. If I was to approach my sixth criterion I needed to find someone who knew their way round an oscilloscope, and it wasn't until the mid-eighties that a bright, if somewhat cranky scuba-diving enthusiast arrived on campus to do a PhD. Milton Keynes is about as far from the sea as it is possible to get in England and Roger Mason's motivation in coming to us was never quite clear to me (perhaps it wasn't to him either, for after four intensive years' research, Roger produced a many-hundred page 'draft' of his thesis, far in excess of what might be required, but finally failed to submit it for examination). Almost as soon as he arrived on campus, he disappeared into a mass of wiring, flashing lights, bleeping tones and tens of metres of multicoloured printouts on endless rolls of chartpaper. Even approaching his workzone was hazardous, as one had to pick one's way through suspended diving gear and disassembled bicycles over a floor awash with chartpaper.*
Nonetheless, after about eighteen months of technological purdah, he emerged, having resolved most of the equipment problems.

Essentially, what we proposed to do was straightforward. We would train chicks on water or methylanthranilate, and then anaesthetise

---

*Such a combination of domestic chaos and high tech. is very common in such labs. The disorder is so great as to astonish the uninitiated, who may well wonder that anything sensible can emerge from the mess. I once visited one of the world's leading PET-scan facilities. In this multimillion-pound complex, tiny samples of water or carbon compounds are placed at the heart of a cyclotron which fires vastly accelerated ion beams at them, creating very short-lived isotopes. The isotopes are pumped up through lead-shielded pipes to a radiochemistry laboratory in which complex syntheses are done in automated equipment designed to produce pure labelled organic chemicals from the fast-decaying isotopes within a few moments of arrival. From here the compounds are transferred to the room in which a subject lies with his head surrounded by the circular array of positron detectors, themselves coupled to bank after bank of computers. Entering this science-fiction world of flashing lights and mazes of piping, one puts on lab coats and disposable shoe-coverings, yet within the lab, benches are scattered with trivial chaos, the detritus of day-to-day life; spatulas and scissors are parked in disused honey jars, the chemists have sited their coffee-maker close to a shielded cabinet, and the cyclotron technician surreptitiously smokes. How can this living disorder coexist with the fantasy world of nanosecond measurements and picogram quantities?

them. The chicks would be maintained on a miniature life-support system in what is known as a stereotactic device. This is a way of holding an anaesthetised animal (by now of course not an animal at all but a 'preparation' which ceases even the pretence of life once the support system is switched off) gently but firmly so that its brain can be exposed and an electrode lowered into it, according to known coordinates. Such electrodes can be of several types; they can be fine glass tubes filled with solutions of drugs or salts to be pumped out close to specific cells, or they can be 'stimulating electrodes' designed to deliver trains of electric pulses – of the sort described in relation to LTP. Or they can be fine metal wires intended simply to record the electrical activity occurring in the cells in their vicinity. We were going to use the latter – extracellular recording electrodes. The question we would ask would be whether there was any difference in the electrical activity of the neurons in the IMHV as a result of training on the methylanthranilate bead.

This sounds straightforward enough, but masks a lot of difficulties. Finding the right anaesthetic, so that one can keep the chick functionally alive for the hours required for the recording is one that proved surprisingly tricky. Interpreting the electrical traces and distinguishing signals from noise is another. As it takes some time to get the system going and prepare the animal for recording, any one experiment can run for many hours, and as a result neuro-physiologists tend, even more than any other lab scientists I know, to be erratic nightworkers, and (at least when they are graduate students) not well cut out for normal social relations.

My part in the experiment was easy. All I had to do was to train chicks on either methylanthranilate or water and hand them over to Roger, who would disappear into the neurophysiology lab with them and emerge many hours later with reams of paper which he would begin to analyse. He was not told which chick was in which group until we had completed the entire first series of experiments (this is our normal lab practice – where possible, especially when two of us are involved in the experiment, to run the work 'blind' until after the data are analysed, to avoid the possibility of unconscious bias). By the time Roger had recorded from sixteen birds, he told me he thought he could detect regular differences between them so large that he could assign them to the two groups even without being given the code. When I challenged him to do so he was right in fourteen out of the sixteen.

What the traces showed was, as we would have anticipated, a steady background buzz of the spontaneous firing of the cells of the IMHV. But superimposed on this background were brief 'bursts' of high-frequency activity, in which whole ensembles of cells were firing in some sort of rhythmic synchrony. This bursting activity was massively – up to fourfold – higher in the methylanthranilate-trained animals than in the controls which had pecked the water bead. The increase in bursting seemed to continue for up to twelve hours after training. It was indeed a bit like an LTP effect, though generated not by the artificial injection of current, but by a behavioural experience.[22] In fact, the analogy may be precise; a couple of years later other researchers were able to produce LTP-like phenomena in IMHV slices stimulated in vitro.[23] To convince ourselves of the specificity of the bursting effect, Roger and I repeated the experiment using the electroshock amnesia approach

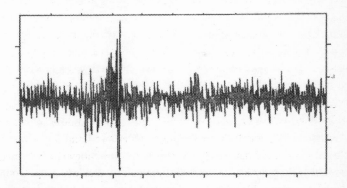

Fig. 10.10
Bursting activity in IMHV neurons
Oscilloscope trace of recording from a group of neurons in the IMHV of an anaesthetised chick. On the vertical axis, calibrations are at 50μvolt intervals; on the horizontal axis, at 20 msec intervals. The trace shows a busy background level of rather low voltage activity punctuated by a burst of high frequency firing of up to 300μv amplitude. This bursting activity increases dramatically on training.

I referred to above; the bursting activity, like the biochemical and structural changes, only occurs in the animals which remember the task.[24]

The end of the story?

Thus to form some sort of representation in the brain of the association between pecking the bead and the bitter taste, such as to result in a lasting change in the chick's behaviour, requires a biochemical cascade of events in a localised region of the forebrain. This cascade results in structural modifications to synapses and dendrites and is reflected in alterations in the electrical properties of the cells, as shown by changes in their spontaneous, rhythmic, electrical activity in the hours after training. Criterion 6 has apparently been met.

So have I solved the problem of how and where chicks make memories? Well, up to a point, Lord Copper, up to a point. But only up to a point. All this biochemistry and neurophysiology and structural change is beautiful; ten glorious years of experiments creating an apparent order out of the seeming chaos of the living world. I don't think I have been fooled by artefacts, or overinterpreted my findings, though it is obvious even to me, let alone a critical outsider, that in fitting the data within a temporal cascade I have not formally proved all the necessary biochemical links; some of my arguments have run dangerously close to falling into the classical trap of assuming that *post hoc* implies *propter hoc*; just because the phosphorylation step precedes the glycoprotein synthesis I cannot automatically assume that the latter depends upon the former. But, give me the benefit of the biochemical doubt, for that is perhaps not, except among professional biochemists, the most important question. Much more relevant is the issue of whether, even without biochemical caveat, memory can really be such a simple, mechanical process, a straightforward linking up of neurons into some novel network in the IMHV, like rewiring a computer? Does this mean that Hebb is right? Are the effects I find specific to the chick – or even merely specific to the young chick remembering about a bitter-tasting bead – or can I legitimately claim that they illustrate some general principles about the mechanisms of memory formation? And shouldn't I anyway be puzzled by the sheer size of the effects we find? Fourfold increases in bursting activity; 60 per cent changes in the numbers of dendritic spines – and all for remembering a little bead? If the chick is going to do this to remember everything that goes on during its lifetime, how will it find enough room in its little brain for all that synaptic machinery?

If I don't try myself to answer such questions, I could be sure that someone would certainly raise them. True, in what I actually do in the laboratory, I am trapped in an artefactual world mediated by machinery. I do not observe nature, as symbolised by my chicks, in an unmediated way. Like all scientific findings, mine are actually nothing but readings on meters, printouts on papers, numbers derived from machines (nothing but pointer-readings, the positivist philosopher and physicist Ernst Mach called such observations back at the beginning of the last century), which I manipulate to extract meaning and which I then endeavour to extrapolate back to stand for, to represent, deductions about the behaviour of molecules, cells and organisms in the real world. Nonetheless, I am fairly unmoved by the current debate in the philosophy and sociology of knowledge about the status of realism and of science. I stand by what I have written above; they are the truth about what I have observed concerning the material universe I study. Set up a lab like mine and run the same experiments, and anyone should be able to come up with the same results, for they do not depend on excessively mysterious skills or tricks, and science is after all, in the words of its most passionately admiring philosophers, public knowledge. But what I have described is interpreted truth – and the public which would share the knowledge is one which also shares my preconceptions (or at least enough of them) about how to interpret it. Nor is it yet the whole truth; presenting the experiments in the way I have chosen, as I said at the beginning of this chapter, has been a logical way to tell a story, even if a story not yet adequately grounded in theory. It has been a rhetorical device then, necessary because, my literary friends tell me, science proceeds by just such rhetoric. But rhetorical nonetheless. Let me open a new chapter, and tell another story.

Chapter 11

Order, chaos, order:
the fifth criterion

The fifth criterion

THROUGHOUT THE EIGHTIES, AS EXPERIMENT AFTER EXPERIMENT SEEMED to fit the cellular cascade so elegantly, I had been haunted by the thought of my fifth criterion:

Removal of the anatomical site at which the biochemical, cellular and physiological changes occur should interfere with the processes of memory formation and/or recall, depending on when, in relation to the training, the region is removed.

Haunted on three grounds; not only did I know that we would have to try to do the experiment, and this would require yet another set of technical skills that would be new to me; I was also inevitably anxious about what we would find; and in addition, for all the reasons discussed earlier, I have always been uneasy aesthetically, perhaps morally, and certainly in terms of the interpretation of any experimental findings, about making lesions. Nonetheless, I couldn't indefinitely avoid the question. In 1988, one of Gabriel Horn's ex-students, Ceri Davies, now working in London, published a paper stating that if he lesioned the left and right IMHV of chicks the day they were hatched, and trained them the next day on the passive avoidance task then, although they pecked the bead, showed the disgust response and in every other way appeared to show normal behaviour, they were nonetheless amnesic, and therefore pecked the bead when they were tested subsequently.[1]* Gabriel himself had already shown something similar with imprinting. This was all as one might have expected from the biochemistry, morphology and neurophysiology of the IMHV after training, but we needed to do the experiment more systematically. I decided to spend a good part of the years 1989 and 1990, during which I had a research grant which would buy me out from a lot of teaching and administrative responsibilities, tackling the question. What follows, then, is the story of these two years – and their aftermath.

What made it possible for us first to reproduce and then to build on Ceri's results was the arrival in the lab of two very disparate post-docs, both psychologists. Teresa (Terry) Patterson came straight from doing her PhD working on chicks with Mark Rosenzweig in Berkeley. Dave Gilbert had a somewhat chequered career since his degree at Birmingham and, research funds being so tight in Britain during the eighties, was at the time I found the funds to go ahead with the lesion work somewhat underemploying his psychology doctorate by working as a freelance windowcleaner. Terry and Dave clashed personally almost from the moment they arrived in the lab, but experimentally and conceptually they made a great team. Between us we designed the experiments so that one of them made the lesions, the second, not knowing which lesions had been made, trained the chicks, and a third, blind to both the former (generally me), tested them.

Lesion-making involves somewhat similar procedures to the

---

*It never fails to amaze me that one can lesion – or ablate; the words mean much the same – such relatively large regions of the chick brain and yet the animal will survive and continue to behave seemingly normally. Irreverent sceptics might wonder what all that brain was doing most of the time; chicken farmers might reply that its only function was to stop the animal running around excessively, as they have known since time immemorial, when the first farmer prepared the first chicken for the pot by cutting its head off.

neurophysiology. The chick is anaesthetised and an electrode (a fine wire) is lowered stereotactically into the brain, so that its tip arrives at a known site. A current is then applied to the electrode so as to raise the temperature at the tip. The heat (or radiofrequency) kills the cells in the vicinity, and the extent of the damage is controlled by varying the strength and duration of the current. The electrode is removed, the scalp stitched and the chick allowed to recover from the anaesthesia overnight. To control for the effects of the operation and the anaesthesia each group of birds contains some which have been 'sham-operated', that is, treated exactly as the others except that no current is passed through the electrode. The operations are delicate, but with skill it is possible to perform as many as a dozen during a working day. The chicks survive the operation and the anaesthesia well, and once they come round the birds with IMHV or LPO lesions seem quite normal, indistinguishable from sham-operated or normal controls. After the behavioural experiment, the animals are killed and the site of the lesion checked under the microscope. As with all our experiments, to be sure of the statistics it is necessary to repeat the entire protocol until we have collected twelve or more chicks in each treatment group. Granted four chick working days a week (the days when the birds hatch), and the need for the sham controls and for verifying the lesion sites, each set of results takes about three weeks to assemble; as something always goes wrong somewhere down the line – a bad hatch, a meeting to go to or whatever – say a month.*

Once we had spent the requisite number of months and made enough mistakes to have the lesioning procedure properly established, the first thing to do was to check Ceri's results. By early 1989 we had confirmed them. There was no doubt about it; chicks with bilateral IMHV lesions showed all the appearances of learning

---

*Long ago I learned by painful experience that when setting up an experiment, you should estimate the maximum time you think it should take, then double it and add a bit for luck – then you will be about right. The other crucial bit of such lab wisdom is that if you are trying to set up a method you've not used before, but just read about in someone else's paper, it tends not to work until at least the third time you try it; somehow and somewhere in the first two attempts you acquire that extra unarticulated fingertip magic ('tacit knowledge', the philosopher and one-time chemist Michael Polanyi called it) which makes the technique yours in practice as well as theory.

the passive avoidance, pecking at the bitter chrome bead on training just like the shams, shaking their heads in the 'disgust response', and backing away, but they simply couldn't remember it when they were tested a few hours later, and pecked the dry bead just as vigorously as those which had tasted it water-coated the first time. So the lesions had no effect on the birds' pecking behaviour, or their sense of taste, or general mobility; all they seemed to do was to stop the chicks remembering the avoidance. The next step was to go beyond what Ceri had done and make unilateral lesions. Again the results were just as we might have hoped. Chicks with left hemisphere IMHV lesions forgot the avoidance, chicks with right hemisphere lesions remembered without difficulty. So far, so good. Exactly as we would have predicted from all the earlier results, and in perfect accord with the fifth criterion, chicks needed their left but not their right IMHV to learn and to remember to avoid the bead. (See figure 11.1.)

Logically, the next thing to do was to check what happened if the lesions were made after the training. The earliest time at which one can do this is an hour later, otherwise, not surprisingly, the anaesthesia itself interferes with the memory formation. To our astonishment, even bilateral lesions made an hour after training did not produce amnesia. So, the first paradox; the chicks need their left IMHVs to acquire the memory for the avoidance, but once they've acquired it (or at least, within an hour of doing so), they don't need their IMHVs any more.[2]

So where has the memory gone? Granted all the earlier biochemical and morphological results, the obvious place to look was LPO. The experiment worked. Bilateral LPO lesions, made an hour after training, indeed resulted in amnesia, although unilateral ones, either of right or left LPO, were without effect. So perhaps in the normal course of events, the memory trace somehow migrates from the left IMHV to LPO after training, and that is why the post-training IMHV lesions don't produce amnesia? This would be a very satisfactory result, because it would explain why we find biochemical and morphological effects in LPO as well as IMHV. The fact that any one LPO, left or right, is sufficient to sustain the memory, so that unilateral lesions don't produce amnesia, would fit with the fact that

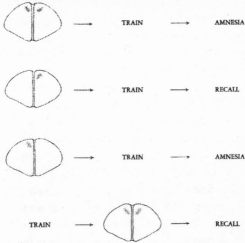

TRAIN → AMNESIA

TRAIN → RECALL

TRAIN → AMNESIA

TRAIN → RECALL

Fig. 11.1
IMHV lesions
In this and subsequent cartoons the shaded areas on the brain diagrams indicate the sites of lessions.

many of the changes we had found were in both left and right LPO, so in some way the 'representation' of the memory for the bead and the avoidance response must be present in both hemispheres.

Never content to leave a simple result alone, we pressed on. What happens if the LPO lesions are made before training? To our surprise, these pre-training lesions were without any effect on memory.[3] (See figure 11.2.)

If the chick doesn't have its LPO when it learns the avoidance, maybe the trace is reorganised in some other way? Perhaps in the absence of the LPO the memory trace simply gets stuck in the IMHV? If this were the case, then in chicks which are given pre-training LPO lesions (which don't produce amnesia), and then trained, post-training IMHV lesions (which also, in the otherwise normal animal, don't produce amnesia) should together summate to produce amnesia. And indeed they did (see figure 11.3). When we got this result, I assumed that it would be the left IMHV which would be vital; Dave bet me otherwise. We did the unilateral lesions, and I was wrong. In the animals without their LPO at the time of

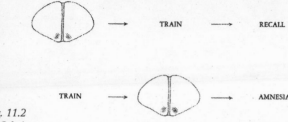

Fig. 11.2
LPO lesions

training, lesioning the left IMHV post-training was without effect; lesioning the right IMHV produced amnesia. The cartoon sequence of figure 11.3 summarises the results.

These results were intriguing. Thinking about them, I decided that everything could be explained on the basis of a slightly mechanical model in which an initial representation of the memory for the bead and the response is made in the left IMHV, and that in the hours after training the memory 'flows' first to the right IMHV and then to the left and right LPO. (See figure 11.4.)

I drew up this scheme purely on the basis of the logic of the lesion experiments; it is the type of drawing which horrifies neuroanatomists, as the obvious question that they ask is whether there are really any direct nervous pathways in the brain between

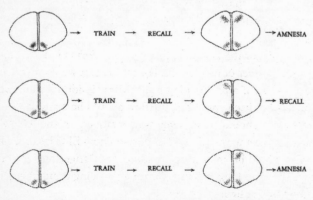

Fig. 11.3
Pre-training LPO lesions, post-training IMHV lesions

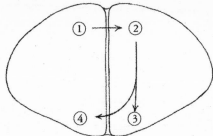

Fig. 11.4
Explaining the data: How memory might flow

IMHV and LPO. A simple 'connectionist' view of memory would have to insist on such direct links. At the time we started the experiments, we didn't know, but it seemed unlikely. We did know that all sorts of pathways from sensory systems, like vision and taste, converged on the IMHV. Indeed, there were even somewhat indirect routes from LPO to IMHV. This is why, after all, the IMHV had come to be regarded as some sort of 'association area' which could integrate many different types of incoming information, for instance, the pairing of the sight of the bead with its taste. By contrast the LPO lies on the 'output' side of the brain, part of the region which controls motor activities such as pecking, but also, perhaps, with 'emotional' types of response. Since we began, the neuroanatomy has become a little clearer; there really are no simple pathways from IMHV to LPO, though there is an indirect route, involving two synaptic connections, through another brain structure called the archistriatum.

Irrespective of the anatomy, this type of information flow model suggests a number of predictions. For instance, it would follow from figure 11.4 that if a pre-training right IMHV lesion is made (which we know is not itself amnestic), the flow should be disrupted, so the memory trace should be in some way 'stranded' in the left IMHV and unable to get to the LPO. Thus post-training LPO lesions, which otherwise would be amnestic, should now no longer be. No sooner thought (well, no longer than a month after being thought anyhow) than tested. (See figure 11.5.)

It worked, and Dave, Terry and I decided to write the definitive

Fig. 11.5

paper showing all the new set of lesions and their confirmation of the model. We quickly mapped out the draft, and just as we sat together to finalise it, one of them – I now can't remember who – suddenly said 'What if we simply make a pre-training right IMHV lesion?' We knew that such a lesion is not itself amnestic, and the model predicts that with such a right IMHV lesion, the memory trace should simply stay in the left IMHV, because there is no route by which it can escape. So in animals with a pre-training right IMHV lesion, a post-training left IMHV lesion should now produce amnesia. This would be the definitive test. We put the paper to one side, and spent the next month doing the lesions. Here's what happened.

Fig. 11.6
But the memory escapes . . .

I had been ready, like a scientific Poirot, to gather all the suspects into the drawing room and reveal the guilty, explaining, with tortuous ingenuity, just how the crime had been committed, how memory worked. Now I couldn't. The experiment was unambiguous – and entirely negative. Nothing; nix; nada; no amnesia. The memory trace couldn't get into the right IMHV, yet it clearly hadn't stayed in the left. Where had it gone?

More than half a century ago the psychologist Karl Lashley wrote a classic paper summing up the results of a decade or so of research on maze learning in rats. He had taught the rats to run complex mazes and then removed various chunks of cerebral cortex to see if he could find the locus of memory storage for the maze. To his surprise, he found there was no specific region whose ablation would completely obliterate the memory; rather it seemed as if the more cortex he took out, the worse the rats behaved, as if what mattered for the memory was simply the quantity of cortex present. On this he based his doctrine of what he called the 'equipotentiality' of the cortex. He called his paper 'In search of the engram', and concluded that memory resided simultaneously everywhere and nowhere in the brain.[4]

Lashley's experiments and his gloomy conclusions were superseded by later work, but the paradox of localisation remains. It serves to remind me, first, that in making IMHV and LPO lesions, we aren't studying the functions of those regions, but rather the functions of the rest of the brain, quick to reorganise itself in their absence. And second, that memory is not to be conceived of as static, 'residing' unambiguously in one site or small ensemble of cells, but must be present in a more dynamic and distributed form. Further, subtle is the brain, and full of devices. Prevent it achieving its goal by one route, and it will find another. If we block the preferred route from the IMHV to the LPO, the chick will find another. The brain doesn't function as a set of simply connected little boxes, but as a richly interacting, functional system. While we must take the neuroanatomy into account, we mustn't be confined by it, because it is clear that the brain is not! To insist on thinking mechanistically, on not heeding my own cautions about the follies of reductionism, means that the experiments are bound to end mired in paradox.

Double waves

The idea that when the normal chick learns there is some sort of flow of information, of memory, beginning in the left IMHV and then migrating to the LPO offers yet another paradox, but also suggests other experiments. The whole burden of the decade of biochemical and morphological work that occupied the last chapter was that, as a result of learning to avoid the bitter bead, there were lasting changes in the chemistry and structure of the cells of the left IMHV and of the LPO. Yet the lesion experiments had seemingly shown that once it has learned the chick doesn't need its left IMHV any more.

So why did the changes in the IMHV persist?

I brooded about this in the evenings, and plugged away with the biochemistry during the day. If the lesion studies were right, early after training there should be biochemical changes in the left IMHV, but as time went on they should begin to develop in the LPO as well. So as well as a flow of 'information' between IMHV and LPO there should be a flow of biochemical change. I had a sudden recollection of some odd results that the Magdeburg group had reported a decade or more back in rats trained on a brightness discrimination task. I checked back their papers and found I had remembered right. They had observed that, after training, there were two distinct waves of increased protein and glycoprotein synthesis, one immediately, and one about six hours, after training, the first in the hippocampus, the second in the cortex. At the time no one had quite known what to make of the data, and the Magdeburg people hadn't followed it up. One or two other bits of data that I'd heard at recent conferences, but hadn't made much sense at the time, began to float into the picture. How to test it?

In the biochem lab we were still struggling with the glycoprotein problem, using the inhibitor of glycoprotein synthesis 2-Dgal. Back when I was working with Reinhard Jork, I had shown that if the 2-Dgal was injected into the brain at any time from two hours before to about two hours after training, it produced amnesia. At three hours afterwards, though, the injection had no effect. I hadn't gone any further out with the time-course, as there hadn't seemed to be much point. But now I could see that it might be worth trying; I repeated the experiment, training chicks in the morning and then injecting separate groups with either 2-Dgal or, as controls, with saline, every hour for the next twelve. The next morning, twenty-four hours after training, I tested them. Figure 11.7 shows what happened.

The first part of the graph simply repeats what we had found earlier: injecting the 2-Dgal around the time of training produces amnesia; injecting it three hours or more later was without effect. *Until I came to the six-hour time point.* Injecting the 2-Dgal then produced a second amnestic wave. So there had to be *two* waves of glycoprotein synthesis in the hours after training, one immediately around the time of training, the other several hours later. Could the first wave be in the IMHV and the second in the LPO? Could it involve different glycoproteins? When, by the time my research leave had run out and I was finalising the text of the first edition of this book, I still didn't know the answer. Now, it looks as if the hunch was right. And since we did that experiment lots of other labs have found the same type of double wave in species as different as rats and fruitflies. But as the next chapter will show, other questions have become more urgent.

Meantime, the neurophysiology was in full flood. John Gigg, the graduate student who had replaced Roger Mason amid the welter of cables and oscilloscopes, now jazzed up with some rather more sophisticated computing gear which took much of the steam out of interpreting the data, was set to map the time course of the 'bursting' on the neurons in IMHV, and look at the LPO as well. John got the

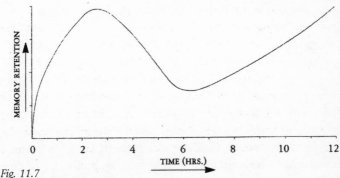

Fig. 11.7
The double wave of memory
In this experiment the amnestic agent 2-Dgal was injected into chicks at various times around the time of training (O) or later, and all the birds were tested for recall 24 hours after training. The birds were amnesic if the agent was injected up to 1–2 hours after training, or around 5–7 hours after training, but not if the injections were at intermediate times or later than 8 hours after training.

time-course of the bursting sorted out at about the same time as we were struggling to interpret the lesion data – and it seemed to fit. At the early times after training, the bursting is in both left and right hemispheres – but by about four to six hours after training the increase is mainly in the right IMHV and at the same time it is coming to a crescendo in both left and right LPO (figure 11.8). So there is a double wave of bursting too. With a bit of juggling the times of increased bursting in figure 11.8 can be superimposed on the times of the glycoprotein double wave of figure 11.7. It all seems to fit with the idea that at least in the normal chick there is a flow of memory from left to right IMHV and thence LPO.

Order out of chaos?

In any ongoing research programme, you tend to wake up one Monday morning with ideas for new experiments, rush off to the lab to start them, and find, weeks or months down the line, that they give ambiguous results which don't fit in with the general picture you have begun to develop. You can decide to dump the experiments, or to publish them in an obscure journal and hope that no one notices that they turn the pre-Raphaelite perfection of

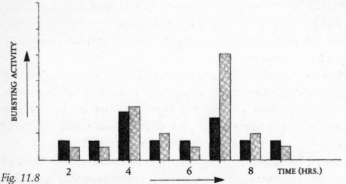

Fig. 11.8
The time-course of bursting
Histograms show increases in bursting activity in left (dark bars) and right (light bars) IMHV at various times after training. Note increase in both hemispheres at around 3-4 hours and much greater increase in right IMHV at 6-7 hours.

your theories into Jackson Pollockish disorder. Over the years the lab has accumulated my fair share of these. Suddenly, though, I have begun to see how they could fit into the picture that these double-wave experiments were revealing. For instance, if instead of training the chick on a bead dipped in the very bitter methylanthranilate, you use a much milder taste – say a weak solution of methylanthranilate, or quinine – the chick will show an avoidance, and remember it for a few hours, but then forget. Suppose that this 'weak' memory involves only a first wave of cellular activity, in the IMHV, which then doesn't get transferred by the more permanent second wave into the LPO? Such weak memories should trigger only the first wave of biochemical change, not the second, and indeed this turns out to be the case.[5] Indeed, it even turns out, from experiments based on the 'conditioned taste aversion' that I described way back in chapter 6, that the mere act of pecking at a dry bead will produce a 'first wave' of glycoprotein synthesis.

The reality is that the chick is all the time observing, noting, exploring, remembering aspects of its environment, because it has to begin with no way of knowing what is important to remember – just like the eidetic memory of human childhood. In the restricted environment of a rather featureless pen, showing it even a neutral tasting bead is a novelty of sufficient importance to be at least noted; the chick pecks, and makes its representation of the bead, probably in the IMHV. If later, as in the conditioned taste aversion experiment, the chick becomes mildly sick, it can only arrive at the sensible, if erroneous, conclusion that it was the bead it pecked that made it sick, if there is still some representation of the experience of pecking the bead somewhere in its brain. And we have shown, in a very simple experiment using the 2-Dgal inhibitor of glycoprotein synthesis, that it needs to be able to synthesise the first wave of glycoproteins to make this representation (figure 11.9).[6]

So far, so good. I can fit together a picture of shorter- and longer-term memory as located in different regions of the chick brain, with a flow of information between them and requiring two distinct waves of cellular activity. But I'm not yet out of the wood. Long-lasting memories leave long-lasting traces in the left IMHV, yet once the

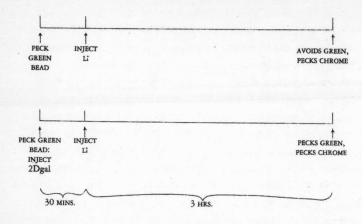

Fig. 11.9
Conditioned taste aversion experiment

memory has been made the chick seemingly doesn't need its left IMHV
any more to remember to avoid the bead. How can this make sense?

When I started to write the first edition of this book I didn't know.
I thought I would have to end this chapter with paradox, with having
turned order into chaos. My first draft concluded:

> This then is the story with which I wish to end the
> present chapter. Take your pick as to interpretation: on
> the one hand a set of experiments designed to meet
> rigorous criteria so as to demonstrate the necessary
> and sufficient cellular processes in defined regions of
> the chick brain which together create the representa-
> tion of a simple associative memory, as a result of which
> the chick's subsequent behaviour changes; on the
> other, a parallel account of memory as an elusive,
> dynamic phenomenon which is sufficiently important
> to the chick that it resists attempts to finally pin it
> down.

It wouldn't do. No one would be happy with that, not me, not
the publisher, not you as the reader. Surely I could do better.

Suddenly, as I was writing, the solution became so blindingly apparent that I could only be amazed at how deeply my thinking had become trapped into a mechanistically reductionist straitjacket. All along I have been talking about the chick pecking at a chrome bead as if this were a simple, unitary experience for the animal. Because we saw the bead as a single unit, just a chrome bead, we assumed the chick did too. But of course there is every reason why it should not. The chick tasting the bitter bead has no way of knowing what is important about it – could it be its colour, its size or shape, the time of day or orientation in which it is presented? Without ways of discriminating between these possibilities, if it is to survive, it has to consider them all. It cannot simply jump to the conclusion that only a bead of this particular configuration is important. So it has to classify what it has seen and experienced.

Suppose, then, that the important features of the bead are its colour, size and shape (there are actually good reasons for thinking that this is indeed the case, which I needn't go into here).[7] When the chick has learned about the bead, any one of these cues will subsequently allow it to avoid it. It is rather like trying to remember a name one has forgotten; you can try to recall it by thinking about the face of the person, by trying the sounds of names that you think are a bit like it, or by working systematically through the alphabet. Each of these is a strategy to seek a different cue as to the forgotten name.

If the chick indeed remembers not 'the bead' but a series of cues as to its characteristics, they don't have to be 'stored' in the brain in one unique site. Perhaps it remembers the bead's colour in one place, its shape in another and so on. We had simply never tested for this possibility; we had merely trained the chick on the chrome bead and tested it again with the same bead. Yet if the colour memory for the bead was stored in IMHV, and other characteristics in the LPO, this would explain why, after training, the chicks no longer need their IMHV to recall a chrome bead, because the other cues are still available via the LPO. Yet there would still be the lasting cellular changes in the IMHV, because they would be forming the 'colour representation' of the bead. We would have escaped Lashley's paradox.

Once I'd thought this through, it was simple enough to test. All we had to do was to train the chick on a bitter yellow bead, and then offer it a choice between the yellow and a novel blue bead. Normal chicks avoid the yellow, but go on pecking the blue. What would the lesioned birds do? If the colour representation was in the IMHV, taking out the LPO pre-training would have no effect, the bird would still learn and remember the discrimination, just as in the earlier experiment of figure 11.2. But taking out the IMHV post-training would now have an effect; the birds should learn the task, but when they were tested they should now avoid both the yellow and the blue beads, because without their IMHVs they would only have the memory cues for shape or size in the LPO, and on these cues they would have no way of discriminating between the beads – all small round objects should be avoided because they taste bitter.

By summer 1991, Terry and I had done this experiment.[8] It worked immaculately. We had found a way out of the reductionist trap and, by coming to respect the brain as an open learning system, had recreated order out of chaos.

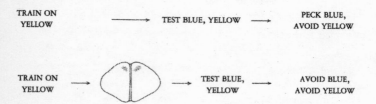

Fig. 11.10
Colour discrimination experiment

Chapter 12

The merits of velcro

TEN YEARS HAVE GONE BY SINCE I WROTE THE CLOSING WORDS TO THE last chapter, and the research world has moved on. Colleagues have left, new students, visitors and post-docs have passed through the lab, bringing fresh visions and insights into our central problem, and I have become ever more conscious, in 20/20 hindsight, of how fragile the certainties of the last two chapters now seem as new themes have gradually taken shape. Those that will dominate this chapter relate first, to the insufficiency of synapses; and second, to the most unexpected consequence of decades of work on the biochemistry of memory, pointers towards repairing the memories lost in the devastating condition of Alzheimer's disease.

Synapses are not enough

The idea that memory is encoded in stably reconfigured Hebb synapses took a beating, it is true, in the last chapter, but I never paused to consider just why and how some things are memorised, while others are not. And to understand this, it is important to recognise that the brain is not a passive information processor, simply receiving inputs from the outside world via the sense organs, selecting and storing them. Reducing the complex routes by which phenomena are perceived, interpreted, analysed and finally learned and remembered to the operation of fixed 'learning modules' innately programmed in a form of 'cognitive architecture' has become a fashionable claim, supported by cognitive neuroscientists and evolutionary psychologists alike, but for all the reasons discussed in chapter 4 (and see note 1) it simply won't do. In day-to-day life we don't approach the world around us as if we were solving crossword puzzles or playing chess. That strategy is fine in the lab, but we won't survive long outside it if we try to operate as if we were mere information processors.

Brains are embedded in bodies, and brain and body are in constant intercommunication. One direction of communication is obvious enough; neural pathways convey commands to muscles, 'control' our every movement. But the body answers back; the rich blood supply of the brain carries an array of signals in the form of hormones, as well as subtle interactions between the immune system and the brain. What and how we remember is not merely monitored but to some extent controlled by these signals, as many hormones cross into the brain and directly affect synaptic activity. Furthermore, the brain itself manufactures a host of signal molecules, neuromodulators and neurohormones. A formidable array of small peptides (growth factors) swim through the gaps between neurons and glia, directing axonal growth and shaping synapses. Another class are the neurosteroids – molecules closely related to other steroid hormones such as cortisol, oestrogen and progesterone.

It is in large part because of these interactions that brains and minds deal in meaning and not simply in information, and why, as I argued in chapter 4, brains are not simply computers made of carbon instead of silicon. For a crucial determinant of what is remembered in real life, as opposed to artificial laboratory situations, is the emotional content of an experience. An elegant experiment by Larry Cahill and Jim McGaugh, at Irvine, in California illustrates this beautifully (we later repeated a modified version of the experiment using the MEG imaging techniques I mentioned in chapter 5, and got similar results). They showed their subjects two series of illustrations, each of which told a similar story. In the first, a mother is taking her child to school, and on the way witnesses an accident. However, neither is hurt and the child arrives safely at school. In the second, the child is hurt in the accident and has to go not to school but to hospital. When Cahill and McGaugh tested their subjects later, the second, more emotional story, was perhaps unsurprisingly the better remembered. The neat follow-up was to repeat the experiment but this time to subjects who had been given the β-blocking drug propanolol, which blocks the effects of the hormone adrenaline and its neurotransmitter analogue noradrenaline. The drug had no effect on memory for the less emotional, more cognitive story, but now the emotion-laden one was remembered no better than the other.[2]

Emotional events produce a surge in hormones, including adrenaline and cortisol. Adrenaline, McGaugh's group has shown, affects neural processing via the amygdala, a brain region close to the hippocampus. While the hippocampus is important in cognitive and spatial learning, the amygdala is crucial in emotional memory – especially for aversive, unpleasant experiences, but also for pleasurable ones. Lesions or damage to particular regions of the amygdala diminish emotional memory, though perhaps not as dramatically as the effects of the hippocampal lesions I discussed in chapter 5. And the amygdala seems to be a key site at which hormones such as adrenaline, produced outside the brain, can modulate neural responses and hence memory.

By contrast, the primary site of interaction for steroid hormones like cortisol is the hippocampus, where the neurons carry receptor sites. Cortisol can bind to these receptor sites, acting as a stimulus for a chain reaction of internal biochemical processes within the cells. Cortisol is sometimes regarded as a stress hormone and its release from the adrenal glands is itself partially controlled by a signal molecule (adrenocorticotrophic hormone, ACTH) coming ultimately from the brain. Levels of cortisol fluctuate according to time of day, but are increased substantially under stressful conditions. Its effects in the brain, via hippocampal and other receptors, are varied, and depend markedly on concentration. Too little cortisol adversely affects neuronal development, but too much, as in chronic stress conditions, can result in premature death of hippocampal neurons.

I hadn't thought much about such interactions until the early 1990s. As a committed European, I had encouraged the European Science Foundation, a Strasbourg-based body, to set up a network of labs working on learning and memory. Our group became the centre, organising conferences, and a full-scale collaborative set of experiments and staff exchanges with four other groups, in Paris, Madrid, Magdeburg and Utrecht. It was under this scheme that our lab was illuminated by the arrival from Madrid of yet another young researcher with golden hands, Carmen Sandi, whose interest was in the effects of stress on learning and memory, though she hadn't worked with chicks before.* Checking through the literature, it turned out that the chick IMHV is packed with receptors for corticosterone – the analogue of cortisol in many non-human species. A drug company was marketing antagonists which could block these receptors, and we soon showed that in the presence of these blocking agents, or of drugs which inhibited corticosterone synthesis, chicks could learn the avoidance response, but would forget it again after a few hours – that is, they couldn't make longer-term memory.[3,4]

The implication is that in order for the pecking at the bitter bead to make a lasting impression on the chick, it is not enough for the bird to see the bead, peck and register the bitter taste. There had to be a whole body stressful response, resulting in corticosterone production which could then feed back into how the brain processed the bead-pecking experience. To check out this interpretation, we measured the levels of corticosterone in the blood, and, sure enough, within five minutes of pecking the bitter bead we found a transient surge of the circulating hormone.[5]

So, if levels of corticosterone help determine whether something

---

*Carmen, as elegant as she is talented and committed as a scientist, has gone on to make an impressive career for herself in Spain, and to write an important textbook on the effects of stress.

is remembered or not, maybe one could enhance memory by increasing corticosterone levels? One of the neat features of the bead-pecking model is that we can use it in two ways. The experiments I have described up till now involve the chick pecking at a bead dipped into the evil-tasting liquid methylanthranilate. This triggers the cascade which I described in the previous two chapters, an initial set of synaptic transients, activation of immediate early genes like fos, and then the 'second wave' of glycoprotein synthesis, some six hours downstream of the bead-peck. But suppose that instead of using such a powerfully unpleasant stimulus as the concentrated methylanthranilate, we used a weaker one – for example quinine, or dilute methylanthranilate? If one does this, the chick will peck, show a disgust response as before, but only remember the experience for about six to eight hours, after which it seems to forget. Ten per cent methylanthranilate may not be very pleasant, but it clearly isn't so nasty or toxic as to worry about for very long, so the memory fades. Train chicks on 10 per cent methylanthranilate, and similar early synaptic transients occur as with strong training, but the 'second wave' of glycoprotein synthesis doesn't occur; the cellular protein synthetic machinery simply isn't activated (see p. 325). Nor, when chicks peck the 10 per cent bead is there the same surge in corticosterone in the bloodstream. However, when we injected an optimum dose of corticosterone just before training on the weak stimulus, the experience seemed to be enhanced, for now the 'second wave' of glycoprotein synthesis occurred and chicks remembered a day later just as well as with strong training.[6]

So one can mimic the training experience physiologically. And, interestingly, we can mimic it behaviourally as well. Remember my description of the training procedure back in chapter 2? The young chicks are put into their pens in pairs, because if they are kept singly they get a little distressed. This provided the trigger for an elegant follow-up to our corticosterone story. Train the chicks on the weak stimulus as usual, and then separate them putting each into a different pen. Result: an increase in blood levels of corticosterone, and enhanced memory for the avoidance. And one final twist in the tail of this experiment: the young male chicks are more fearful than the females, their corticosterone levels are higher, and their memory is stronger![7]

These findings helped redirect part of my research. If we could manipulate the strength of memory by such external interventions, perhaps there might be scope to discover other substances that might function as memory enhancers – 'smart drugs' – as they are known colloquially. I had always been sceptical about the claims made for such substances, claims that pack the shelves of health food stores and fill internet websites with dubious products. Indeed, back in the 1980s I had done a thorough survey of the smart drug literature and products for the Consumers' Association, and found little stronger than snake oil available. But just maybe there could be something here? Corticosterone presumably functions by stimulating the biochemical pathways that lead to the synthesis of the glycoproteins that are involved in remodelling synapses. But other agents might act earlier, by stimulating the synaptic transients themselves. One substance claimed to act as a cognitive enhancer, if not the elixir of life, is the neurosteroid, structurally related both to corticosterone and to oestrogen, dehydroepiandrosterone (DHEA). We found that DHEA too strengthens memory for the weak learning task, and affects males and females differently.[8] In the first edition of this book I wrote that I had never found a sex difference in the chicks' behaviour. Well, I clearly hadn't looked hard enough, for these male/female differences now stared me in the face.

By contrast, other substances such as the secreted neuromodulator Brain-Derived Neurotrophic Factor (BDNF) affected memory retention in a sex-blind way.[9] The nice thing about such experiments is they make possible a neat double test of my criteria: blocking BDNF synthesis, or DHEA or corticosterone receptors, results in amnesia for strong training, whilst injecting corticosterone or DHEA, or BDNF, enhances memory for the weak training task. But I will come back to the implications of developing such memory enhancers later in this chapter. Just now I want to take up the biochemical story begun long ago in chapter 2 and in more detail in chapter 10, and trace it to its surprising conclusion: unexpected hope for a treatment for Alzheimer's disease.

Velcro molecules and Alzheimer's disease

Ten years ago, I finished the biochemistry section of chapter 10 with an expression of ignorance; memory formation seemed to need the synthesis of glycoproteins but I didn't know which of the many glycoproteins were involved or what function they might serve. The first years of the 1990s saw us painstakingly analysing complex mixtures of proteins, attempting to separate them. I won't bore you with the technology, but the upshot was that we were able to identify two of the crucial actors, called NCAM (Neural Cell Adhesion Molecule) and NG-CAM (Neuronal and Glial Cell Adhesion Molecule). Both these molecules are embedded in synaptic membranes, with a 'tail' that sticks into the interior of the cell (the cytosol) and a region that projects out of the membrane into the space which separates the pre- from the post-synaptic side. These extracellular domains, as they are called, are sequences of amino acids with sugars bound onto them (that's the glyco-addendum to the protein). The sugar units are sticky, and can clutch one another so that the extracellular domain of a cell adhesion molecule (CAM) protruding from the pre-synaptic side can stick to the matching region sticking out from the post-synaptic side. CAMs thus function a bit like velcro, holding the synapse together. The idea that they might have a central role to play in both helping to shape the nervous system during development and remoulding it in response to new experience dated back into the 1980s, and was developed by the immunologist Gerald Edelman and the molecular neurobiologist Jean-Pierre Changeaux, but we now had clear proof of their role in learning and memory.[10]

Blocking the synthesis of NCAM with the anti-sense technique I mentioned in chapter 10 resulted in amnesia in our chicks; they could learn but not remember for more than a few hours – the NCAM seemed to be important during the second wave of protein synthesis.[11] We were able to find an antibody which specifically bound to the external domain of NCAM, blocking its function. Injecting the antibody during that second wave period also resulted in amnesia. Our collaborators in other labs also found that if antibodies to NCAM were injected into rats about six hours after a training trial they would become amnesic.[12]

It was another collaborator, Ciaran Regan, working in Dublin, who developed most clearly the model of how the NCAMs might work their magic. He realised that NCAM, for instance, comes in a number of slightly different flavours. In particular, the juvenile form, present during development carries long polymeric sugar chains built of sialic acid (polysialic acid or PSA). The heavy polysialylation makes the NCAM slippery, so cells and synapses do not adhere. There is an enzyme, polysialylase, which removes the excess sialic acid, and in this, mature form, the NCAM achieves its Velcro-like quality. This happens during development as neurons migrate from the sites in the brain where they are born to their ultimate residence perhaps several centimetres distant. Ciaran proposed that during the process of forming long-term memory, mature NCAM could be sialylated once more, thus de-adhering pre- and post-synaptic sides and enabling synapses to grow or shift into new configurations. Then the NCAM would be desialylated again, fixing the synapses in their new patterns. He was able to show that in rats trained on a version of the passive avoidance task, such changes in sialylation really did occur,[13] and although we haven't proved it for the chicks, it seems a pretty convincing argument.

But of course nothing in biology is quite so simple. The adhesion molecules turn out to have other functions as well as sticking cells together. One has to do with helping signal molecules cross the cell membrane and enter the neuron. Advances in analysing protein structure led to the understanding that different regions of the external domains of the CAMs were involved in the different functions, so one was more involved in the signalling, another in the adherence. One of our collaborators, Melitta Schachner, then based in Zurich, gave us two artificially synthesised fragments derived from the external domain of Ng-CAM, designed to stick to and inactivate the different regions. When we looked at their effect on memory formation, we found that one produced amnesia only when injected at around the time of training, the other only when injected about six hours downstream of training. It looked as if we could begin a molecular dissection of the CAMs in relation to their role in memory formation.[14]

But while we were happily at work decoding the molecular

mechanisms involved in remodelling synapses, progress in other areas of neuroscience and medicine continued to impinge on my thinking. In particular, more and more evidence was accruing concerning the cascade of molecular events that leads to the widespread death of neurons that is the consequence of Alzheimer's disease (AD). By the 1990s, the increasing world-wide incidence of Alzheimer's disease, largely as a consequence of people living longer, was being brought loud and clear to the attention of politicians and medical research funders. AD is a progressive disease of which the early diagnostic signs are lapses of memory, associated with apathy, depression and agitation. As the disease progresses the neurons become filled with 'tangles' of fibres of a protein called tau, and the spaces between them with 'amyloid plaques', – masses of congealed fragments of a peptide called beta-amyloid. Soon neurons, especially in the hippocampus, begin to die, the brain shrinks and a slow but seemingly inevitable deterioration sets in. There is a passionate debate among researchers on AD as whether 'the' primary problem in the disease is the tangles ('tauists') or beta-amyloid ('βAPPtists')

There is no single cause of AD; a few years ago there was a flurry of interest in the possibility that excess aluminium was a problem, because the amyloid plaques contained high concentrations of the metal, and many of us traded in our aluminium cooking pans; but it turns out that the plaques precede the accumulation, so it couldn't be the cause. There is a also small number of very early onset cases associated with a specific genetic defect. Advances in human genetic analysis also led to the identification of a few genetic risk factors – that is genetic modifications that increase the chance of getting the disease. And there are environmental risk factors, like concussion. Furthermore, there is a sort of 'use it or lose it' factor – the more educated you are and the more intellectually demanding your life is, the less likely you are to get Alzheimer's. But the major predictor is age, followed by sex – women are more likely to suffer from the disease than men, even when age is factored out. Oestrogen is mildly protective against the onset of AD (perhaps because of its relationship to the neurosteroids I mentioned above), and of course oestrogen levels decrease dramatically in women post-menopause.

The fact that hippocampal cells are affected early may be one reason why memory loss is one of the earliest features of AD, and also led to the early attempts at therapy. One of the major neurotransmitters in these cells is acetylcholine, and nearly all the drugs presently licensed for treating the disease are aimed at boosting the levels of this transmitter by blocking one of the enzymes that breaks it down. Unfortunately they aren't very effective and people often have strong adverse reactions to them. But the intense focus of the genetic and biochemical studies of the last decade have provided some clues as to the molecular sequence of events that leads to the accumulation of the amyloid plaques. The broken peptide fragments of beta-amyloid are derived from a protein called, unsurprisingly, the amyloid precursor protein, APP. Like the CAMs, APP is a transmembrane protein, present in synapses, with an intracellular tail and a long extracellular region (Figure 12.1). APP is rapidly synthesised and broken down in day-to-day life. The enzymes responsible for the breakdown cut off portions of the extracellular region into peptide fragments, and this is clearly important in the normal functioning of the APP as the fragments seem to have signalling functions. However, in some cases the enzymes cut the protein in the wrong place, producing the 42-amino-acid long beta-amyloid, which then congeals into the plaques.

So when, towards the end of the 1990s, I had begun to clarify the role of the CAMs in memory formation in our chicks, the neuroscience world was buzzing with questions about the biochemical cascade which led to the formation of the plaques. So what might be the normal function of APP? Like the CAMs it was an adhesion molecule and had some signalling functions. I wondered if it were necessary for memory formation. By this time my closest collaborator in the lab was a molecular biologist, originally from Belgrade, Radmila (Buca) Mileusnic. Buca and I had worked together on and off since we first met in the late 1970s, and she had made regular brief research visits to our group. As former Yugoslavia disintegrated and the Milosevic-led Serbian military atrocities accumulated, she and her husband, as members of the small Serbian opposition movement found life increasingly intolerable, and Buca threw up her Belgrade

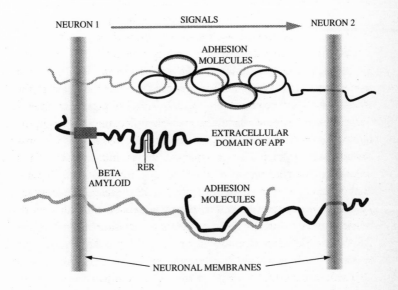

Fig. 12.1
Cell adhesion molecules link pre- and post-synaptic neurons. The figure also shows the amyloid precursor protein APP, and the location of the beta-amyloid segment which forms plaques in Alzheimer's disease and the small peptide segment RER which we have found to rescue memory.

professorship and came to Milton Keynes, first on a grant and later as a permanent member of the academic staff, and we were working together on the CAMs. We decided to see what would happen if we blocked APP function with antibodies. The answer – early onset amnesia. The same happened if we stopped APP from being synthesised, using anti-sense. Normal APP function is therefore necessary for conversion of short- to longer-term memory. Could this be the cause of the memory loss in AD?[15]

An exciting finding but perhaps not altogether surprising, as others had found not dissimilar effects. But then I began wondering whether it might be possible actually to rescue the memory loss. Different parts of the extracellular domain of APP are known to have different functions – in particular there is a stretch, a few amino acids long, which is known to promote growth of dendrites in cell cultures. Little peptides derived from this sequence are commercially available. I am not quite sure now quite why I focused on this particular possibility – perhaps it was just a wild shot experiment – but we bought some of the pentapeptide RERMS (Arginine-Glutamate-Arginine-Methionine-Serine). We blocked APP synthesis with the anti-sense and injected RERMS just before or just after training. As if by magic, memory was restored. In another experiment, we injected beta-amyloid, which results in amnesia, and again RERMS rescued the memory. We trained the chicks on the weak, 10 per cent methylanthranilate task, and RERMS boosted memory just as we had found with BDNF and corticosterone earlier.

We were worried. How specific was this magic molecule? We made the reverse peptide sequence, SMRER as a control. To our embarrassment, it too restored memory. Fortunately a different sequence, RSAER, was without effect. Why did both forward and reverse peptide sequences work though? Look at the sequences, and it becomes obvious. Both contain the tripeptide sequence RER, a palindrome, that reads the same forwards and backwards. And as I write these sentences in October 2002, we have shown that RER is indeed the active sequence. This tiny peptide binds to the neuronal membranes, substitutes for the missing APP, and prevents beta-amyloid from doing its evil work.

Could it be that after decades of studying the cellular and molecular processes of memory formation as an exercise in fundamental science, of increasing human knowledge about how brains work, that we may have stumbled on a clue as to a possible treatment for AD? Too early to tell, for there is a lot that needs to be done to turn a laboratory finding into potential clinical use. We need to show that RER is effective in other memory models – in for instance rats or mice. We need to be able to 'protect' the peptide so that it will not break down but can be administered peripherally – by injection or inhalation – a drug that you have to inject directly into the brain is not much use! We've started along this road and things are looking promising. If all that works there would be the long task – not for us but for some pharmaceutical company – of all the necessary toxicity and clinical trials. I will give the latest update on all of this when this book is in proof – but if it works at all, it doesn't seem a bad way to end a research career, and my chicks will finally have brought some hope to people suffering from an appalling disease.

Chapter 13

Interlude: laboratories are not enough

IT NEARLY ALWAYS STARTS THE SAME WAY. AN EARLY MORNING FLIGHT from Heathrow leaves you, several time zones later, standing outside a strange airport, in disorienting late afternoon sunshine, waiting for a taxi – or if you are lucky the labelled coach – to the conference hotel. The feeling of gloom which you have been trying to suppress ever since you boarded the plane that morning is beginning to turn to anxiety. Why have you come? You won't know anyone; the programme which looked so exciting when you got the first announcement nearly a year ago now seems, when you come to look at it more closely on the plane, stale in the subjects you know about and incomprehensible in those you don't. The conference location, which appeared exotic in the travel brochure, is a noisy city in the heat of summer or a plastic holiday resort slightly out of season so as to get good hotel rates. And anyhow your timetable looks to be so crowded that there won't be time to see anything outside the meeting.

Check in at the hotel reception, dump your bags in a bedroom which could have been cloned from any one of the half-dozen or so other meeting venues you've already hit this year, and make for the Conference Centre and Registration. If you are in the United States, it's likely to be going on in the same hotel you are staying at; if in mainland Europe, a custom-built Palais de Congrès; if you are out of luck and this year's meeting is United Kingdom-based, it will be a rather run-down university block and your cloned hotel will have downgraded itself into a distinctly tatty student hall of residence. In any event you will queue in front of makeshift trestle tables behind which harassed secretaries and students drafted in for the occasion (or at the plusher events professional conference organisers neatly turned out in car rental uniforms) search through their documents to check if you are an invited speaker or have paid your registration fee, and finally present you with a stuffed plastic briefcase, courtesy of some drug company or local Syndicat d'Initiative.

The next wise move is to sit down somewhere and unload the briefcase of its inevitable complimentary copies of unwanted journals and advertising brochures, leaving yourself just the meeting programme and the book of abstracts. Mostly one is too mean to throw away the spare pads of paper and pens embossed with the insignia of biochemical manufacturers, so they get hoarded until you return home, when they will eventually get lost or ditched. Interleaved among all this dead weight for recycling will be a name tag and a clutch of envelopes containing tickets or heavily gilt invitations for the receptions, dinners and private junketing that accompany all such jamborees. Inevitably, the first of these events, the conference mixer or opening party, will be starting almost immediately, so skip back to the hotel, drop the briefcase, unpack your toothbrush because you'll be too knackered to do it later (but don't put on a clean shirt; you will need one tomorrow and the standards of sartorial elegance at these events are no higher than the quality of the canapés), pin your name tag on your lapel, and head for the drinks.

Your earlier feelings of gloom tinged with anxiety will by now have expanded into the blind panic of the launch into the unknown, but suppress all such sensations; you are engaged in a crucial part of the scientific enterprise. It is easy to see this as no more than the academic tourism which Ashley Montagu once characterised as the 'leisure of the theoried classes', and which has formed the backdrop to dozens of campus-based novels ever since. But there is more to it than that. Within minutes of joining the party in its ornate city hall, hotel ballroom or poolside venue you will no longer be hunting distractedly for a glass of mediocre wine and peering anxiously at illegible badges to discover at least someone whose name you vaguely recognise, but talking intimately with people you saw at the previous congress, and the one before that, who on this basis alone qualify as old friends. You've begun the business of the meeting, and what happens at the parties and receptions is as much part of science as the imparting of the latest research findings in the lecture theatres.

Internalism and externalism

It is with this business of science – science as a social activity – that the present chapter is concerned. Looking back on this book as it has developed up till now, I find that, having begun by trying to unpick my own biography as a way of understanding the many meanings of research on 'memory' within the laboratory, my text has moved firmly back into the cognitive domain, a detective story of the hunt for the cellular processes of memory which has perforce become stripped of much of its social framework. There is nothing in my account to which a strict Popperian, still less a Kuhnian philosopher of science could take exception. After all, much of it has been presented in almost classical Popperian vein; research as a matter of conjectures and refutations, of setting up hypotheses and testing them. There has even, for Kuhnians, been a paradigm within which I have been working – that of memory as coded for by the strengthening of synaptic connections according to Hebbian rules – and, in the last experiments, a paradox that seems in a very small way to foreshadow what Kuhn would regard as a 'revolution' in science – that is, a phenomenon for which the old theory could not account.*

Classical philosophy of science would be happiest to leave the account there, and so, in some ways would I. It is of course so much easier, as a working researcher, to tell the story like this; it permits

---

*In its purest form, now admittedly superseded even by Karl Popper himself, and certainly by his followers, science was seen as a matter of the making and testing of hypotheses. No such test could ever prove a hypothesis true in some absolute sense; however, it could falsify the hypothesis. Crucial experiments – those that characterise good science

a modicum of vanity, allows for the understated heroism of scientific achievement, and avoids social difficulties. The problems that I have so far permitted myself to identify have been essentially internal ones, those of false clues or inadequate experiments – and even most of these became sanitised away in the ultimately progressivist account of the last three chapters. The problems with which my self-portrait has shown me struggling have been of two kinds – conceptual and technological. The first, the conceptual ones, involve the week-by-week wrestling with the sheer intractability of the biological phenomena I am studying, the inherent cussedness of nature, the need to judge whether an experiment which gives a result I hadn't expected is a freak to be discounted, or is a genuine signal from the world outside demonstrating that my simple models are fatally flawed. To the outsider, it is these conceptual problems which

– are for Popperians those which, if a hypothesis fails to pass, mean that it must be abandoned. Thomas Kuhn, in his famous book *The Structure of Scientific Revolutions*, first published in 1962, challenged this Popperian view. Most normal science, he argued, is not about testing hypotheses but merely about puzzle-solving within a given understanding of the world – an understanding that Kuhn called a paradigm. Further, if research results appear to contradict a hypothesis, it isn't usually abandoned – the results or their interpretation may be challenged, or the hypothesis may be modified in order to accommodate seeming anomalies. However, the time comes when the accumulation of anomalies becomes too much for a hypothesis to bear, and a period of 'revolutionary' science ensues in which a new hypothesis is formulated which better explains both the old and the new anomalous data. Kuhn's view of science became especially popular amongst sociologists of science, because it opened a door for the 'social' to enter the natural sciences. What made a paradigm attractive, for instance; and why did scientific revolutions occur? Kuhn left such questions as internal to science, or perhaps to the psychology of the individual scientist, but the sociologists could see that there was more to it than that, and that the social order external to the laboratory might influence what went on inside. This enabled them to recapture the insights into the social functions and relations of science that an earlier generation of Marxist scientists had already found and it presaged their eventual move into a hyper-reflexive stratosphere within which scientific knowledge of the natural world itself dissolved into a miasma of social relations.[1] Philosophers accepted that Kuhn had outpointed Popper, but sought an alternative way that avoided his sociological conclusions and retained the validity of the internalist account of science. In particular, Imre Lakatos suggested that science proceeded by way of 'research programmes'. A programme could be progressive, in which case it went on suggesting puzzles that could be solved within it and seemed to remain fertile of insights about the world, or it could be degenerate, in which case it threw up increasing anomalies. In this sense the Hebbian, associationist programme is progressive in that it can accommodate both Kandel's cellular alphabets of memory and long-term potentiation as a model mechanism, but the Garcia experiments and my lesion studies, described in chapter 11, are anomalies which may mean that the Hebbian programme is becoming degenerate, or, in Kuhnian terms, we need a revolution.

are likely to seem the most interesting. But for those of us directly engaged in making the experiments it is the second class of problems – the technological ones – with which we find ourselves struggling most frequently in everyday practice. What we want to measure is at or below the sensitivity and resolving power of the equipment we have; an assay simply doesn't exist and we have to invent it; more mundanely, the damned cryostat has broken down again and we can't get a service engineer to fix it till next week, or there aren't enough eggs in today's hatch to run the experiment we planned. The conceptual engagement has to wait for the seminar discussion or for when we finally write the paper describing our results.

To go further than this in exploring the non-cognitive aspects of science runs the risk of transgressing the bounds of decency, of moving into the terrain inhabited by sociologists of science, who can examine natural scientists' practices critically but with the degree of comfort afforded by the fact that they are talking about other people's work, not their own. Bruno Latour has become the archetypical figure here, in such books as *Laboratory Life*, and *Science in Action*.[2] His deconstructionist accounts of science began with his experience, as a post-doctoral anthropologist, spending a year as a partially participant observer in a Californian laboratory working on the identification and isolation of a neurohormone. The account he gave of this in his first book, and the theoretical extrapolations that have followed, have influenced a generation of sociologists of science.*

To move into this territory would present few problems if I were intending voluntarily to retire from lab work, but I'm not quite ready for this yet. I have to try to tread the dangerous edge of reflecting on my own practice without thereby disqualifying myself from continuing it. To confine my account too sharply would still be to

---

*It has been left to Hilary Rose to point out the extent to which, by failing to subject his own methods to critical analysis, Latour fails to carry through his project, despite the scale of what she has described as the 'Latourisation' of the social studies of science.[3] For sure the whole field of the social studies of science has moved on since I wrote these footnotes, just as has my own, but to go further in discussing it would take me far outside my current themes.

miss the social nature of science in action which is one major part of my concern in this book.

The conference with which this chapter began is but one of the half dozen or more I will go to in a year. They may be the biggies in which a session on learning and memory is but a single half-day symposium among many other simultaneous sessions during the week-long programme, or specialist intensive weekends for twenty-five people focused on a single theme. Within neuroscience the memory mafia – the key players worldwide – probably amounts to some couple of hundred in all, with perhaps a hundredfold wider penumbra of interested observers (when in the 1990s I compiled a register of all the labs working in the field in Europe, I counted 93). The small size of the key group means that we are likely to run across each other several times a year at successive meetings.

Why go? Granted that all research is eventually published, and one can read the experimental details more easily than listen to someone lecturing about their work, why not just stay at home and read the journals? It is not simply the lure of foreign travel, though most of us are likely to respond more eagerly to an invitation to a meeting in Nice in May than to one in Glasgow in November.[4] Partly it is of course to keep up with the gossip of our particular global village; the hatched, matched and dispatched of our profession. To be missing from a meeting is to have ceded your place within the charmed circle, to run the risk of falling out of fashion, as serious a fate for the ambitious researcher as for the would-be member of the glitterati who misses a party in London, Paris or New York.

But most seriously because, despite all the marvels of electronic publishing, from the time when a research paper is submitted to the time it is printed will be at least eight months. Then of course one has to find it – to notice the significant title from a quick scan of the contents pages of the journals in a visit to the library (or much more likely these days, through a keyword search on the net or an online journal subscription – library visits are becoming as redolent of ancient, more leisurely times as are fountain pens and manual typewriters and most students seem rarely to abandon their computer screens). So rapidly does research move on that, for those in the know, the result may be already almost obsolete by the time it appears (not for nothing is the scientific literature referred to as archival). Going to the conference, hearing the results first-hand and talking direct with the person who has made the experiment short-circuits that process. Reading a paper is always something of an abstraction – can you really believe the experiments reported in the dry formalisms which the scientific style demands? Talking direct with the person who has done the research (and still better, drinking with them) is a wholly different experience – it may increase your confidence in their results – or it may sharply diminish it. Like the difference between chess and poker, it is yet another demonstration that to regard research as a purely cognitive activity is to mistake the ideology of science for the reality of what we do. As a result it is a rare meeting from which I don't return to the lab with a batch of ideas, both about new experiments to do, but also about ones I might have been contemplating but now know I shouldn't bother with.

There are two lessons from this story, and both point to aspects of the competitive nature of the way science is done. It might seem surprising that knowledge of the natural world can become obsolete. Yet it may happen simply because the experiments which the published paper suggest to you have already been done by someone else, or a new and improved technique developed, within the intervening eight months, so that working with the old technique or redoing the already done experiments won't interest anyone. It may even be that by the time the paper appears its data have been disproved. In science, we are continually being told, there are no prizes for coming second. Of course, this attitude is – or ought to be – a bit silly, because there are enough experiments to be done, enough to find out about the world, to keep us all busy and cooperative indefinitely far into the future. But the competitive pressure, to be first, to publish, to be recognised, to lead rather than follow fashion, is very powerful in the grant-dependent, soft money world of Western science, with most researchers employed on short-term contracts and great pressure exerted for quantifiable results –

that is, measurable numbers of papers published – within a four-year grant cycle.*

Such pressures can even affect the work of the conferences themselves; in very active research fields people can be reluctant to discuss their own work in any detail until it has been safely written up and accepted for publication by a journal in case someone hears what they are doing and steals a march by rushing off and repeating the experiment (journals try to diminish such priority disputes by announcing the dates at which a paper has been received and accepted for publication). It has become sufficiently common practice to take a camera to meetings and photograph the slides or the posters for some conference organisers to impose a camera ban.

The second implication is even more important. The charmed circle of the memory mafia is drawn overwhelmingly from North America, Europe and Japan. Would-be researchers in peripheral labs in Latin America or India cannot attend the conferences, and they receive the journals, even when they can afford them, irregularly and late. Thus they are condemned always to follow, never to lead, the fashion. In science as in industry the production processes of metropolitan capitalism, by which knowledge becomes a commodity whose value diminishes with time and use, constantly marginalise the third-world research economies.[6]

A similar failure of the production system of science underlay the relative inadequacy of the research enterprise in the command economies of eastern Europe and the Soviet Union prior to the collapse of 'actually existing socialism' in the late 1980s. Travel was made difficult if not impossible, so that the exchange of ideas at conferences between researchers from East and West was virtually prohibited; there were few subscriptions to Western scientific journals

---

*The increasing numbers of examples of fraud in science, exhaustively catalogued by Congressional committees in the United States stand as evidence to the extent and consequences of this pressure, but also in Europe. A classic example is the Dingell hearings which in 1991 led to the Nobel Laureate and Rockefeller President David Baltimore being forced publicly to withdraw a paper he had co-authored five years previously because the forensic evidence conclusively demonstrated that the lab books on which it was based had been tampered with to give misleading data. Philip Kitcher has exhaustively chronicled the whole wretched affair.[5]

and some (such as *Science* and *Nature*) were censored before distribution. And the business of rigidly centralised planning of research prevented the inspired laboratory anarchy which has been so instrumental in the rapid development of biochemically based research in the last decades. If I come back from a conference with the idea for an experiment requiring some new chemical, drug or isotope, I can telephone a supplier and have the substance in the lab within forty-eight hours, in time to start the experiment while the idea is still hot. The atrophied command economies of eastern Europe and the Soviet Union seemed to treat doing science rather as they did heavy industry, and made such flexibility impossible; experiments, and their chemical and equipment requirements, had to be specified far in advance, orders could take months to fulfil and the rapid shifts of gear and approach which make for flexible research have been impossible.

Facts and resources

So, going to the conferences, keeping in the know and ahead of the game, has become as integral a part of doing successful science as the experiments themselves. As Latour argues, the 'work' of research takes many different forms. The establishing of a scientific 'fact' is not just a matter of designing and running experiments; these are but part of a complex process which involves activities both upstream and downstream of the laboratory bench. To make an experiment presupposes a functioning laboratory, equipped with machinery, technical staff, the funds to buy laboratory consumables. The researcher at the bench, post-doc or student, takes these for granted as part of the necessary research environment, the upstream element of which it is someone else's responsibility to provide so that the researcher, who is the 'real scientist', can get on with the work of 'doing' the science. If you run short of a reagent or isotope, you reorder it; if a machine breaks down, you ask the lab superintendent to call the service engineer; if you need a small piece of apparatus made, you cajole the electronics workshop along the corridor from the lab. None of this is 'doing science'; it is part of the irritating necessity of preparing yourself for the experiment – the real thing.

For the laboratory chief, however, establishing and maintaining this research environment becomes practically a full-time job, in which, Latour argues, the student or post-doc may be no more than just another part of the necessary lab equipment. From this perspective it is the chief who is doing the 'work' of science. Thus, whether what constitutes science is the work at the bench or the organising of the resources and the setting of the programme, all depends on your point of view. More and more, since the first edition of this book appeared, my own time is taken up with money, an almost obsessive issue for most British-based researchers these days. How to find the funds for the salaries of the post-docs, the grants to the students, the capital costs of new equipment, the consumable budget to buy the isotopes and the reagents. Designing an experiment means asking whether we can afford it as well as whether it will give us the answers we seek (this is part of the down side of the market approach which eastern Europeans are slowly beginning to recognise as they recoil from their command economies). The money for my chicks comes from a mix of places: the university; the United Kingdom and European Research Councils; exchange programmes. There are charitable trusts, drug companies, the military.*

---

*The military remain major players in the funding game. Their interests are very different from those of the Research Councils, Foundations or drug companies. The United States Airforce and Navy and ARPA have an ongoing enthusiasm for neural modelling, the computer games of connectionism. Especially because of the current concerns with allegations concerning Iraqi chemical and biological weapons, and the anthrax scares in the United States, all branches of the United States military (and to a lesser extent the British), are interested in neural mechanisms and the action of neurotransmitters for what they can reveal about actual and potential chemical weapons – and they are not small spenders. I remain, however, unwilling to take such money. It isn't that my research is of no military interest. Nor is it a simple statement about morality, for I am neither a pacifist nor excessively naïve. If the military want the products of my research, they can obtain it from the published literature whether they pay me or not. If you live in a militarised society, as we do, this is hard to avoid. However, if you want to be credible as a public critic of that militarism, better not take their cash at the same time. Many neuroscientists who do accept military money take it with the argument that because what they do is basic research it is of no real military relevance and cannot be used directly – for instance to develop new types of chemical or biological weapons. People believe, of course, what they want to believe, but I am inclined to think that military contractors know what they are about.[7]

Every grant has to be applied for separately, budgeted separately; a lab runs like a small business and a major portion of my time is spent trying to raise the money and manage it once I get it. So is doing all this grant writing and organising not science at all? Ask most people employed as 'scientists' and they will say no – the ideology demands that real science is limited to what you do at the bench or with your animals, the rest is a necessary evil. Latour wants to rescue the scientific manager from such unceremonious ideological dumping. But I notice that I didn't choose to describe a day at the computer or on a committee to begin this book with – and I only sing in the lab.

All this helps explain why to tell the chick story of the last chapters in its Popperian mode would be to misunderstand a fundamental aspect of scientific work. I cannot pretend that we can effect all things possible, that there are no financial limits on what we do or that the little question of how we can raise the money does not constantly constrain our choice of experimental question and technique. Much of what we – what I – do is at best inspired adhoc-ery. Certainly, I have a game plan; my obsession with memory has continued for far too many years now for that not to be the case. But it is a plan which is constantly being diverted by what we can and what we can't get money for, by new technological possibilities and their limitations, by chance reading of papers in someone else's field, by the rush of ideas to the head that comes from going to a worthwhile conference, and above all by the recruitment of colleagues, visitors to the lab, students and post-docs who come with their own ideas and experience to feed into the work of the group. The order that the last three chapters imposed is in some ways a historian's order, describing a battle after it has been won and lost. It doesn't feel like that in the thick of it, though it may read that way later in the textbooks. And when I wrote the first two I had no idea that within ten years our work would have turned into a hunt for a potential treatment for Alzheimer's. Indeed rereading some of the drafts from that first edition, I can see how sceptical I then was about such a prospect. History, just as much as scientific papers, is in this sense a fraud – or at least an art-form, and finding the right funding is as important and integral a part of doing successful science as is finding the right organism.

Spreading the word

These, then, are the upstream elements of the research; the *matériel* I must assemble before I can even contemplate running an experiment. Downstream, the messages our chicks give us must be incorporated into a package of knowledge that bears our stamp, is identified as the work of our lab. Packages are posters, papers, reviews, conference lectures, invited seminars – sometimes even books.* Latour, in his discussion of how scientific facts come to be established, repeatedly adopts military metaphors. All those necessary features of the research environment described above, from the funding of the research and the development of technologies powerful enough to achieve the desired goals, through to the publication and dissemination of results, he describes as the 'allies' that need to be recruited before the 'fact' becomes a fact. Whether one likes his metaphor or not, he is absolutely right that the job of doing science is not complete without the paperwork. Just to have discovered something for oneself is not enough. The public nature of science demands that the discovery be made part of the common stock of the literature of science, but the vast size of that literature means that it is not enough simply to launch one's small piece of knowledge into the wide world and wait patiently for acclaim to follow; modern marketing methods demand more sophisticated approaches if it is not to sink without trace.

The simplest and weakest form of package is a poster at a conference. Most conferences, as well as the plenary lectures and symposia included in their programmes, have such sessions, in which you are given a 1.5 x 1.5 metre pinboard and a designated time at

---

* Does this book count? I don't know; I would want it to, but I suspect that as a form of scientific communication it is perhaps a sort of meta-package. Even today professional scientists remain a bit sceptical about the status of attempts to make our work accessible outside the charmed circle.

which your poster can be displayed. On it, you can put up a few sheets describing the aims of your experiment, some of the data and its implications. Granted that yours will be one of hundreds – perhaps thousands at the big meetings – of such posters, you need to think carefully about the design, for here, more than anywhere else in science, McLuhan rules: the medium is the message. The term 'poster' is after all not accidental – you are in an important way advertising here. Students know this very well; they sometimes seem to spend more time discussing poster designs and getting the best possible images to include on them than they do on the actual experiments.

Few people can get the funds to come to the conference unless they at least give a poster, and the sessions are therefore the place for graduate students to make their first tentative entry into scientific existence; for distinguished elder colleagues who haven't been given a plenary spot at the meeting to summarise decades of work and in doing so make clear they still exist; and for outsiders desperate to convey the secret of the universe to the waiting world by means of a few incomprehensible scribbled formulae. Round the posters mill the conferees, sometimes reading them with interest, more frequently being volubly harangued by the poster-givers who stand like stall-keepers by their boards, and most often using the session as an extension of the conference mixer – a giant cocktail party at which to catch up with old acquaintances. Poster sessions are the ultimate manifestation of science as an open institution – rarely refereed and only lightly controlled, they have become the researchers' equivalent of Beijing's Democracy wall.

But just because they aren't refereed, the posters don't really count. A 'proper' publication has to be in the form of a paper published in one of the hundreds of possible neuroscience journals. Where and how to publish becomes an important strategic decision. As any part of the work of the lab nears publishable form (cynics speak of 'MPUs' – Minimum Publishable Units) the debate among the authors begins. Publication has many aims other than the simple 'objective' recording of a scientific finding. It shows the flag for a laboratory and a research group. But the group is not an anonymous collective; it is an assemblage of individuals, each with careers to make. So whose names go on the paper, and in what order they appear, becomes important. There is no tidy solution to the problem, because people's objective interests are not the same. Many very competent technicians see their laboratory work as a steady job without feeling much intellectual involvement in the end product and are therefore relatively indifferent to whether their name is on the paper or not – for them it matters much more that they have decent working conditions and the prospect of promotion. For researchers on short-term contracts, the situation is very different and the names on the paper – and the order they appear in – are all-important. A strong publication record will help determine whether, as a student or post-doc, you get the next job, but if your name is but one amongst many your claim to the paper will be diluted. For the laboratory chief, to be seen to be still publishing is essential if one is to have any chance of obtaining the next grant. (In this sense the downstream fate of the research observation helps determine the flow of upstream resource.)*

Simple publication of course is not enough; it matters where you publish and how many other researchers subsequently cite the paper you have written. To get into either *Nature* or *Science*, the two major general scientific international weeklies, is especially prized. As they are virtually obligatory reading – or at least scanning – for all lab scientists, publication guarantees you will reach the widest audience. But only a small proportion of the papers submitted to *Nature* or *Science* are accepted; the gatekeepers – the editors and their panel of referees – operate rather rigid exclusion principles. Failing these, there are literally hundreds of more specialist journals with varying degrees of international prestige. Before submitting it, you need to think carefully about the flavour of the particular MPU in question – how

---

*Another seemingly increasingly common phenomenon is to muscle in on someone else's paper. I once contacted a Japanese biochemist, requesting he supply me with a small sample of a substance to test for possible amnestic effect in the chicks. The normal practice would be to acknowledge the gift by thanking the donor at the end of the paper; however, in this case he replied that he was willing to supply the chemical only provided he was listed as a co-author on any paper that resulted from its use. I suppose I wouldn't have minded doing so if the experiments had worked out; in the Latourian sense we became allies, but I can't think it would have done either of us much good.

strong is the paper; will it interest biochemists more than clinicians or behaviourists; is this a journal the Americans will read?

Why so many journals? It is not only the diversity of subjects, but competition, which keeps them all going. As soon as it is founded, each 'learned society' – the Biochemical Society, the Physiological Society or whatever – comes under pressure from its members to publish a journal. Because each major scientific nation has its own learned societies, there are parallel journals in each subject area. As international societies are formed, these too want their own journals, so there are European Journals of This-and-that, and International Journals of The-other. But the greatest pressure comes from commercial publishers who begin to recognise potentially profitable markets, and move in with their own variants of the title. Now none of the big academic publishing houses can afford to be without a competitive title in any given subdivision of the neurosciences. Granted that it is possible to cover costs for a monthly journal with a circulation of no more than a few hundred and it is always possible to scrape together an editorial board of scientists with more or less competence in the field who will serve for free for the prestige of seeing themselves listed on the front cover, such journals can scarcely fail to attract submitted papers and to provide profits for their publishers. The numbers of neuroscience journals ceasing publication or going broke in the last decades can probably be numbered on the fingers of a single hand. And so 'the literature' proliferates; eventually, even the weakest of papers is likely to find a published home, though whether it will ever be read or referred to again is a different matter.*

---

*A high proportion of papers, even in the most prestigious journals, are it seems, never cited again. With increasing emphasis on the productivity of research, and efforts being made by the fund-givers not merely to count publications – bibliometrics – but also to assess the quality of these publications, by the number of citations they receive. A new index has been devised, called an 'impact factor', to describe the quality of any journal and the papers it publishes. Jobs, grants and promotions are beginning to depend not merely on publishing, but on publishing with a high impact factor. At the same time the proliferation increases the chance of your paper being lost; if you want to be read and noticed, it becomes necessary to say the same thing often and in many different places. This is one of the reasons why so much research is published in the form of MPUs – it means one can split the work up and publish part in one journal, part in another, reaching slightly different audiences each time, and so spread the word.

No wonder that the 'literature' of science is in chaos and libraries in despair. There are serious proposals to abandon all pretence at publishing papers in traditional journals, and instead to store all publishable data on the web. Indeed some journals already publish supplementary details of methods or gene or protein sequences this way. Finally, perhaps only the titles or abstracts will be published in the traditional form. Such a prospect was forecast decades ago, long before it became technically feasible, by the visionary Marxist crystallographer J. D. Bernal.[8] The resistance is still strong however, from publishers and researchers alike. The publishers, because their livelihood will diminish; the researchers, because while each is likely to agree that it is the most efficient way to discover what others are doing, almost no one will want to pass up the prospect of seeing their own work, the visible proof of their creative existence, in real print.

Writing the record

I've left till the end the small matter of the content of any possible paper. The experiments of the last three chapters were originally published as more than seventy research papers, each no more than fifteen pages long, and written in the rigidly prescribed, almost sonnet-like form which the journals require, together with reviews and conference abstracts.[9] The style and voice of a scientific paper are like no other piece of writing that I know. Absolutely characteristic is the almost obligatory use of the passive voice. The active, participating observer or experimenter disappears: animals are observed, brains are removed, tissue is homogenised and centrifuged; gels are visualised. Equally obligatory is a condensed form of prose in which nouns and verbs are transformed into adjectives to produce long word strings which save printing space but make comprehension very hard. It used to be the case that only other researchers, and occasionally philosophers and historians, read our papers, and they by and large seemed to take such conventions for granted; more recently the Eng. Lit. critics have seized upon scientific papers as texts for analysis, and have had a field day working over the significance of this arid style.[10]

The gist of their argument is that the research paper, which purports to be providing a neutral description of some aspect of the natural world and its properties, should instead be regarded as a narrative about that world, which uses particular rhetorical devices to convince the reader of the validity of its claims. The passive voice is one of these devices which, by removing the active participating experimenter, gives the text apparent authority and credibility; to doubt it requires that you doubt, not an individual scientist, but nature itself. If I report that I have observed chicks behave in such-and-such a way, I may be mistaken or biased. If it is the chicks which 'were observed' (by whom?) behaving, you can blame only them, not the now disappeared observer.

Within the formal construction of the paper, the starting point is always a brief Introduction, setting the background and purposes of the experiments. This is intended to cite previous work which has led up to the present experiments, and set out the accepted framework for what follows. My papers tend to begin with a statement to the effect that

Day-old chicks will peck at bright objects but rapidly learn to distinguish bad-tasting ones (references). This is a particularly good model system in which to study the cell biology of memory (references). When chicks are trained on this task a cascade of cellular processes occur, in particular . . . (references). Other workers have suggested that process X may be involved in e.g. long-term potentiation (references). Therefore this paper reports a study of process X in the chick (references).

Such a format establishes our own priority in the work, relates what we are doing to other currently interesting approaches to the study of memory, and 'trails' the salient findings of the present paper. Granted that a busy reader may only glance at the Introduction to a paper before deciding to pass it by, it is important that the trailer be made intriguing, without overselling – a delicate matter.

There follows a Methods section, which is supposed to contain enough detail to enable anyone else who has the same facilities as your own lab to set up and repeat the experiments. In fact, it rarely does; one may think that a paper one writes describes one's methods in precise detail, but when a novice comes to try to replicate them they turn out to be written in a dense code, interpretable only by reference to dozens of other earlier papers and probably requiring several years of initiation. Research students, when starting out, get this initiation by serving as apprentices in working labs and soon learn the trick of writing their own Methods sections. Viewed from the outside, however, such conventions are as arcane as those of any medieval guild. What is more, there are many such guilds; molecular neurobiologists and neurophysiologists may work with the same animal and similar research questions, but neither is likely to understand or be able to carry through the other's methods. We are all the prisoners of our techniques, and strong hermeneutic traditions are concealed within the seeming democracy of science.

Nonetheless, for those in the trade, the Methods section is generally easy to write. What follows is the meat of the paper, the Results, and this is much harder. What to include and what to omit? How to present one's data, to extract logical sense from a series of experiments which may have taken over a year to run; which contain many replications but with minor variations; to try to remember just why you ran this particular experiment in what must have seemed the right way at the time but now appears very cackhanded; what to do about an anomalous point on a graph which looks untidy – should one repeat the experiment to try to eliminate the anomaly, or live with it, or even commit a small sin of omission and drop the point entirely? The Results section is not a neutral presentation; it doesn't descibe the story of a year in your research life in a chronological way, but imposes a particular post-hoc rationalisation on it. All data, as Darwin knew long ago, is collected for or against some hypothesis, and so must be presented to tell a story. The rhetoric of neutrality and objectivity is in fact deployed precisely for the purpose of persuading the reader to believe the results, to convince that all possible alternative explanations for the data being presented have been considered and ruled out by control experiments. Presentation becomes all-important. Should the data be given as a table, or as a graph or as a series of histograms? Which shows the results in their most convincing light? Does the story seem to hold water as one reads critically through the draft? If you were a hostile referee, could you pick holes in it? Is there some crucial control experiment you have missed? It is as much a piece of pleading as any case presented by counsel in court, calling and cross-examining witnesses to convince the jury. No wonder so many researchers who love the lab hate the slog of writing up their work for publication, for the act of creating this work of art may force them for the first time to articulate what they are doing.

Once the Results are done, it is downhill all the way, for the final section (bar the credits and references) is Discussion. Here for the first time one is allowed to become a little more discursive, to draw out the implications of the experiments, to set them into context or to try to develop a small-scale theory to explain them. Don't spend too long about it though, for journal space is precious and referees will certainly get impatient and ask you to cut down on the speculation – just give us the facts, officer, just the facts, and leave us to draw our own conclusions. Don't omit, however, to cite respectfully other researchers in the field, especially anyone who might be called upon to act as a referee for the paper, or they are likely to be offended and become, with the privilege of anonymity afforded by the refereeing process, excessively critical.

As a finale there comes the almost obligatory phrase pointing a vague way forward:

To test this possibility however, more research is needed . . .

or, if one wants to stake a more positive claim to territory:

Experiments are now under way in our laboratory to test this possibility.

That's it; the paper is done; send two copies off to your chosen journal and wait.

Two months later you'll get the referees' comments (try not to let them upset you; the criticisms may seem trivial or misguided, or even ill-intentioned, but you will get your own back when you in turn referee others' papers; that is the joy of science's peer review system). With a fair wind you should be able to deal with the comments, turn the paper back and after another three months or so it will appear in print. At last your little brick in the edifice of science is mortared into place, your immortality in the infinite archive assured. You will have satisfied your funding agency that they have backed publishable work. Colleagues will write to you to ask for reprints; some may cite you in papers of their own – living proof that you exist in the community of science – and bibliometry may even prove that you have become for a brief while a citation classic. You may even end up passing the Warhol test and becoming famous for fifteen minutes – at least among the few hundred in the memory mafia, and then only enough to have your fares to the next conference paid. If you want more than this, and to be recognised in the street, you will need to appear on television.

Chapter 14

Memories are made of this

IT IS TIME TO PULL THE THREADS TOGETHER. MEMORIES ARE PUBLIC records of past events, more or less transformed to meet current ideological needs, as when revisionist historians rewrite the past of Nazi Germany – or, in the reverse direction, of cultural revolutionary China. Memories are collective acts of recovery of lost experience, as when Black Americans re-discover (re-member, as Toni Morrison describes it, emphasising that it is an active, not a passive process) their roots in slavery, or when feminists restore the records of those women scholars whose names masculinist histories have systematically erased. Memories are the fictions of novelists for whom symbolic episodes provide keys to unlock the mystery of who and what one is, from Proust's madeleine to Atwood's cat's eye. Memories are snatches of songs once heard. Memories are the technological metaphors of a computer age. Memories are the transmuted re-creations of our own childhood and dead parents, our continued efforts to make coherence of our own lives, to synthesise past and present so as to face the future. Memories are the easing of muscles and sinews into long-unpractised rhythms when, in middle age and after decades of travelling in motorised comfort, we again mount a bicycle.

And are memories also the rejection of a once-bitter bead by a day-old chick? Can decoding the intimate mechanisms of this process illuminate any of the multiple meanings that such a potent word as memory has in our daily lived experience? The easy answer may seem to be no; that the identity of a word, memory, does not imply the identity of the many varieties of phenomena we use that word to describe. You may have been intrigued – even convinced – by the account of the findings of my lab in the preceding chapters, and yet reluctant to concede that they can have much significance in helping come to terms with your own subjective experience of personal memory. This chapter will reject such an easy answer, but my rejection will not be itself an easy one. Yes, I will conclude, the rejection of a once-bitter bead is also memory, and the study of even these small memories reveals important truths about ourselves as humans.

There are many books that have been – that could be – written about memory. As Gayle Greene has put it, discussing what feminists have described as 'the hard work of remembering':

> All writers are concerned with memory, since all writing is a remembrance of things past; all writers draw on the past, mine it as a quarry. Memory is particularly important to anyone who cares about change, for forgetting dooms us to repetition . . .[1]

But this book is about memory seen from the – perhaps peculiar – perspective of the neuroscientist. Its central focus has been a double one: the brain processes that are in increasingly well-understood ways responsible for memory (even, I would maintain, *are* memory, as written in the language of biochemistry and physiology) – and the scientific, experimental processes involved in identifying and interpreting them. Before the previous chapter's brief glance away from the lab bench towards the social processes of acquiring and disseminating scientific knowledge, I had brought my account to the frontier of biochemical, cellular understandings of these brain processes – face-to-face with the paradoxes that the experiments have begun to reveal and their potential for alleviating some at least of the distresses and diseases of the ageing process. Although I will in this chapter attempt to confront the larger questions, I will continue to insist that the biological details of what goes on inside a chick brain when it pecks a bitter bead must form as much a part of our understanding of memory as does the novelist's quarry.

The arguments against this insistence are of three types: methodological, epistemological and ontological. The methodological arguments are simple. If chick memory exists, they claim, either it is a global and unanalysable property of the chick brain as a system, or, if it is a molecular property, then the molecular processes involved are likely to be small and subtle, too refined for the coarse methods of the biochemist. Therefore to seek for the manifestation of memory in molecular and cellular changes is to approach the problem with the wrong tools and at the wrong level of analysis. This is an argument internal to neuroscience, and the whole of the last part of this book has been an effort to answer it. If you aren't convinced, you never will be, and I won't say any more about it now.

Much more interesting are the arguments from epistemology and ontology. The epistemological argument says that there are many different types of knowledge about the world. I can say what I like about the chick; poets, psychoanalysts and sociologists can say what they like about the human, but these are different forms of knowledge each with equal status. Even if you allow me the word memory to describe the behaviour of my chicks, this can tell us nothing of relevance about the peculiarities of the subjective, autobiographical, episodic and semantic features of our own complex human memories. I therefore have no right to privilege the reductionist knowledge of the laboratory over the experiential knowledge of human life outside the lab. The ontological argument is more extreme. By contrast to my claim that the world is a unity, it asserts that there are many different kinds of world-stuff. It is not that chicks and humans do not contain similar molecules, but whatever I may find out about the molecular processes going on during learning and memory says nothing about the content and meaning of those memories to their possessors; these are inscribed in a different and radically incommensurate language, which it is the job of other disciplines than neuroscience to explore. I can spend a lifetime researching the physiology of walking, but this will not explain why someone gets up from their chair at a particular moment and strides across the room. Such an explanation, it is claimed, has to be framed in the language of mind, intentionality, psychology. This is the modern version of the Cartesian split. Or, in an equally favoured computer metaphor, investigating the molecular processes of memory merely (*sic*) describes the hardware necessary for memory to occur, but such a study says nothing about the content of memory, the software or program.

In my view both the epistemological and the ontological arguments are designed to perpetuate the divorce between biological and personal/social explanations of the world. They may be invoked to privilege the personal, to protect us from a crassly reductionist biology – at its worst, vulgar sociobiology – but such attempts at privilege merely serve to harden the resolve of biological reductionists, and to encourage the fragmentation of our understanding of what can only ultimately be understood as a unitary world.

Post-modernism, epistemology and ontology

Nonetheless, it is hard to deny that such arguments have their attractions. When I go home at the end of my day in the laboratory, I re-enter the several worlds of human memories; my private world with its web of present personal relations built on past shared experience, and the public world with its endless media chronicles of human misery and folly. Understanding and functioning in these worlds, it seems, demands radically different epistemologies to the controlled reductionism of the lab. Am I then an irretrievably fractured person, entering and re-entering these different worlds of meaning? The post-modernist movement, in literature, in philosophy, in politics, regards such a fracturing of identity as inevitable. We all, whether scientist or not, inhabit multiple personae, given by class, race, gender, sexual orientation and individual experience. For each persona there is a new twist to the world's kaleidoscope, offering a different 'reality'. Why should I worry if I abandon my lab persona and its reductionist epistemology when I close the door of the animal house or switch off the centrifuge? I should reconcile myself to living in ambiguity. I might even learn to enjoy it. Should one not then simply accept the fact of multiple epistemologies, taking them as an inevitable feature of the casual barter of day-to-day existence? Thinking of memory in terms of membrane phosphorylation, gene activation and cell adhesion molecules is fine during the working day, but in the evening over dinner, and in the night in bed, we remember other things in other ways, and effortlessly invoke social or psychological explanations for our conduct and that of others.

I have no option but to accept that we do indeed all live with such different epistemologies; when I try to remember the name of the person who has just announced himself on the phone to me I don't consciously do so in terms of protein phosphorylation or neuronal bursting. But I have no difficulty in accepting that these processes are going on as I make my memory and that in some way which I only partially yet understand they can be translated into that memory. To work within different epistemologies is not to concede that the world is irretrievably fractured, merely that, in our present state of understanding, it is the best that we can do. We can envisage a world made whole again, but it remains still a promised land visible at best just on the horizon.

What I do reject utterly is that vociferous version of post-modernism that goes even further, not merely questioning the legitimacy of science's insistence that it is the epistemology which gives true knowledge of the world, but challenging its claims to have any legitimacy at all. To such extreme philosophy, science's seemingly inevitable reductionism, its effort to extract fixed categories from the seamless web of the world, to go behind appearances in the search for universals, must fail intellectually even if it avoids the Samson-like fate of bringing the world down with it. This view is reflected in the anti-rationalism, anti-science, of a wide swathe of contemporary intellectual and political movements, from literary criticism and history and philosophy of science to the animal rights movement in the streets. Despite the popularity of this way of thinking as millennial intellectual fashion retreats from classical rationality towards a gothic romanticism, abandoning Voltaire and the search for Truth in favour of Nietzsche and an orgy of supernatural horror videos starring anything from gremlins to ninja turtles, it nonetheless carries the seeds of its own destruction.

No one who has come so far with me in this book is likely to have much patience with such a totalising rejectionism. Nonetheless, having conceded the possibility – even, at present, the necessity – of different epistemologies I have to deal with the ontological argument. There remains a problem about the ontological status that I claim for the laboratory phenomena I am studying. I am not referring here to that philosophical tradition which argues that the only true knowledge can be of my own interior world, of percepts and the present instant of consciousness which illuminates them, but to something more troubling. The world outside the lab, the world of war, famine, injustice, the begging homeless in the street, is real enough even though it be a world that we know by its surface appearances. Since the days of Galileo, the task of scientific method has been seen as to seek the unchanging immutable qualities which are believed to lie beneath such surfaces, the abstractions, for physicists, of mass, energy, force, number. Such abstractions, for neuroscientists, include molecules, electrical fields, units of behaviour . . . By contrast with the world of instantly perceived objects, my laboratory, although it seems real enough, is devoted to the production and isolation of such artificial phenomena, unnatural, created and given meaning only by my actions, legitimated though they be by three hundred years of scientific history and the collective effort of will and imagination of millions of scientists across the globe.*

---

*This emphasis on the artefactual nature of the modern laboratory, in which scientists deal with constructs that have no natural existence outside the laboratory environment where they are created, was first advanced more than thirty years ago, in the context not of biology but of high-energy physics, by Jerry Ravetz in his influential book *Scientific Knowledge and its Social Problems*.[2]

The fragments that I have extracted from the seamless web of the life of my chicks, pecking or avoiding beads, shaking their heads or backing away, peeping and twittering, are abstract generalisations that I have drawn from many hundreds of thousands of individual acts by individual birds that I have observed. By what right do I define and classify these abstractions as 'real', distinct units of behaviour? Am I not simply imposing my own meanings, a personal epistemology – or at best one shared by other behavioural scientists – on the multitudinous disorder of the world?

The short answer again is no. The order I strive to impose is not one that I can vary by random will; the theoretical models that one builds of the world as a natural scientist have constantly to be tested by practice; that is, they suggest experimental tests and can – sometimes at least – be falsified. Injecting an inhibitor of protein synthesis into a chick before training it either does or does not result in amnesia. In the early stages of an experiment one can argue with the phenomenon – that it is a chance event, or not unequivocal or whatever. Repeat the experiment enough times, and with enough precautions, and one is no longer able to disagree that the phenomenon occurs; all one can do is to debate the reasons why; in this sense, despite the philosophers and sociologists of science, most practising scientists, most of the time, are more-or-less naïve realists. Scientific knowledge is in this sense public knowledge – provided always of course that all members of that public share a common agreement about what constitutes avoiding or pecking at the bead, and that these may be defined as remembering and forgetting.

Memories are made of what?

So let me summarise what we now know happens when memories are made, and then get on to the much more interesting issue of why they happen and what they mean. Of course, there are differences of opinion between neurobiologists about the intimate biochemical details – differences at least as important to us, and perhaps as unimportant to the rest of the world, as the angels on a pinpoint whose number haunted medieval scholastic debates. Nonetheless certain important general neurobiological principles have emerged from the last decades of experimentation, on chicks, on the hippocampus, on Aplysia, and on many other experimental models which I have not found space to mention here. When an animal learns – that is, when it confronts some novel environment, some new experience which requires it to change its behaviour so as successfully to achieve some goal – specific cells in its central nervous system change their properties. These changes can be measured morphologically, in terms of persistent modifications to the structure of the neurons and their synaptic connections as observed in the light or electron microscope. They can be measured dynamically, in terms of localised, transient changes in blood flow and oxygen uptake by the neurons during the processes of learning or of recall. They can be measured biochemically, in terms of a cellular cascade of processes which begins with the opening of ion channels in the synaptic membranes and proceeds by way of complex intracellular signals to the synthesis of new proteins which, inserted into the synaptic and dendritic membranes, generate these morphological changes. And they can be measured physiologically, in terms of the changed electrical properties of the neurons that also result from their altered membrane structures.

Hydra-like, each established experimental result breeds a dozen new questions. And, much though answering each day's questions may tell me about neurobiology, what can they really say about memory? Is my manic pursuit really any different from that of any other obsessive, reaching eagerly for the morning's crossword puzzle or heading for the Vegas fruit machines?

I know, from having worked steadily through the criteria I discussed in the earlier chapters, that the cellular processes I have described are necessary for memory, in the sense that if they are prevented from occurring my experimental subjects cannot remember; and they are specific at least in the sense that blocking them doesn't seem to affect any other aspect of the chick's behaviour except memory – at least no other aspect that I can observe.

Although I cannot yet say that they are sufficient, because there may be many other processes occurring that I do not suspect or have not been able yet to study, I have no doubt that in this sense I am indeed studying the processes that generate the engram, the memory trace within the brain. Further, the types of morphological change we have found and that I discuss in Chapter 10, the changes in synaptic connections due to growth of new dendritic spines, the increases in synapse numbers and dimensions – changes analogous to those found in other labs and in other experimental paradigms – make theoretical sense. At the simplest interpretation, they lead to a view of the brain as a Hebbian-type mechanism, in which new information is coded by changes in local synaptic strengths, along the lines of figure 6.1.

The experimental models and the neurobiological tools now exist to define these cellular processes with increasing certainty and precision, and it is certain the fine detail will become increasingly transparent in coming years. If the history of biology is anything to go by, we will begin by discovering seemingly universal principles, and then realise that the intense variability of the biological world means a seemingly infinite regress of apparently minor differences, to which we will struggle to give meaning; we will oscillate between marvelling at simplicity and struggling with complexity.

For instance, just which neurons in which parts of the brain are involved in the registration of memory will vary with what is being learned. We don't yet know how universal the biochemical changes that are being found are to the different types of memory catgorised by the psychologists with their taxonomy of procedural and declarative, episodic and semantic. Other types of distinction, say between modalities (verbal versus visual) may also turn out to show differences at the cellular level. At the very least, in different brain regions and cell ensembles different neurotransmitters will be involved, and there are likely to be other variations in important detail in the biochemical and cellular learning strategies adopted by different species.

However, I feel confident in claiming that the general principles of the cellular mechanisms involved in animal learning are no

longer in dispute. I am even prepared to assert the central importance of the cell adhesion molecules in the development of longer
term memory. So much is clear. I have spent most of my research
lifetime investigating the minutiae of these processes and I see no
end in sight of happy and productive play amongst the cells and
molecules, further unravelling their intimate relationships. Each
laboratory day means a new experiment, the results often depress-
ingly negative or frustratingly ambiguous – and on rare occasions
the fierce private sense of joy at a micro-hypothesis tested
triumphantly. I know of no other sensation quite like this sense of
joy, at the same time intensely cognitive but deeply emotionally
satisfying. Critics of science – especially the new feminist critics –
have analysed the ways male scientists speak of these cognitive
triumphs, with their dominating metaphors either military or sexual.
We researchers commonly speak, it appears, of winning battles
against nature, of dominating her, unveiling and penetrating her
deepest secret recesses. These metaphors, from philosopher Francis
Bacon in the seventeenth century to physicist Richard Feynman in
the twentieth, make a depressing catalogue.[3] They are not metaphors
I share. My enthusiasm for the logical elegance of the experimental
sequence I describe in chapter 11, and for what seemed, by the end
of that sequence, to be a solution to several mysteries, is intense,
tempered though it is with the knowledge that to achieve it I had
to perform operations on young chicks that I would wish it were
possible to avoid. Yet I am convinced, and I hope I have convinced
you, that this is not the dubious pleasure of domination, military
or sexual. Rather it is that of understanding, of bringing order out
of confusion, of discovering yet another way in which, to quote
Einstein, God is not playing dice with the universe. That the
outcome, hinted at in chapter 12, may indeed help in the creation
of an aid for sufferers from Alzheimer's disease, is an immensely
satisfying bonus, but it isn't in all honesty where I started, and it
would be disingenuous of me to try to justify everything I have done
in the light of that potential goal. So still the question remains, what
can such understanding of the ordered world of molecular processes
tell me about the rich experiential world of memory?

Memories are more than information

The several variants of the Hebbian synapse as a memory mechanism are models entirely in accord with the types of information processing approach to memory favoured by today's exponents of artificial intelligence, the theorists of parallel distributed processing and neural nets discussed in chapter 4. 'Information' in this sense can indeed be stored in interconnecting webs of neurons – nets – by alterations in synaptic weights. Such a conclusion is not a merely mathematical device. As I mentioned in that chapter, computers that learn and change their output as a result of experience in this way have been built.[4]

And yet, there is something ultimately unsatisfactory about such a view of the brain as an information storing and processing device. I've argued after all that brains deal in meaning rather than information, that affect is as important as cognition. And experimentally, the limitations of the simplistic information processing/neural network view are shown by the series of lesion experiments I described in chapter 11. The seemingly paradoxical result of the first set of these experiments was that regions of the brain which were necessary for memories to be made, and which showed lasting cellular changes as a result of that learning, after an hour or so ceased to be necessary for recall. The solution to this paradox was the recognition that memories are dynamic and dispersed, located in different ways in different parts of the brain.

Let me expand on the example I offered in that chapter. Think of the (for me, all-too-) familiar experience of trying to remember the name of a person one has met but whose name is somehow lost, though one has the feeling that it is 'there' somehow if only one can locate it. Most of us faced with this problem, unless we have earlier used some personal mnemotechnic device, run through a variety of strategies, conjuring up the person's face, the context in which we first met them, a feeling that their name sounded like or rhymed with or began with . . . something that we know is nearly but not quite right. That is, the memory of the person is somewhere in our head, classified in different ways; the hope is that, by running through several possible classifications we will eventually catch hold of – retrieve – the elusive name. There is no reason why such different classifications should each involve the same region of the brain or even the same sets of cells, though presumably there must be some communication between them.

Just so with the chick; if it cannot remember the bead by its colour, it will do so by its shape. Thus even for seemingly simple memories in the chick, and presumably much more so for complex human memories, a Hebbian model is not enough. At best, cellular associationism is telling us only part of the memory story, that part which is about the processes of learning; what is missing from this story is that memory is not only about learning, but also about subsequently recalling that memory, retrieving it. And it is here that the hard work of re-membering comes in, for chicks as well as humans.

The obstacle to our thinking about this process lies in our reliance on the technological metaphors of office management as ways of thinking about biological memory. Our imagination is dominated by computers and filing systems. Memories become items of 'information', to be 'stored', 'classified', brought out of store on demand and later refiled. Thinking about memory in this way, instead of in terms of human or biological meaning has come about as a result of the marriage of the neurobiologist's enthusiasm for Hebbian synaptic models and the 'bottom-up' school of neural modelling discussed in chapter 4. Whilst the information-processing metaphor dominates the language and thinking of much of present day neuroscience – typified by the manifesto for a theoretically committed reductionist 'computational neuroscience' by philosophers like Patricia Churchland[5] – such enthusiasm is not universally shared, even within neuroscience, still less outside it.

I have already referred to the immunologist Gerald Edelman's critique of the information processing metaphor, in his books on memory and consciousness.[6] The critique is based on the premise that the development of the nervous system and its capacity to change its properties as a result of experience are to be seen as processes of continuous selection of pre-existing groups of neurons and their synaptic connections in response to environmental challenges or constraints. This is the process Edelman calls, by explicit analogy with evolution by natural selection, 'neural Darwinism'. Although catchy, I do not find his phrase apt. Darwinian evolution is a process of preservation of favoured genotypes as a consequence of differential survival and reproduction of phenotypes. Neuronal ensembles do not survive and reproduce in this way – indeed they don't even replicate. Evolution and selection are poor metaphors to describe the processes of interaction, feedback, stabilisation and growth of cells and synapses occurring during development – and indeed throughout an entire lifetime. What is to be welcomed, however, is Edelman's insistence on just that dynamic and developmental nature of biological processes which the computer analogy suppresses – better caught linguistically in Susan Oyama's term 'the ontogeny of information', a phrase which in addition emphasises meaning as the antithesis of information.[7]

A more experimentally-based critique of simple associationism and its metaphors has come from the Berkeley neurophysiologist Walter Freeman. Based on his studies of the electrophysiological properties of the rabbit olfactory cortex – that is, the region of the rabbit's brain which responds to and can learn to discriminate odours – he rejects the information-processing model with its reliance on permanent changes in synaptic weights within a network. Analysing multiple simultaneous recordings from different regions of the cortex during and following odour learning, they argue that memories are represented in terms of fluctuating dynamic patterns of electrical activity across the entire brain region, fluctuations from which the application of chaos theory can extract pattern and order.[8]

Re-membering and forgetting

Without dwelling on the mathematics here, the emphasis of these alternative ways of thinking is very much upon memory, like other aspects of brain function, being an emergent property of the brain as a dynamic system rather than a fixed and localised engram. But perhaps the case is put most clearly by the cognitive psychologist Endel Tulving (whose work on the taxonomy of human learning was discussed in chapter 5):

> There is nothing wrong in principle with the idea that the information that is necessary for remembering something is recorded in a particular site in the brain . . . [however] . . . the concept of the engram has mesmerised brain scientists into acting as if there was nothing more to the problem of memory and the brain than the engram and its characteristics, including its location in the overall structure . . . the engram is an unfinished thought about memory, at best only one-half the story of memory . . . A biological memory system differs from a mere physical information-storage device by virtue of the system's inherent capability of using the information in the service of its own survival . . . The Library of the Congress, a piece of videotape or a Cray supercomputer . . . could not care less about their own survival. So anyone who is interested in memory, but looks only at the storage side of things, is essentially ignoring the fundamental distinction between dead and living storage systems, that is, ignoring the essence of biological memory (p.89).[9]

In this interview, Tulving goes on to argue that cognitive psychology does better that neurobiology in its understanding of memory by its recognition that 'memory storage' is not the problem; while synaptic studies may reveal the processes of learning, the 'problem of memory' is that of the retrieval of memories. Indeed, it was precisely this point which led me, in chapter 5, to emphasise the significance of Lionel Standing's demonstration that human recognition memory, as opposed to recall memory, seemed to be essentially unsaturable.

What though, of 'the engram'? Tulving offers a more radical challenge yet than Edelman or Freeman in his suggestion that although there may be what he calls physical – that is biochemical – changes in the brain in association with learning, a permanent engram as such does not exist as a durable 'physical' change, but only comes into existence when activated (he calls it 'ecphorised') by the act of remembering. He concludes his interview with a parody of a once better known phrase:

'Students of memory of the scientific world, unite in the study of the myriad aspects of the essence of biological memory, unite in the study of the interaction between the processes of storage and recall.' (p. 94)[9]

I suspect that Tulving is deliberately teasing his interviewer here – not in his insistence on the importance of retrieval, but in doubting that there is in the brain some physical instantiation of the engram which remains after learning and even when it is not 'ecphorised'. The argument hinges on just what is meant by 'physical'. For sure, all the biochemical and physiological phenomena I catalogued in chapters 10 and 12 are transient in the sense that the increases in protein synthesis, neural bursting activity and so forth do not persist beyond a few hours after the learning experience. But those few hours have permanently altered the brain, if only by shifting the number and position of a few dendritic spines on a few neurons in particular brain regions. The spatial map of cells and their connections in the brain – certainly as much of a physical change, in Tulving's sense, as is the imposing of a magnetic trace on a cassette tape – has been lastingly altered. In the limited sense that the music on the tape only exists when the cassette is subsequently played, the engram only exists when it is 'ecphorised' – and the memory is retrieved.

Tulving is right; as we come to understand the synaptic mechanisms of learning, it is retrieval rather than learning which will become 'the' question for memory research. For to remember is much more than simply to extract a file from a computer store. It is, in its dictionary meaning, to 'bring to mind', to 'think of again', 'to recollect' – terms which, as Greene points out (pp. 297–8),[1] suggest a connecting, an assembling, a bringing together of things in relation to one another. Virginia Woolf's metaphor is characteristically richer than that of the computer modellers; for her memory is a 'seamstress' who runs 'her needle in and out, up and down, hither and thither.'[10]

Woolf's metaphor, Greene's emphasis on the work of re-membering, and Tulving's critique come close together here, because above all they emphasise that memory – declarative memory, that is – is not some passive inscription of data on the wax tablet or silicon chips of the brain, but an active process. Furthermore, this view points to something the psychoanalysts have long emphasised; that forgetting can be more than mere erasure of stored information, as in wiping a disc clean. It is also an active process. I have already made the point in connection with the filtering process of short-term memory. Information stored in such a memory need not be transferred to a more long-term store – and indeed there is a biological necessity that much of it must be filtered out if we are not to collapse with memory overload. But once in long-term store, what happens to memories that are not re-membered? The story of Mr Goss, and the experiments of Standing, tell us that they need not be lost even if they cannot be found; they may be difficult to access because we can't find a key to how we have classified them, because we need a cue to recover them; because the work of re-membering is for whatever reason just too hard. We don't have to go all the way with orthodox psychoanalysis in its argument that the reasons for not re-membering are to do with blocking out experiences that are difficult for us to assimilate to agree that often this must be the case. Just as re-membering is hard work, so too may be forgetting. However, often the reasons for such forgetting may be more trivial. And we know, from the evidence of artifical memorisers down the ages, that we can train ourselves to re-member better. An important clue may come from the study of procedural memory here. Procedural memories, unlike declarative, do not seem to become forgotten in the same way, suggesting that they are both learned and re-membered by a very different mechanism from declarative ones. Perhaps this is because memories that involve procedural rather than declarative modes – such as riding a bicycle – are not confined simply to the brain but involve whole sets of other bodily memories, encoded in muscles and sinews?

What we as neurobiologists don't know – don't, I believe yet have a clear way of thinking about – is just how re-membering may occur as a physiological or biochemical process. It is relatively easy to design an experiment in which to measure the biochemical changes occurring when an animal learns in response to new experience. To ask what happens when that memory is revived is not so easy if only because in order to show us that it is remembering the animal has to perform some task. By contrast, human subjects can be asked specifically to remember some experience, and the activity of brain regions engaged in such remembering observed using imaging methods like fMRI or MEG – indeed the 'virtual shopping' experiment I described in chapter 5 was an experiment about re-membering, not about learning and making memories.

This isn't to say people haven't tried to do more biochemical or physiological experiments, and indeed there is a literature running back over several decades that attempts to approach the question. Social amnesia is strong in the biological sciences though, as I mentioned way back in chapter 3, and so when in 1999 the New York based group led by Joe LeDoux published a paper implying that re-membering activated the same or a similar biochemical cascade as that during initial learning, there was a surge of interest.[11] But their results were not unambiguous. Learning, memorising some new experience, involves protein synthesis, and, as I have discussed in earlier chapters, if that synthesis is blocked by an inhibitor around the time of training, then the animal can learn but cannot make long-term memory, so it soon forgets. Similarly, as I describe in chapters 11 and 12, if the protein synthesis is blocked during the 'second wave' a few hours downstream of the training, once again the animal becomes amnesic. However, if the inhibitor is given at any other time, before or after training, then the memory is unimpaired. So, the research community concluded, protein synthesis is necessary for long-term memory, but once the memory is made, then it is in some way 'fixed', immune to pharmacological assault.

What Karim Nader, Joe LeDoux and their colleagues did was to teach their rats that a particular environment was dangerous. As expected, injecting them with a protein synthesis inhibitor around the time of training made them amnesic whilst injecting the inhibitor twenty-four hours later had no effect. However, if just after they had injected the inhibitor twenty-four hours after training they gave the rat a reminder of the experience by putting it back into what was previously the dangerous environment, it was once more amnesic when they tested it later. The interpretation was that reminding the rat – recreating for it the context of the earlier experience – reactivated the memory, and the reactivation evoked the same type of biochemical cascade as initial learning. Re-membering then is not a passive process, simply opening up some closed computer file, but the active re-making of the memory, a process described as reconsolidation.

If this interpretation is valid, the implications are quite important. The experiment suggests that memories are highly dynamic and unstable records. Each time you seem to be 'remembering' an event, you are actually not re-membering the event itself, but the last time you re-membered it. Perhaps this could explain some of the furore that has erupted over the last decade concerning patients going through psychotherapy who claim to re-member long forgotten episodes of sexual or physical abuse – so-called 'recovered memory'.[12] Whilst some of these apparently recovered memories may be true records of what actually occurred, others transparently are not, and some have argued that they were in some way 'implanted' during the psychotherapeutic process – hence 'false memory syndrome'.[13] The battle over claims about true and false memories has ricocheted through courtrooms and the case notes of psychiatric social workers. My point here is that because of the dynamism of memory and the active nature of re-membering, such apparent memories may be biologically real for the individual, in that they correspond to traces in the brain, even though those traces in the brain may have been induced not by a real event but by the later implanting of a false memory. Historical 'reality' and biological 'reality' are not one and the same.

However, the interpretation of the LeDoux experiment is not uncontroversial. As I said, social amnesia had itself blotted out the prior history of experiments of this sort. Those with a longer memory of the field were able to point to experiments showing that, yes, combining a reminder with an inhibitor or transmitter antagonist often resulted in amnesia – but the amnesia was temporary, and after a few hours memory returned. The problem then is not a failure of reconsolidation, but a temporary block on retrieval – the drug prevented access to an otherwise unimpaired memory.[14] Our own intervention into this debate was begun by Kostya Anokhin, in experiments he began in Moscow and followed up in our labs. Train chicks on the bitter bead, and twenty-four hours later remind them by showing them the bead again. If he injected protein synthesis inhibitors, or antagonists to glutamate receptors just before or just after the reminder and tested the chicks a few hours later, they were indeed amnesic and happily pecked the bead. But if he postponed the test for a day or so, the previously amnesic birds seemed to have recovered their memory and avoided the bead once more. Thus unlike the effect of the inhibitors on initial training, which is to produce a permanent amnesia, the amnestic effect is transient. This points to a failure of retrieval rather than of reconsolidation, and follow-up experiments that we have been doing over the last year or so confirm that at least for our chicks, although re-membering certainly renders the memory trace dynamic and labile once more, it does not result in a simple recapitulation of the molecular and cellular processes involved in initial learning.[15] LeDoux does not agree, and experiments in the various labs are still producing conflicting results as I write. A fascinating if somewhat polemical debate is underway, as there are clearly interesting consequences for any theory of re-membering involved.

Truth to tell, the upshot is that we still haven't the slightest idea of just how re-membering occurs, how a simple clue can evoke the sequential memory of an entire scene. In fact, we are not much better off than St Augustine 1,800 years ago. Meantime, psychotherapists can share a field day with the novelists in describing the phenomena of memory, of consciousness and the unconscious, and in analysing the strategies of human memory from Simonides' theatres onwards, without us being able even to define the corresponding neurobiological tasks. If memory is to serve the purpose I have claimed for it, of becoming the Rosetta stone in the search for the translation rules between brain and mind, neurobiology and psychology, then the next decades need to go a long way further towards solving this particular mystery.

Levels of meaning

Meanwhile, what does it mean to reject an information-processing model for a more dynamic understanding of memory as a biological process? In part, the question is one about the correct level at which we might expect to find 'memory' in the brain. 'Level' is a much used and somewhat slippery concept in science in general, and neurobiology in particular. Sometimes it refers to what I have in this book called language – for instance, people speak of phenomena at 'the biochemical level' or the 'physiological level'. Sometimes it contrasts theory and practice (the 'conceptual' and 'experimental' levels). Sometimes it just means 'scale' (synapse versus neuron versus network versus brain – Churchland uses it like this). And it is also used in evolutionary and developmental senses.

I have emphasised that we can find learning-associated changes at all 'language' levels, morphological, biochemical, physiological. None of these 'levels' is in any way more fundamental than any other; as I have argued throughout this book, each can be translated into the other. Aspects of the unitary phenomenon of the change in cellular properties and connections that occurs in learning can be meaningfully described in any of my proferred languages, but the full understanding of the process demands that we use them all. This is because the three languages also offer us distinct dimensions of understanding. Morphology maps in space; biochemistry describes composition; physiology is essentially dynamic, describing events occurring in time. To understand memory we need at least these three dimensions.

But perhaps three are not enough, for still implicit in them is the idea of localisation, the idea that Tulving – or for that matter Freeman – criticises. If we take Churchland's sense of level as scale, just how many neurons, or synapses, can be said to be required to contain or represent the memory? Even if we reject the idea, discussed in chapter 9, of a cellular alphabet of memory, connectionist theories would imply that the site of the memory was a discrete albeit distributed ensemble of cells. The fact that in all the biochemical experiments on memory one can find measurable changes occurring in relatively large brain regions (hippocampus, IMHV or whatever) says that such an ensemble must be rather large, or the changes would be below the sensitivity of our measuring instruments. And the lesion experiments say that it, the engram, is not confined to a single brain region. But I want to go further than this, and to argue that in an important sense the memory is not confined to a small set of neurons at all, but has to be understood as a property of the entire brain, even the entire organism.

How can I justify such a paradoxical conclusion after having spent the whole of chapters 10 and 11 explaining just how we can indeed localise the cellular changes of learning to small brain regions? The point is that the sites of change are not equivalent to the sites of the property that they change. Come back to the tape recorder analogy. When I record a piece of music on the tape, its engram is a change in the magnetic properties of the tape, but to extract those properties, to retrieve the music, it is not enough to have the tape, I have to have the tape head, the electronics and loudpeakers that constitute the entire instrument. This is what I mean by saying that identifying the site of storage, the engram, is not the same as uncovering the mechanism or site of retrieval. The level at which memory has to be understood is thus the level of the system as a whole.

Further, whilst the 'system' that comprises the tape recorder and its tape is fixed and inanimate – dead – the essential feature of live biological systems is that they develop and change with time. The tape on which I am playing Miles Davis tracks as I write this text will play the same music however often I run it through whatever machine, merely changing its quality and slowly degrading with time. Not so the brain 'tape'. Because, as a living organism I live in a world of meaning and not simply information, the same Davis track will sometimes thrill and sometimes sadden me, depending on my present mood and past history. These present moods and past history are part of my own development, and any of my own brain engrams must be replayed, and can only achieve meaning within, these contexts. And moods are as much a matter of physiology – of hormones and immune system – as they are of brain processes, as I tried to illustrate in the experiments of chapter 12.

How far can I take my argument that the meaning of memory resides in the system rather than in the loci of change without coming to the conclusion that my life's research is after all no more than doing elaborate crossword puzzles? To argue this way is not, absolutely, to go down the Cartesian, or software/hardware route. In the very first chapter of this book I introduced my own metaphors, of translation and the search for a Rosetta stone, to describe the task that I believe we face in moving between the languages of the brain and those of mind, between those of molecular events and those of meaning. I would not want to depart from that metaphor here. It is just that what needs to be translated to understand memory, as we move between the languages of brain and of mind, is not simply a small fraction of the brain's activity, but all of it, and of the body too, including of course, but not limited to, that set of cells whose modified connections incarnate the engram.

Humans are animals too

In all that has gone before, in this chapter and in earlier ones, I have oscillated rather freely between discussing memory in humans and in the animals which are my day-to-day experimental subjects. For all the reasons that I have discussed in chapter 2 and again in chapters 6 and 7, I find this a natural and easy transition to make. But I am aware that not everyone shares this degree of tranquillity, and many people may indeed be quite uneasy about it. There seem to be two main types of reason. One says that humans and their brains are so much more complex than chicks – or even monkeys – that no valid extrapolations can be made from non-humans to humans. The second says that what we can investigate in experimental animals are behaviours in response to contingencies of reinforcement. Human memory does not seem to require such reinforcement; much of our re-membering does not manifest itself in behavioural change at all but includes verbal and visual memories of intense subjectivity. So what are the grounds for assuming that these forms of learning are subject to the same biological constraints as those that determine whether a chick learns to avoid a bead?

It is obvious that I cannot provide an answer to such arguments that would convince those determined not to believe me, but who are instead committed to the view that there is a radical breach between human activity and non-human animal psychology. All I can do is to argue for evolutionary continuities of the sort that I have spelled out in detail in chapter 7 and have continued to emphasise throughout this book. At the cellular and biochemical level neurons from the human brain are virtually indistinguishable from those of any other vertebrate; there are no obviously unique human brain cell types or even brain proteins, and the physiological and organisational properties of non-human mammalian and human brains seem very similar. Regions of the human brain that are known to be involved in memory formation are analogous to the similar (homologous) regions in non-human mammals – notably the hippocampus. All the biochemical mechanisms known to occur in non-human animal brains seem to operate in humans too. Of course it isn't possible to show that protein synthesis inhibitors block memory formation in humans – but electroconvulsive shock therapy certainly does, just as it does in animals – and memory not merely for behaviourally manifested actions but also the 'higher' forms of specifically human verbal and visual memory. I thus see no good reason to oppose the proposition that when engrams are formed in the human brain, their formation employs broadly the same types of biochemical mechanism as is the case in other vertebrates.

Improving memory?

But will learning about how chicks make memories say anything of interest about human memories? Is there more here than simply adding to our knowledge of the world? Certainly the designators of the Decade of the Brain and the Japanese Human Frontier Programme, with their emphasis on funding research into the mechanisms of memory seem to think so – as of course do the military and drug companies the industrial world over. Apart from the hope that understanding biological memory will improve the design of silicon computers, or even offer the prospect of biological computers based on ensembles of neurons in culture, the biology of neural plasticity and memory is relevant across the whole of human life, from the development of memory and learning ability in young children to the disabling deteriorations of accident, disease and later life.

The potential benefits seem obvious, even if I am somewhat cautious about the increasingly strident claims made for the development of drugs which may rectify loss or weakness of memory – the so-called nootropics. I am reasonably optimistic that it will be possible to develop therapies against memory loss in diseases like Alzheimer's and even to delay its progression.

But I am much more sceptical about the possibility of the use of the nootropics to 'improve' memory in healthy people, for all the reasons spelt out in chapter 5 and re-emphasised here. Remembering and forgetting are biologically balanced activities in a healthy person; to insist, even were it theoretically possible, that what should be forgotten is instead remembered is to run the risk of ending up like Funes or Shereskevskii – fates not to be recommended. But I am not sure that such a prospect is even theoretically possible, if we accept the argument about the nature of re-membering and for-getting and their relationships to the engram which has been the thrust of this chapter. Re-membering will always remain hard work, without much by way of short chemical cuts. Rather than offering a chemical fix to anxious students revising for examinations or flustered business executives who can't remember the name of the person they are talking too, perhaps we should prescribe them a course of Cicero or Ad Herrenium, offer everyone the chance to build their own private memory theatre?

The uniqueness of humans

I have insisted on the commonality between human and animal memory. But there is of course more to it than this, for there are profound ways in which human memory is very different from that of non-human animals. The first, and perhaps the least important, is that because humans are the only speaking animals (say what you will about parrots, or devotedly taught chimpanzees), we must presume that we are the only ones to possess a verbal memory. Such a verbal memory means the possibility of learning and remembering without manifest behaviour. Insofar as we can even begin to imagine what it could be like not to have such a verbal memory, it must mean that our powers of memory are overwhelmingly richer than those of other animals. Whereas procedural memory dominates the lives of non-human animals, it is declarative memory which profoundly shapes our every act and thought. Nonetheless, I see no reason to believe that in principle the cellular mechanisms which have been found to operate in non-human declarative memory should not operate in human verbal memory too. The richness of our linguistic recall may be biologically no more mysterious than the capacity of a homing pigeon to navigate precisely over hundreds of kilometres or a dog to distinguish and remember thousands of different odours at almost infinitesimally low concentration.

What does much more to distinguish our specific human from non-human memory is our social existence, and the technological facility which has created a world in which memories are transcribed onto papyrus, wax tablets, paper or electronic screens; that is, a world of artificial memory. It is artificial memory which means that whereas all living species have a past, only humans have a history. Although the biological mechanisms of each human's individual memory may be the same as that of our fellow vertebrates, artificial memory is profoundly liberatory, transforming both what we need to and what we are able to remember. The multimedia of modern memory devices free us of the necessity to remember vast areas of facts and processes, liberating, presumably, great numbers of neurons and their synapses to other purposes.

And it is the existence of artificial memory too, which makes possible the third great difference between human and non-human memory, the importance to all our lives of what I have called collective memory. Where there is no artificial memory, each individual animal lives in its own unique and personal set of memories, memories which begin with its birth and end with its death and can represent only its own experience. Each human, just as each non-human animal, experiences the world and remembers it uniquely, yet artificial memory presents the same picture, set of words, video image to many hundreds, thousands or millions of us, resetting, training and hence limiting our own individual memories, creating instead a mass consensus about what is to be remembered and how it is to be remembered.

Thus whilst for each of us the experience of collective memory is an individual biological and psychological one, its existence serves purposes which transcend the individual, welding together human societies by imposing shared understandings, interpretations, ideologies. It is no wonder that at any time the dominant social group endeavours to impose its own interpretation of this collective memory on the rest of society. Think about Britain in the last half century, and you may recall images ranging from the great student uprisings and the liberatory music of the 1960s to the wars and terror bombings of the 1990s and this new decade. For very few of us are these images and their interpretation our own; each has been in some measure manufactured in the effort to create a certain type of social cohesion and viewpoint about the world and how we could and should live in it.

If this is hard to think about in the context of our own society, we might ask instead what part such imposed collective memories play in the conflicts between, say, Israelis and Palestinians, or the much bruited 'clash of civilisations' between the Christian West and Islam, whose passions are rooted in images which run back through hundreds of years. The present events are incomprehensible unless we take into account this collective memory.

Memories of this type are no part of our biology, yet they dominate our lives. Which is why each new social movement needs to begin with the hard work of creating its own collective memories. Socialism struggled to recreate the submerged memories of working-class people, black movements have rediscovered their roots, feminists the suppressed history of women. These collective memories, whether imposed from above as ruling ideologies, or forged from below by the struggle of emerging social movements, are the means whereby we re-member the past, our history, and therefore they both guide our present actions and shape our futures. Nothing in biology in general, or in our own human life in particular, makes sense except in the context of memory, of history.

Do we need to understand the intricate cellular and biochemical processes which have been the major theme of this book in order to help make such sense? I believe we do. Let me revert to a more domestic example. Understanding the biochemistry of cooking and the physiology of digestion will surely never reduce the enjoyment of the meal to 'mere' biology – but it undoubtedly enriches and improves both our cooking and our eating.

Those dimensions of understanding which depend on social ambience, on the company in which we eat, on our own subjective states, are never thus reducible even though they too have their biological correspondents. A commitment to a belief in the onto-logical unity of the biological and social dimensions of our world never reduces the social to the biological, never privileges one type of explanation over another, but continues to search for ways of learning the translation rules between the two languages.

The search for the Rosetta stone and the effort to decode it which have formed the themes of this book are for me ways of integrating my day-to-day activity as a neuroscientist, with my own intense personal memories of early childhood, of air-raid shelters or of birthday parties. At its best, research on memory may help heal the split in our lives between subjectivity and objectivity, reduce the fractures in our own personae. As we face the challenge of the new millennium in an increasingly fragmented world, this goal seems not abstract but urgent.

But psychobiology and neuroscience are never going to replace the equally hard work of the novelist or poet in exploring this subjectivity, in re-membering and re-creating the foreign country which is the past. Here, to end with, is the playright Brian Friel, working at this re-membering in the closing meditation of his hero, now adult, recalling his childhood in the play *Dancing at Lughnasa*:

> And so, when I cast my mind back to that summer of 1936, different kinds of memory offer themselves to me. But there is one memory of that Lughnasa time that visits me most often, and what fascinates me about that memory is that it owes nothing to fact. In that memory atmosphere is more real than incident and everything is simultaneously actual and illusory. In that memory, too, the air is nostalgic with the music of the thirties. It drifts in from somewhere far away – a mirage of sound – a dream music that is both heard and imagined; that seems to be both itself and its own echo . . .[16]

References

Chapter 1

1 Sorabji, R. *Aristotle on Memory*, Brown University Press, Providence, 1972.
2 St Augustine, *Confessions*, translated by R. S. Pine-Coffin, Penguin, Harmondsworth, 1961.
3 For the debate on realism in science, see:
   Bhaskar, R. *A Realist Theory of Science*. Leeds Books, 1975.
   Lawson, H. and Appignanesi, L. *Dismantling Truth*, Weidenfeld and Nicolson, 1989.
   For the new sociology/anthropology of science:
   Barnes, B. and Shapin, S. *Natural Order*, Sage, 1979.
   Haraway, D. *Primate Visions,* Routledge, 1989.
   Latour, B. *Science in Action*, Open University Press, 1987.
   Latour, B. and Woolgar, S. *Laboratory Life*, Sage, 1979.
   Mulkay, M. *The Word and the World,* Allen and Unwin, 1985.
   Novotny, H. and Rose, H. A. *Countermovements in the Sciences*, Reidel, 1979.
4 For the development of Hilary Rose's and my analyses see:
   Rose, H. A. and Rose, S. P. R. *Science and Society*, Penguin, 1969.
   Rose, H. A. and Rose S. P. R. *The Political Economy of Science* and *The Radicalisation of Science*, both Macmillan, 1976.
   Rose, S. P. R., Lewontin, R.C. and Kamin, L. *Not in our Genes,* Penguin, 1984.
   Rose, S. P. R. *Lifelines: Biology, Freedom, Determinism,* Penguin, 1997.
   Rose, H. A. *Love, Power and Knowledge,* Polity, 1994.

Chapter 2

1 Kevles, Daniel J. *The Baltimore Case: A Trial of Politics, Science and Character*, Norton, New York, 1998.

Chapter 3

1 Bergman, I. *The Magic Lantern*, Hamish Hamilton, 1988, 2.

2 McIlwain, H. *Biochemistry and the central nervous system*, 1st edn, Churchill 1955.

3 For a perspective on Hydén see my tribute: Rose, S. P .R. (2000), Holger Hydén and the biochemistry of memory, *Brain Res. Bull.*, 50, 443.

4 Rose, S. P. R. Preparation of enriched fractions from cerebral cortex containing isolated, metabolically active neuronal and glial cells, *Biochemical Journal*, 102, 1967, 33–43.

5 Rose S. P. R., Cragg B. G. Changes in rat visual cortex on first exposure to light, *Nature*, 215, 1967, 253–7.

6 The best account of Lorenz's involvement with the Nazis can be found in Müller-Hill, B. *Murderous Science, Elimination by Scientific Selection of Jews, Gypsies and Others, Germany, 1933–45*, Oxford University Press, 1988.

7 Bateson, P. P. G., Horn, G. and Rose, S. P. R. Effects of an imprinting procedure on regional incorporation of tritiated lysine into protein of chick brain, *Nature*, 223, 1969, 534–5.

Chapter 4

1 Ong, W. *Orality and Literacy*, Methuen, 1982, 146.

2 Cicero, *De Oratore, II, lxxxvi*, 351–4; quoted in Yates, F. *The Art of Memory*, Penguin, 1966, 17–18.

3 Quoted by Patten, B. M. The history of memory arts, *Neurology*, 40, 1990, 346–52.

4 Yates, F. *The Art of Memory*, Penguin, 1966.

5 Spence, J. D. *The Memory Palace of Matteo Ricci*, Viking, 1985.

7 Rose, H. A. and Rose, S. P. R. *Science and Society*, Penguin, 1969.

8 Draaisma, D. *Metaphors of Memory: A History of Ideas about the Mind*, Cambridge University Press, 2000: also Haken, H., Karlqvist, A. and Svedin, U. (eds.), *The Machine as Metaphor and*

Karlqvist, A. and Svedin, U. (eds.), *The Machine as Metaphor and Tool*, Springer, Berlin, 1993.

9 Wall, P. D. and Safran, J. Artefactual intelligence, in *Science and Beyond*, eds. Rose, S. P. R and Appignanesi, L., Blackwell, 1986, 115–30.

10 Many of the references to this debate will be found in the two books edited by myself and Hilary Rose: Rose, H. A. and Rose, S. P. R., *The Political Economy of Science* and *The Radicalisation of Science*, both Macmillan, 1976.

11 Needham, J. *Science and Civilisation in China,* Cambridge University Press (continuing series of volumes).

12 Rose, S. P. R., *Lifelines: Biology beyond Determinism,* Penguin, 1998.

13 Descartes, R. *Philosophical Works*, 1:116.

14 Rose, H. A., Learning from the new priesthood and the shrieking sisterhood: debating the life sciences in Victorian England, in *Reinventing Biology: Respect for Life and the Creation of Knowledge,* eds. Bitke, L. and Hubbard, R., Indiana University Press, Bloomington, 1995, 3–21.

15 Descartes, R. *Les Passions de l'ame*, 1649, quoted in Dudai, Y., *The Neurobiology of Memory*, Oxford University Press, 1989.

16 From the reference to Brain in *The New Book of Knowledge,* vol 2 BIB-CHIC – I was brought up on this encyclopedia.

17 Hodges, A. *Alan Turing: The Enigma of Intelligence*, Counterpoint, 1983.

18 Papert, S. One AI or many? in *The Artificial Intelligence Debate*, ed. Graubard, S. R., MIT Press 1988, 3–4.

19 Lighthill, J. *Artificial Intelligence,* Science Research Council, 1973.

20 Rumelhart, D. E., McClelland, J. L. and the PDP Research Group, *Parallel Distributed Processing,* MIT Press, 2 vols. 1986.

21 Churchland, P. S. *Neurophilosophy: Towards a Unified Science of the Mind–Brain,* MIT Press, 1986.

22 Churchland, P. S. and Sejnowski T. J. Perspectives on computational neuroscience, *Science,* 242, 1988, 741–5.

23 Minsky, M. *The Society of Mind,* Heinemann, 1987.

24 Boden, M. in *Science and Beyond*, eds. Rose, S. P. R. and Appignanesi, L., Blackwell, 1987, 103–14.

25 Grand, S. *Creation: Life and How to Make It,* Weidenfeld and Nicolson, London, 2000.

26 Aleksander, I. *How to Build a Mind,* Orion, London, 2000.

27 Penrose, R. *The Emperor's New Mind,* Oxford University Press, 1990.

28 e.g. Wegner, D. M. *The Illusion of Conscious Will,* MIT Press, 2002. But there are dozens more.

29 Crick, F. H. C. *The Astonishing Hypothesis: The Scientific Search for the Soul,* Simon and Schuster, London, 1994.

30 Changeux, J.-P. and Ricoeur, P. *What Makes us Think?* Princeton University Press, 2002.

31 Searle, J. R. *Minds, Brains and Science,* Harvard University Press, 1984.

32 Edelman, G. *Neural Darwinism; The Theory of Neuronal Group Selection; Topobiology;* and *The Remembered Present,* Basic Books, 1987, 1988, 1989.

33 Griffith, J. S. *A View of the Brain,* Oxford, 1967.

34 All these examples of memory metaphors occur in the current model-builders' lexicons. See for instance Delacour, J. and Levy J. C. S. (eds.)' *Systems with Learning and Memory Abilities,* Elsevier, 1988.

35 See the references to this debate in chapter 1.

36 Austen, J. *Mansfield Park,* 1816.

37 Frame, J. *An Autobiography,* The Women's Press, 1990.

38 Borges, J. L. Funes the memorious, in *Fictions,* Calder, 1965.

39 Atwood, M. *The Handmaid's Tale,* Virago, 1987.

40 Bindra, D. Metaphors, computers and the brain, Annual meeting of the Royal Society of Canada, mimeo, 1980.

Chapter 5

1 Luria, A. R. *The Mind of a Mnemonist,* Cape, 1969.

2 Haber, N. R., Eidetic images, *Scientific American,* 220, 1969, 36–40.

3 Ebbinghaus, H. in *Memory: A Contribution to Experimental Psychology,* Dover, 1964, 5.

4 James, W. *Principles of Psychology,* Holt, 1890, 646, 648.

5 Miller, G. A. The magic number seven, plus or minus two, *Psychology Review,* 9, 1956, 81–97.

6   Hacking, I. *Rewriting the Soul: Multiple Personalities and the Sciences of Memory*, Princeton, 1998.

7   Russell, W. R., *Brain, Memory, Learning*, Oxford, 1959, 69–70.

8   Standing, L. Remembering ten thousand pictures, *Quarterly Journal of Experimental Psychology*, 25, 1973, 207–22.

9   Baddeley, A. The concept of working memory: a view of its current state and probable future development, *Cognition*, 10, 1981, 17–23.

10 Squire, L. R. *Memory and Brain*, Oxford, 1987.

11 Tulving, E. *Elements of Episodic Memory*, Oxford, 1983.

12 The history of neuroscience is still in its infancy, though there is now a journal devoted to it. See, e.g. Star, S. L. *Regions of the Mind: Brain Research and the Quest for Scientific Certainty*, Stanford University Press, 1989.

13 Scoville, W. B. and Milner, B. Loss of recent memory after bilateral hippocampal lesions, *Journal of Neurology, Neurosurgery and Psychiatry*, 20, 1957, 11–21.

14 Milner, B., Corkin, S. and Teuber, H. L. Further analysis of the hippocampal amnestic syndrome: 14 year follow-up study of HM, *Neuropsychologia*, 6, 1968, 215–34.

15 Squire, *Memory and Brain*, 178–9.

16 Penfield, W. and Perot, P. The brain's record of auditory and visual experience, *Brain*, 86, 1963, 595–696, quoted by Squire, *Memory and Brain*.

17 Maguire, E. A., Frackowiak, R. S. J. and Frith, C. D. Recalling routes around London: activation of the right hippocampus in taxi drivers, *Journal of Neuroscience*, 17, 1997, 7103–110.

18 Brautigam, S., Stins, J. F., Rose, S. P. R., Swithenby, S. J. and Ambler, T. Magnetoencephalographic signals identify stages in real life decision processes, *Neural Plasticity*, 8, 2001, 241–53.

Chapter 6

1   Zeki, S. *A Vision of the Brain*, Blackwell, 1993.

2   Synaptic stabilisation is a major theoretical issue in neurobiology, and its mechanism and function has been made the basis of broad theories of brain function both by the Parisian molecular

neurobiologist Jean-Paul Changeaux and the New York immunologist Gerald Edelman. See, for instance: Changeaux, J.-P. and Danchin, A. Selective stabilisation of developing synapses as a mechanism for the specification of neuronal networks, *Nature*, 264, 1976, 705–12; Changeaux, J.-P. *Neuronal Man*, Pantheon, 1985; Edelman, G. *Neural Darwinism*, Basic Books, 1987.

3 The founder of the team which began this work has reviewed its thirty-year history; Renner, M. J. and Rosenzweig, M. R. *Enriched and Impoverished Environments*, Springer, 1987.

4 Held, R. and Hein, R. *Journal of Comparative Psychology*, 56, 1963, 872–6.

5 Singer, W. Developmental plasticity – self-organisation or learning? in Rauschecker, J. P. and Marler, P. (eds.) *Imprinting and Cortical Plasticity*, Wiley, 1987, 171–6.

6 Amongst useful accounts, from very differing perspectives, of the history of Russian (Soviet) psychology, see Graham, L. R. *Science and Philosophy in the Soviet Union*, Knopf, 1972; Joravsky, D. *Russian Psychology: A Critical History*, Blackwell, 1989; Luria, A. R. *The Making of Mind: A Personal Account of Soviet Psychology*, Harvard University Press, 1979.

7 Skinner, B. F. *Beyond Freedom and Dignity*, Cape, 1972. Chomsky, N. *Language and Mind*, Harcourt Brace and World, 1968.

9 Pinker, S. *The Language Instinct*, William Morrow, 1994.

10 Deacon, T. *The Symbolic Species: The Co-Evolution of Language and the Brain*, Norton, 1997; Karmiloff-Smith, A. Why babies brains are not Swiss army knives, in Rose, H. and Rose, S. (eds.) *Alas Poor Darwin: Arguments Against Evolutionary Psychology*, Cape, 2000, 144–56.

11 Hebb, D.O. *The Organisation of Behaviour*, Wiley, 1949, 62–3.

12 Tanzi, E. I fatti e le induzioni nell'odierna istologia del sistema nervosa *Rivista sperimentale, Freniat. Med. leg. Alien. Ment.*, 19, 1893, 419–72.

13 Kohler, W. *The Mentality of Apes*, Routledge and Kegan Paul, 1925.

14 MacPhail, E. M. *Brain and Intelligence in Vertebrates*, Oxford, 1982.

15 Garcia J., Ervin, F. R. and Keolling, R. Learning with prolonged delay of reinforcement, *Psychonomic Science*, 5, 121–2.

16 Barber, A., Gilbert, D. and Rose, S. P. R. Glycoprotein synthesis is necessary for memory of sickness-induced learning in chicks, *European Journal of Neuroscience,* 1, 1990, 673–7.

Chapter 7

1 Rose, S. P. R. *Lifelines*, Penguin, 1998.

2 Gould, S. J. *The Structure of Evolutionary Theory,* Harvard, 2002.

3 Webster, G. and Goodwin, B. *Form and Transformation: Generative and Relational Principles in Biology,* Cambridge University Press, 1996.

4 Kerkut, G., Emson, P. and Walker, R. J. Learning in the lower animals, in Ansell, G. B. and Bradley, P. B. *Macromolecules and Behaviour*, Macmillan, 1973, 241–57.

5 Menzel, R. Searching for the memory trace in a mini-brain, the honey bee, *Learning and Memory*, 8, 2001, 53–62.

6 Benzer, S. Genetic dissection of behavior, *Scientific American*, 229 (12), 24–37; Dudai, Y. Genetic dissection of learning and short term memory in Drosophila, *Annual Review of Neuroscience*, 11, 1988, 537–63; Weiner, J. *Time, Love, Memory*, Faber, 1999.

7 Young, J. Z. *A Model of the Brain*, Oxford, 1964.

8 Carew, T. J., Walters, E. T. and Kandel, E. R. Differential classical conditoning of a defensive withdrawal reflex in *Aplysia californica*, *Science*, 219, 1983, 397–400.

9 Alkon, D. L. *Memory Traces of the Brain*, Cambridge, 1987.

10 Alport, S. *Explorers of the Black Box*, Norton, 1986.

Chapter 8

1 Grey Walter, W. *The Living Brain*, Duckworth, 1953.

2 Hyden, H. Changes in brain protein during learning, in Ansell, G. B. and Bradley, P. B. (eds.), *Macromolecules and Behaviour*, Macmillan, 1973, 3–26.

3 Dingman, W. and Sporn, M. B. The incorporation of 8-azaguanine into rat brain RNA and its effect on maze learning by the rat: an enquiry into the biochemical basis of memory, *Journal of Psychiatric Research*, 1, 1961, 1–11.

4 Davis, H. P. and Squire, L. R. Protein synthesis and memory: a review, *Psychological Bulletin*, 96, 1984, 518–59.

5 See, e. g. Ziman, J. *Public Knowledge*, Cambridge University Press, 1968.

6 This point is strongly made by the anthropologist of science, Bruno Latour; Latour, B. *Science in Action*, Open University Press, 1987.

7 Horn, G., Bateson, P. P. G. and Rose, S. P. R. Experience and plasticity in the nervous system, *Science*, 181, 1973, 506–14.

8 Stent, G. *Paradoxes of Progress*, Freeman, 1978.

9 Mekler, I. B. Mechanism of biological memory, *Nature*, 215, 1967, 481–4.

10 Griffith, J. and Mahler, H. R. DNA ticketing theory of memory, *Nature*, 223, 1969, 580–2.

11 Conrad, M. Molecular information structures in the brain, *Journal of Neuroscience Research*, 2, 1974, 233–54.

12 Friedrich, P. Protein structure: the primary substrate of memory, *Neuroscience*, 35, 1990, 1–17.

13 McConnell, J. V. Memory transfer through cannibalism in planaria, *Journal of Neuropsychiatry*, 3 (supp. 1), 1962, 42–8. For a review of the early enthusiasms for these experiments, see Corning, W. C. and Ratner, S. C. (eds.), *Chemistry of Learning*, Plenum, 1967; less enthusiasm can be found in Corning, W. C. and Riccio, D. The planarian controversy, in Byrne, W. L. (ed.), *Molecular Approaches to Learning and Memory*, Academic Press, 1970, 107–50.

14 Babich, F. R., Jacobson, A. L., Bubash, S. and Jacobson, A. Transfer of a response to naive rats by injection of ribonucleic acid extracted from trained rats, *Science*, 149, 1965, 656–7.

15 Cameron, D. E., Sved, S., Solyom, L., Wainrib, B. and Barik, H. Effects of ribonucleic acid on memory deficit in the aged, *American Journal of Psychiatry*, 120, 1963, 320–5.

16 Ott, J. and Matthies, H.-J. Some effects of RNA precursors on development and maintenance of long-term memory: hippocampal and cortical pre- and post-training application of RNA precursors, *Psychopharmacologia*, 28, 1973, 195–204.

17 Watson, P. *War on the Mind*, Hutchinson, 1978.

18 Byrne, W. L. and 22 others, Memory transfer, *Science*, 153, 1966, 658–9.

19 Ungar, G., Desiderio, D. M. and Parr, W. Isolation, identification and synthesis of a specific behaviour-inducing brain peptide, *Nature*, 238, 198–202. Stewart, W. W. Comments on the chemistry of scotophobin, *Nature*, 238, 1972, 202–9.

20 Johnston, A. N. B., Clements, M. P. and Rose, S. P. R. Role of brain-derived neurotrophic factor and presynaptic proteins in passive avoidance learning in day old domestic chicks, *Neuroscience*, 66, 1999, 1033–42.

21 Glassman, E. The biochemistry of learning: an evaluation of the role of RNA and protein, *Annual Review of Biochemistry*, 38, 1969, 387–400.

22 Bateson, P. P. G. Are they really the products of learning? in Horn G. and Hinde, R. A. (eds.), *Short Term Changes in Neural Activity and Behaviour*, Cambridge University Press, 1969, 553–66.

23 Experiments showing that the effects of protein synthesis inhibitors on memory could be mimicked by injecting large quantities of some of their constituent amino acids were made by one of my earliest graduate students, John Hambley, on his return to Australia, before his sadly premature death: Hambley, J. and Rogers, L. J. Some neurochemical correlates of permanent learning deficits associated with intracerebral injections of amino acids in young chick brain, *Proceedings of the International Society for Neurochemistry*, 6, 1977, 359.

24 Rose, S. P. R. What should a biochemistry of learning and memory be about? *Neuroscience*, 6, 1981, 811–21.

25 Rose, S. P. R. Early visual experience, learning and neurochemical plasticity in the rat and the chick, *Philosophical Transactions of the Royal Society, Series B*, 278, 1977, 307–18.

26 Brenner, S. Nobel Lecture 2002.

Chapter 9

1 Rose, S. P. R. What should a biochemistry of learning and memory be about? *Neuroscience*, 6, 1981, 811–21.

2 McGaugh, J. L. Time-dependent processes in memory storage, *Science*, 153, 1964, 135–8.

3 Dudai, Y. *Memory from A to Z: Keywords, Concepts and Beyond*,

Oxford, 2002; Dudai, Y. *The Neurobiology of Learning and Memory: Concepts, Findings, Trends,* Oxford University Press, 1989.

4 Tully, J. and deZazzo, T. Dissection of memory formation. From behavioral pharmacology to molecular genetic, *Trends in Neuroscience*, 18, 1995, 212–17.

5 Kandel, E. R. and Schwartz, J. H. *Principles of Neural Science*, Elsevier, many editions.

6 Kandel, E. R. From metapsychology to molecular biology: explorations into the nature of anxiety, *American Journal of Psychiatry*, 140, 1983, 1277–93.

7 Allport, S. *Explorers of the Black Box,* Norton, 1986.

8 Kupfermann, I., Castellucci, V., Pinsker, H. and Kandel, E. R. Neuronal correlates of habituation and dishabituation of the gill-withdrawal reflex in *Aplysia, Science*, 167, 1971, 1743–5.

9 Rayport, S. G. and Schacher, S. Synaptic plasticity in vitro: cell culture of identified Aplysia neurons mediating short-term habituation and sensitisation, *Journal of Neuroscience*, 6, 1986, 759–63.

10 Carew, T. J., Hawkins, R. D. and Kandel, E. R. Differential classical conditioning of a defensive withdrawal reflex in *Aplysia californica, Science*, 219, 1983, 397–400.

11 Goelet, P., Castellucci, V. F., Schacher, S. and Kandel, E. R. The long and the short of long-term memory – a molecular framework, *Nature*, 322, 1986, 419–22.

12 Morris, R. G. M synaptic plasticity, neural architecture and forms of memory, in McGaugh, J. L., Weinberger, N. M. and Lynch, G. (eds.), *Brain Organisation and Memory: Cells, Systems and Circuits,* Oxford University Press, 1990, 52–7.

13 Colebrook, E. and Lukowiak, K. Learning by the *Aplysia* model system: lack of correlation between gill and motor neuron responses, *Journal of Experimental Biology*, 135, 1988, 411–29.

14 Bailey, C. H. and Chen, M. Morphological alterations at identified sensory neuron synapses during long-term sensitization in *Aplysia,* in Squire, L. R. and Lindenlaub, E. (eds.), *The Biology of Memory,* Schattauer Verlag, 1990, 135–54.

15 Rankin, C. H. and Carew, T. J. Development of learning and memory in Aplysia, II: habituation and dishabituation, *Journal of*

Neuroscience, 7, 1987, 133–43; Carew, T. J., Marcus, E. A., Nolen, T. G., Rankin, C. H. and Stopfer, M. The development of learning and memory in *Aplysia*, in McGaugh, J. L., Weinberger, N. M. and Lynch, G. (eds.), *Brain Organisation and Memory: Cells, Systems and Circuits*, Oxford University Press, 1990, 27–51.

16 Bourtchouladze, R. *Memories are Made of This*, Orion, 2002.

17 Bliss, T. V. P. and Lømo,T. Long-lasting potentiation of synaptic transmission in the dentate area of the anaesthetised rabbit following stimulation of the perforant path, *Journal of Physiology*, 232, 1973, 331–56.

18 Levy, W. B. and Steward, O. Temporal contiguity requirements for long-term associative potentiation/depression in the hippocampus, *Neuroscience*, 8, 1983, 791–7.

19 Matthies, H.-J. (ed.) *Learning and Memory: Mechanisms of Information Storage in the Nervous System*, Pergamon, 1986.

20 O'Keefe, J. and Nadel, L. *The Hippocampus as a Cognitive Map*, Oxford University Press, 1978.

21 Rolls, E. Functions of neuronal networks in the hippocampus and of backprojections in the cerebral cortex in memory, in McGaugh, J. L., Weinberger, N. M. and Lynch, G. (eds.) *Brain Organisation and Memory: Cells, Systems and Circuits*, Oxford University Press, 1990, 184–210.

22 Anokhin, P. K. *Biology and Neurophysiology of the Conditioned Reflex and its Role in Adaptive Behaviour*, Pergamon Press, 1974.

23 Vinogradova, O. Registration of information and the limbic system, in Horn, G. and Hinde, R. A. (eds.), *Short-term Changes in Neural Activity and Behaviour*, Cambridge University Press, 1969, 95–140.

24 Graham, L. R. *Science and Philosophy in the Soviet Union*, Knopf, 1972.

25 Rose, S. P. R. (ed.) *Against Biological Determinism*, Allison and Busby, 1982.

26 Lynch, G. and Baudry, M. The biochemistry of memory: a new and specific hypothesis, *Science*, 224, 1984, 1057–63.

27 Lisman, J., Schulman, H. and Cline, H. The molecular basis of CAMKII function in synaptic and behavioural memory, *Nature Reviews Neuroscience*, 3, 2002, 175–90.

28 Wittenberg, G. M., and Tsien, J. Z. An emerging molecular and cellular framework for memory processing by the hippocampus, *Trends in Neuroscience*, 25, 2002, 501–5.

29 Morris, R. G. M., Anderson, E., Lynch, G. S., and Baudry, M. Selective impairment of learning and blockade of LTP by an NMDA receptor antagonist, AP5 (1986), *Nature*, 319, 1986, 774–6.

30 Morris, R. G. M. Long-term potentiation and memory, *Philosophical Transactions of the Royal Society*, B 358, 2003, 643–7.

Chapter 10

1 Bateson P. P. G. The characteristics and context of imprinting, *Biological Reviews*, 41, 1966, 177–220.

2 Gibbs M. E. and Ng, K.T. Psychobiology of memory: towards a model of memory formation, *Biobehavioural Reviews*, 1, 1977, 113–36.

3 Rose, S. P. R., Gibbs, M. E. and Hambley, J. Transient increase in forebrain muscarinic cholinergic receptors following passive avoidance learning in the young chick, *Journal of Neurochemistry*, 5, 1980, 169–72.

4 For this and others of Medawar's essays, see Medawar, P. B. *Pluto's Republic*, Oxford University Press, 1982; Medawar, P. B. *The Limits of Science*, Oxford University Press, 1985.

5 Kossut, M. and Rose, S. P. R. Differential 2-deoxyglucose uptake into chick brain structures during passive avoidance training, *Neuroscience*, 12, 1984, 971–7; Rose, S. P. R. and Csillag, A. Passive avoidance training results in lasting changes in deoxyglucose metabolism in left hemispheric regions of chick brain, *Behavioural and Neural Biology*, 44, 1985, 315–24.

6 Eccles, J. C. *Evolution of the Brain: Creation of the Self,* Routledge, 1989.

7 Gould, S. J. *The Mismeasure of Man*, Norton, 1981; Harrington, A. *Medicine, Mind and the Double Brain*, Princeton University Press, 1987.

8 Andrew, R. J. (ed.) *Behavioural and Neural Plasticity: The Use of the Domestic Chick as a Model,* Oxford University Press, 1991.

9   Nottebohm, F. Brain pathways for vocal learning in birds: a review of the first ten years, *Progress in Psychobiology and Physiological Psychology*, 9, 1980, 85–124.

10 Ali, S. M., Bullock, S. and Rose, S. P. R., Phosphorylation of synaptic proteins in chick forebrain: changes with development and passive avoidance training, *Journal of Neurochemistry*, 50, 1988, 1579–87. Burchuladze, R., Potter, J. and Rose, S. P. R., Memory formation in the chick depends on membrane-bound protein kinase C, *Brain Research*, 535, 1990, 131–8.

11 Salinska, E. J., Chaudhury, D., Bourne, R. C. and Rose, S. P. R. Passive avoidance training results in increased responsiveness of synaptosomal voltage and ligand gated channels in chick brain synaptosomes, *Neuroscience*, 93, 2000, 1507–14.

12 Anokhin, K. V., Mileusnic, R., Shamakina, I. and Rose, S. P. R. Effects of early experience on c-fos gene expression in the chick forebrain, *Brain Research*, 544, 1991, 101–7.

13 Purves, D. *Body and Brain: A Trophic Theory of Neural Connections*, Harvard University Press, 1988.

14 Patel, S. N. and Stewart, M. G. Changes in the number and structure of dendritic spines, 25 hr after passive avoidance training in the domestic chick *Gallus domesticus, Brain Research*, 463, 1987, 168–73.

15 Stewart, M. G. Morphological correlates of long-term memory in the chick forebrain consequent on passive avoidance learning, in Squire, L. R. and Lindenlaub, E. (eds.), *The Biology of Memory*, Schattauer Verlag, 1991, 193–215.

16 Benowitz, L. and Magnus, J. G. Memory storage processes following one-trial aversive conditioning in the chick, *Behavioural Biology*, 8, 1973, 367–80.

17 Rose, S. P. R. and Harding, S. Training increase [3H] fucose incorporation in chick brain only if followed by memory storage, *Neuroscience*, 12, 1984, 663–7.

18 Patel, S. N., Rose, S. P. R. and Stewart, M. G. Training-induced spine density changes are specifically related to memory formation processes in the chick *Gallus domesticus, Brain Research*, 463, 1988, 168–73.

19 Anokhin, K. V. and Rose, S. P. R. Learning-induced increase of immediate early gene messenger RNA in the chick forebrain, *European Journal of Neuroscience*, 3, 1991, 162–7.

20 Burchuladze, R., Potter, J. and Rose, S. P. R. Memory formation in the chick depends on membrane-bound protein kinase C, *Brain Research*, 535, 1990, 131–8.

21 Rose, S. P. R. and Jork, R. Long-term memory formation in chick is blocked by 2-deoxygalactose, a fucose analogue, *Behavioral and Neural Biology*, 48, 1987, 246–58.

22 Mason, R. and Rose, S. P. R. Lasting changes in spontaneous multiunit actvity in the chick forebrain following passive avoidance training, *Neuroscience*, 21, 1987, 931–41.

23 Bradley, P. M., Burns, B. D. and Webb, A. C. Potentiation of synaptic responses in slices from the chick forebrain, *Proceedings of the Royal Society of London, Series B*, 243, 1991, 19–29.

24 Mason, R. and Rose, S. P. R. Passive avoidance learning produces focal elevation of bursting activity in the chick brain: amnesia abolishes the increase, *Behavioral and Neural Biology*, 49, 1998, 280–92.

Chapter 11

1 Davies, D. C., Taylor, D. A. and Johnson, M. H. The effects of hyperstriatal lesions on one trial passive avoidance learning in the chick, *Journal of Neuroscience*, 8, 1988, 4662–6.

2 Patterson,T. A., Gilbert, D. B. and Rose, S. P. R. Pre- and post-training lesions of the intermediate medial hyperstriatum ventrale and passive avoidance learning in the chick, *Experimental Brain Research*, 80, 1990, 189–95.

3 Gilbert, D. B., Patterson, T. A. and Rose, S. P. R. Dissociation of brain sites necessary for registration and storage of memory for a one-trial passive avoidance task in the chick, *Behavioural Neuroscience*, 105, 1991, 553–61.

4 Lashley, K. S. In search of the engram, *Symposia of the Society for Experimental Biology*, 4, 1950, 454–82.

5 Bourne, R. C., Davies, D. C., Stewart, M. G. and Cooper,W. Cerebral glycoprotein synthesis and long term memory

formation in the chick following passive avoidance training
depends on the nature of the aversive stimulus, *European Journal
of Neuroscience*, 3, 1990, 243–8.

6 Barber, A. J., Gilbert, D. B. and Rose, S. P. R. Glycoprotein
synthesis is necessary for memory of sickness-induced learning
in chicks, *European Journal of Neuroscience*, 1, 1990, 673–7.

7 Vallortigara, G., Zanforlin, M. and Compostella, S. Perceptual
organisation in animal learning: cues or objects? *Ethology*, 85,
1990, 89–102.

8 Patterson, T. A. and Rose, S. P. R. Memory in the chick: multiple
cues, distinct brain locations, *Behavioural Neuroscience*, 106, 1992,
465–70.

Chapter 12

1 Rose, H. and Rose, S. *Alas Poor Darwin: Arguments Against
Evolutionary Psychology*, Cape, 2000.

2 Cahill, L. and McGaugh, J. L. Mechanisms of emotional arousal
and lasting declarative memory, *Trends in Neuroscience*, 21, 1998,
294–9.

3 Sandi, C. and Rose, S. P. R. Corticosterone enhances long-term
retention in one day old chicks trained in a weak passive avoidance learning paradigm, *Brain Research*, 647, 1994, 106–12.

4 Loscertales, M, Rose, S. P. R. and Sandi, C. The corticosteroid
synthesis inhibitors metyrapone and aminoglutethimide block
long-term memory for a passive avoidance task in day-old chicks,
Brain Research, 769, 1997, 357–61.

5 Sandi, C. and Rose, S. P. R. Training-dependent biphasic effects
of corticosterone in memory formation for a passive avoidance
task in chicks, *Psychopharmacology*, 133, 1997, 152–60.

6 Sandi, C., Rose, S. P. R., Mileusnic, R. and Lancashire, C.
Corticosterone facilitates long-term memory formation via
enhanced glycoprotein synthesis, *Neuroscience*, 69, 1995, 1087–93.

7 Johnston, A. N. B. and Rose, S. P. R. Isolation-stress induced
facilitation of passive avoidance memory in the day-old chick,
Behavioral Neuroscience, 112, 1998, 1–8.

8 Migues, P. V., Johnston, A.N.B. and Rose, S.P.R.

Dehydroepiandosterone and its sulphate enhance memory retention in day-old chicks, *Neuroscience*, 109, 2001, 243–51.

9   Johnston, A. N. B., Clements, M. P. and Rose, S. P. R. Role of brain-derived neurotrophic factor and presynaptic proteins in passive avoidance learning in day old domestic chicks, *Neuroscience*, 66, 1999, 1033–42.

10   Rose, S. P. R. Cell adhesion molecules and the transition from short- to long-term memory, *J. Physiol (Paris)*, 90, 1996, 387–91.

11   Mileusnic, R., Lancashire, C. and Rose, S. P. R. Sequence specific impairment of memory formation by NCAM antisense oligo-nucleotides, *Learning and Memory*, 6, 1999, 120–7.

12   Roullet, P., Mileusnic, R., Rose, S. P. R. and Sara, S. J. Neural cell adhesion molecules play a role in rat memory formation in appetitive as well as aversive tasks, *Neuroreport*, 8, 1997, 1907–11.

13   Doyle, E., Nolan, P., Bell, R. and Regan, C. M. Intraventicular infusions of anti-neural cell adhesion molecule in a discrete post-training period impair consolidation of a passive avoidance response in the rat, *Journal of Neurochemistry*, 59, 1992, 1570–73.

14   Scholey, A. B., Mileusnic, R., Schachner, M. and Rose, S. P. R. A role for a chicken homolog of the neural cell adhesion mole-cule L1 in consolidation of memory for a passive avoidance task, *Learning and Memory*, 2, 1995, 17–25.

15   Mileusnic, R., Lancashire, C. and Rose, S. P. R. APP is required during an early phase of memory formation, *European Journal of Neuroscience*, 12, 2000, 4487–95.

Chapter 13

1   See Rose, H. and Rose S. (eds.) *The Political Economy of Science*, Macmillan 1976.

2   Latour, B. *Science in Action*, Open University Press, 1987.

3   Rose, H. *Love, Power and Knowledge*, Polity, 1994.

4   A point once made, in the heyday of the radical science movement, by French physicist Jean-Marc Levy LeBlond, in a letter explaining why he was refusing, on principle, to go to a meeting in an idyllic setting in Provence to which several of us had not too guiltily succumbed.

5 Kevles, D. J. *The Baltimore Case: A Trial of Politics, Science and Character*, Norton, 1998.

6 Ciccotti, G., Cini, M. and De Maria, M. The production of science in advanced capitalist society. In Rose, H. and Rose S. (eds.)*The Political Economy of Science,* Macmillan, 1976, 32–58.

7 Evans, R., Butler, N. and Gonçalves, E. *The Campus Connection*, Campaign for Nuclear Disarmament, 1991.

8 Bernal, J. D. *The Social Functions of Science,* Routledge, 1939.

9 The style has been subject to what should have been lethal criticism by the Nobel prizewinning immunologist Peter Medawar. See e.g. Medawar, P. B. *Pluto's Republic,* Oxford, 1982.

10 Gross, A. G. *The Rhetoric of Science*, Harvard University Press, Cambridge, Mass, 1990.

Chapter 14

1 Greene, G., Feminist fiction and the uses of memory, *Signs: Journal of Women in Culture and Society*, 16, 1991, 290.

2 Ravetz, J. R. *Scientific Knowledge and its Social Problems*, Oxford University Press, 1971.

3 E.g. Merchant, C. *The Death of Nature: Women, Ecology and the Scientific Revolution*, Wildwood House, 1982.

4 Graubard, S. R. (ed.) *The Artificial Intelligence Debate*, MIT Press, 1988.

5 Churchland, P. S. *Neurophilosophy*, MIT Press, 1986.

6 Edelman, G. *Neural Darwinism*, Basic Books, 1987.

7 Oyama, S. *The Ontogeny of Information*, Cambridge University Press, 1985.

8 Freeman, W. J. *How Brains Make Up Their Minds.* Orion, 2000.

9 Interview with Endel Tulving, *Journal of Cognitive Neuroscience*, 3, 1991, 89.

10 Woolf, V., quoted by Greene in note 1.

11 Nader, K., Schafe, G. F. and LeDoux, J. F. Fewer neurones require protein synthesis in the amygdala for reconsolidation after retrieval, *Nature*, 406, 2000, 722–26.

12 Schachter, D. L. *Searching for Memory: The Brain, the Mind and the Past,* HarperCollins, 1997.

13 Loftus, E. and Ketchum, K. *The Myth of Repressed Memory: False Memories and Allegations of Sexual Abuse*, St Martin's Press, 1994.

14 Sara, S. J., Retrievals and reconsolidation: towards a neurobiology of remembering, *Learning and Memory*, 7, 2000, 73–84.

15 Anokhin, K. V., Tiunova, A. A. and Rose, S. P. R., Reminder effects – reconsolidation or retrieval deficit? Pharmacological dissection with protein synthesis inhibitors following reminder for a passive avoidance task in young chicks, *European Journal of Neuroscience*, 15, 2002, 1759–65.

16 Friel, B. *Dancing at Lughnasa*, Faber, 1990.

Index

Page numbers in italics refer to illustrations and those followed by 'n' to footnotes.

and protein synthesis 210
in rats 163–4
in vertebrates 204
synaptic connections 65, 208, 254
in brain development 162
and Hebb 174, 175
indeterminacy 100
synaptic gap 190
synaptic terminals *59*
Szent-Gyorgyi, Albert 46

Tanzi, Eugenio 175–6, 177
tau 337
Tauc, Ladislav 202, 249
taxonomy of memory 136, 137–8, 139, 230–1, 371
technological metaphor 78–81
technology 69, 70, 108–9
and biology 79, 108
and collective memory 10
and laboratories 10–11
see also computers
temporal lobe *142*, 143, 148, 149, *151*
thalamus 147
theatres of memory 14, 74, 76–8
thiamine 128
time course 122–6, *280*, 288–301
see also long-term memory; short-term memory
Tonagawa, Susumu 244
Tower of Hanoi 146
Tully, Tim 247–8, 291
Tulving, Endel 138, 376–7, 382
Turing, Alan 88, 90
Turing Test 90, 99, 105, 137n

Unconditioned reflex 167
Ungar, Georges 221–4, 225–6, 233
United States of America 3, 15, 21n, 35, 46, 48, 52, 343–4, 350n
and behaviourism 169
PhD system 55
psychology 171

Van Leeuwenhoek, Antoni 62
velcro molecules and Alzheimer's disease 335–40
vertebrates 65, 182, 198, 204–5, 231, 258–9, 385
Vietnam 10, 281n
Vinogradova, Olga 270
visual cortex 160, 161, 166
vivisection 83
see also animal rights
Voltaire 368
voluntary memory 123
von Frisch, Karl 199
von Neumann, John 88, 89
von Neumann bottleneck 92
Vygotsky, Luria: *Mind of a Mnemonist* 169n

Walker, Alice 9
Wannsee 70, 71
wasps 33
water maze 212
Watson, John B. 46, 169, 170, 217, 218
wax tablets 80, 108, 112, 378, 387
Webster, Gerry 186n
Western culture
dualism 81
fragmented 7, 13
and mnemotechnics 72
Whittaker, Victor 64
Wiener, Norbert 88, 99
Wiesel, Elie 70
Wilson, Edward 184n
Woolf, Virginia 377–9
working memory 231, 269

Yates, Frances: *The Art of Memory* 74, 77–8
Young, John Zachary 160, 201, 202

Zaïrean bard 70, 72, 108
Zeki, Semir 161